Microwave Heating as a Tool for Sustainable Chemistry

Sustainability: Contributions through Science and Technology

Series Editor: Michael C. Cann, Ph.D.
Professor of Chemistry and Co-Director of Environmental Science
University of Scranton, Pennsylvania

Preface to the Series

Sustainability is rapidly moving from the wings to center stage. Overconsumption of non-renewable and renewable resources, as well as the concomitant production of waste has brought the world to a crossroads. Green chemistry, along with other green sciences technologies, must play a leading role in bringing about a sustainable society. The **Sustainability: Contributions through Science and Technology** series focuses on the role science can play in developing technologies that lessen our environmental impact. This highly interdisciplinary series discusses significant and timely topics ranging from energy research to the implementation of sustainable technologies. Our intention is for scientists from a variety of disciplines to provide contributions that recognize how the development of green technologies affects the triple bottom line (society, economic, and environment). The series will be of interest to academics, researchers, professionals, business leaders, policy makers, and students, as well as individuals who want to know the basics of the science and technology of sustainability.

Michael C. Cann

Published Titles

Green Chemistry for Environmental Sustainability
Edited by Sanjay Kumar Sharma, Ackmez Mudhoo, 2010

Microwave Heating as a Tool for Sustainable Chemistry
Edited by Nicholas E. Leadbeater, 2010

Sustainability: Contributions through Science and Technology

Series Editor: Michael C. Cann

Microwave Heating as a Tool for Sustainable Chemistry

Edited by
Nicholas E. Leadbeater

CRC Press
Taylor & Francis Group
Boca Raton London New York

CRC Press is an imprint of the
Taylor & Francis Group, an **informa** business

Cover image created by Nicholas E. Leadbeater and Sarah Louise Upjohn.

CRC Press
Taylor & Francis Group
6000 Broken Sound Parkway NW, Suite 300
Boca Raton, FL 33487-2742

First issued in paperback 2017

© 2011 by Taylor and Francis Group, LLC
CRC Press is an imprint of Taylor & Francis Group, an Informa business

No claim to original U.S. Government works

ISBN-13: 978-1-4398-1269-3 (hbk)
ISBN-13: 978-1-138-11198-1 (pbk)

Library of Congress Cataloging-in-Publication Data

Microwave heating as a tool for sustainable chemistry / editor, Nicholas E. Leadbeater.
  p. cm. -- (Sustainability)
"A CRC title."
Includes bibliographical references and index.
ISBN 978-1-4398-1269-3 (hardcover : alk. paper)
  1. Environmental chemistry--Industrial applications. 2. Chemical processes. 3. Microwaves--Industrial applications. 4. Heat--Transmission. 5. Sustainable engineering. I. Leadbeater, Nicholas E.

TP155.2.E58M53 2011
660'.28--dc22   2010026996

Visit the Taylor & Francis Web site at
http://www.taylorandfrancis.com

and the CRC Press Web site at
http://www.crcpress.com

Contents

Series Preface

Sustainability is rapidly moving from the wings to center stage. Overconsumption of nonrenewable and renewable resources, as well as the concomitant production of waste has brought the world to a crossroads. Green chemistry, along with other green sciences and technologies, must play a leading role in bringing about a sustainable society. The *Sustainability: Contributions through Science and Technology* series focuses on the role science can play in developing technologies that lessen our environmental impact. This highly interdisciplinary series discusses significant and timely topics ranging from energy research to the implementation of sustainable technologies. Our intention is for scientists from a variety of disciplines to provide contributions that recognize how the development of green technologies affects the triple bottom line (society, economy, and environment). The series will be of interest to academics, researchers, professionals, business leaders, policy makers, and students, as well as individuals who want to know the basics of the science and technology of sustainability.

Michael C. Cann
Scranton, Pennsylvania

Preface

After arriving home hungry following a long day in the laboratory or the office, we all know that the fastest way to heat up last night's leftovers is to use a microwave oven. Since Percy Spencer first noticed that candy bars melt when close to radar sets, thus leading to the development of the first domestic microwave oven in 1947, the technology is now pretty much in every home. Dow Chemical Company filed a patent in 1969 in which they documented carrying out chemical reactions using microwave energy, and it was in 1986 that the first reports appeared in the scientific literature showing that microwave heating can be used in organic chemistry. Since these early days, the use of microwave heating as a tool in preparative chemistry has transitioned from a curiosity to mainstream, both in industrial and academic settings. Perhaps the main driving force behind this is the short reaction times that are often possible when using microwave heating. Alongside this, chemists have found that product yields can improve. The development of scientific microwave apparatus has been instrumental (quite literally) in the advance of the field. There is now a range of equipment available for performing chemistry on milligrams as well as kilograms of material. Advantages over domestic microwave ovens include accurate measurement of parameters such as temperature and pressure as well as, most importantly, safety. Household microwaves are great for heating food but are not designed for synthetic chemistry, as many of us who started out working with them found out firsthand.

Alongside the development of microwave heating for preparative chemistry has come the somewhat controversial topic of "microwave effects." In an attempt to rationalize the short reaction times and different product distributions observed when using microwave as opposed to "conventional heating," a range of theories has been suggested, some of which, if true, would require rewriting the laws of science. When comparisons are made under strictly identical conditions, the general observation is that, be it in a microwave or an oil bath, heating is just that—heating. However, the operational ease with which reactions can be performed makes microwave heating a very valuable addition to any preparative chemistry laboratory. No longer do you have to work in high boiling point solvents with messy oil baths and lengthy reaction times in order to obtain high yields of your target molecule.

This book will showcase the application of microwave heating in a number of areas of preparative chemistry, a theme running through it being sustainability. Looking at the online resource, Wikipedia, *sustainability* is defined as "the capacity to endure." Within the chemistry community, sustainability is becoming front-and-center as evidenced by the fact that at the end of 2009 two of the largest chemical societies, the American Chemical Society (ACS) and the Royal Society of Chemistry (RSC), agreed to collaborate to promote chemistry's role in a sustainable world. In addition, the topic for the Spring 2010 ACS National Meeting was "Chemistry for a Sustainable World." So how then can microwave heating be used as a tool for sustainable chemistry? There are some clear-cut examples shown in this book: microwave heating for making biodegradable polymers and efficient battery materials, for teaching the chemists of tomorrow the concepts of green chemistry, and for use in conjunction with water as a solvent, to name but a few. Running through every chapter there is the general theme of microwave heating being an easy, rapid, effective way to make a wide range of molecules. While not every transformation shown may be classed in itself as "sustainable," the overall drive of chemists to develop cleaner, greener routes to their target compounds is undoubtedly being facilitated by the incorporation of microwave heating into their toolkits. Returning to the Wikipedia definition, microwave heating and all the advantages it brings definitely shows that it has the capability to endure, and I firmly believe it will increasingly become the heating method of choice in the laboratory.

All the authors of the chapters in this book are steadfast microwavers. I want to offer my heartfelt thanks to each and every one of them for being willing to take the time and energy to contribute their wealth of knowledge in compiling their chapters. When you read them, I think you will sense their enthusiasm for microwave chemistry, their excitement over where the field has come from, and their passion for seeing it develop in the future.

I am also indebted to the sustainability series editor Mike Cann and to Taylor & Francis, especially the chemistry acquisitions editor Hilary Rowe, for giving me the opportunity to gather a team of people and put this book together. Unlike a microwave reaction, the book has taken a bit of time to reach completion. This is totally my fault and I thank the publishing team (especially Pat Roberson, my project coordinator, and Tara Nieuwesteeg, my project editor) who have been very accommodating of my requests for "just a bit more time."

I would not be editing this book, nor would I be so deeply involved in microwave chemistry, if it were not for the students who have worked in my research group over the last 10 years since I first took in that microwave oven from home to "try something." Their enthusiasm, good ideas, and willingness to "give it a go" when I suggest something is greatly appreciated. Alongside this, we have been incredibly fortunate to have close relationships with the major microwave manufacturers. Their willingness to give us access to nice new shiny equipment and take back broken things for repair has been instrumental (quite literally again) to our development of new chemistry. Finally, I must thank my wife for her patience and willingness to "just let me get on with it" throughout the editing stage of this book.

I hope you enjoy and learn from the contents here, and I close this preface with the words of two of my graduate students over the years. First, Jason Schmink, who says "… go out and try even your craziest idea in the microwave. It will take, after all, just a few short minutes of your time!" Second, Maria Marco, who established our group motto: "Get a life. Get a microwave!"

Nicholas E. Leadbeater

Contributors

Ping Cao
Progenra Inc.
Malvern, Pennsylvania

Mauro Ianelli
Milestone s.r.l.
Sorisole, Italy

Nicholas E. Leadbeater
Department of Chemistry
University of Connecticut
Storrs, Connecticut

Cynthia B. McGowan
Department of Chemistry
Merrimack College
North Andover, Massachusetts

Jonathan D. Moseley
AstraZeneca Process Research and
    Development
Avlon Works
Bristol, United Kingdom

Gregory L. Powell
Department of Chemistry and
    Biochemistry
Abilene Christian University
Abilene, Texas

Jason R. Schmink
Department of Chemistry
University of Connecticut
Storrs, Connecticut

Robert A. Stockland, Jr.
Department of Chemistry
Bucknell University
Lewisburg, Pennsylvania

Steven L. Suib
Department of Chemistry
University of Connecticut
Storrs, Connecticut

Grace S. Vanier
CEM Corporation
Matthews, North Carolina

1 Microwave Heating as a Tool for Sustainable Chemistry
An Introduction

Jason R. Schmink and Nicholas E. Leadbeater

CONTENTS

1.1 MICROWAVE HEATING

The microwave region of the electromagnetic spectrum is broadly defined as that with wavelengths ranging from 1 m down to 1 mm (Figure 1.1). This corresponds to frequencies of between 0.3 and 300 GHz. Since applications such as wireless devices (2.4 to 5.0 GHz; U.S.), satellite radio (2.3 GHz), and air traffic control operate in this range, regulatory agencies allow equipment for industrial, scientific, and medical (ISM) use to operate at only five specific frequencies: 25.125, 5.80, 2.45, 0.915, and 0.4339 GHz.

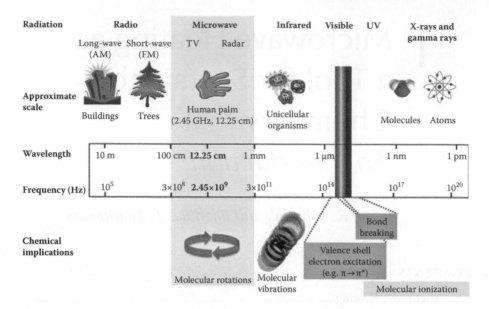

FIGURE 1.1   Regions of the electromagnetic spectrum with approximate scale as well as chemical implications for selected wavelength regions.

Domestic microwave ovens operate at 2.45 GHz (12.25 cm wavelength), and this same frequency has also been widely adopted by companies manufacturing scientific microwave apparatus for use in preparative chemistry, with only a few exceptions.[1]

Microwave heating is based on the ability of a particular substance such as a solvent or substrate to absorb microwave energy and effectively convert the electromagnetic energy to heat (kinetic energy). Molecules with a dipole moment (permanent or induced) attempt to align themselves with the oscillating electric field of the microwave irradiation, leading to rotation. In the gas phase, these molecular rotations are energetically discrete events and can be observed using microwave spectroscopy.[2] However, in the liquid and solid phases, these once-quantized rotational events coalesce into a broad continuum as rotations are rapidly quenched both by collisions and translational movement.

Molecules in the liquid or gas phase begin to be rotationally sympathetic to incident electromagnetic irradiation when the frequency approaches 10^6 Hz.[3] Conversely, above a frequency around 10^{12} Hz (infrared region), even small molecules cannot rotate an appreciable amount before the field changes direction. The optimal frequency at which a molecule turns incident electromagnetic radiation into kinetic energy is a function of many component parts, including the permanent dipole moment, the size of the molecule, and temperature. However, for most small molecules, the relaxation process is most efficient in the microwave region (0.3–300 GHz) of the electromagnetic spectrum.

The interaction of microwave energy with a molecule can be explained by analogy to baseball or cricket. During the swing, the batter or batsman can be said to be "rotationally excited" and can deliver some amount of rotational force to the incoming pitch (delivery in cricket). At the point of impact, the rotational energy is rapidly converted into translational energy of the ball. Similarly, one water molecule excited rotationally by incident irradiation can strike a second molecule of water, converting rotational energy into translational energy. Under microwave irradiation, a large number of molecules are rotationally excited and, as they strike other molecules, rotational energy is converted into translational energy (i.e., kinetic energy) and, as a consequence, heating is observed (Figure 1.2).

Since microwave heating is dependent on the dipole moment of a molecule, it stands to reason that more polar solvents such as dimethylsulfoxide, dimethylforma-mide, ethanol, and water better convert microwave irradiation into heat as compared to nonpolar ones such as toluene or hexane. Previous efforts have been undertaken to quantify relative microwave absorptivities[4] and correlate this with the dielectric constant (ε'), dielectric loss (ε''), or a combination of both, termed loss tangent or loss angle (tan $\delta = \varepsilon''/\varepsilon'$). The dielectric constant describes the polarizability of a molecule in the microwave field, while the dielectric loss expresses the efficiency with which a molecule converts the incident electromagnetic irradiation into molecular

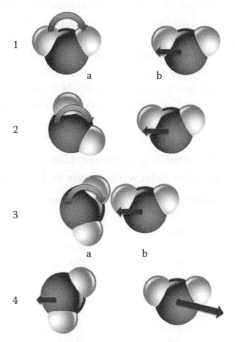

FIGURE 1.2 Microwave heating. Panels 1–3 show a molecule a that has been rotationally excited by microwave irradiation being approached by a second molecule b. Upon impact (panel 3), the rotational energy of molecule a is converted to the translational movement of molecule b. In panel 4, note the increase in translational vector magnitude, the consequence of which leads to an increase in molecular collisions (kinetic energy). This concept is not so unlike that of baseball or cricket players about to strike a ball and impart their rotational energy to the ball in the form of translational energy, hopefully enough to vault the ball over the outfield fence or to score a "six."

rotation, and hence heat. The loss angle (tan δ) is a measure of reactance (resistance in a capacitor) of a molecule.[5] The easiest way to understand this concept is to examine the extremes. A material that has tan $\delta = 0$ is completely transparent to microwave irradiation, and incident irradiation passes through with its path unchanged ($\delta = 0$). For a perfectly absorbing material, tan $\delta = \infty$; $\delta = \pi/2$ radians or 90°. Here, the material under irradiation shows complete resistance to the incident irradiation. Practically speaking, materials with tan δ approaching 1 are very strong microwave absorbers. For instance, ethanol (tan $\delta = 0.941$) or ethylene glycol (tan $\delta = 1.350$) are both exceptional absorbers of microwave irradiation at 2.45 GHz (see Table 1.1).

While the dielectric loss or tan δ value of a molecule can be used to assess microwave absorbance, the use of any single parameter drastically oversimplifies the issue of "efficient" microwave heating. A number of other factors contribute to this. Attributes such as specific heat capacity and heat of vaporization of the substance, as well as the depth to which microwave irradiation can penetrate into the sample, can sometimes have a larger impact on heating rate than its respective dielectric loss or loss tangent.[6] In addition, dielectric loss and dielectric constant are functions of both irradiation wavelength as well as temperature, specific heat changes as a function of temperature, and heat of vaporization changes as a function of pressure. These can all affect microwave absorptivity individually and in combination. Room temperature water, for instance, is most microwave absorbent at approximately 18 GHz, but as temperature increases, so does the optimum frequency at which water converts microwave irradiation to heat. Generally, however, when synthetic microwave chemists speak of "good" or "bad" microwave absorbers, implied is a 2.45

TABLE 1.1
Dielectric Constant (ε'), Dielectric Loss (ε''), and Loss Tangent (tan δ) for Selected Solvents at 2.45 GHz

Solvent	Dielectric constant (ε')	Dielectric loss (ε'')	Loss tangent (tan δ)
Water	80.4	9.89	0.123
Ethanol	24.3	22.9	0.941
DMSO	45	37.1	0.825
DMF	37.7	6.07	0.161
Acetonitrile	37.5	2.32	0.062
Acetone	20.7	1.11	0.054
DCM	9.1	0.382	0.042
THF	7.4	0.348	0.047
Ethyl Acetate	6	0.354	0.059
Toluene	2.4	0.096	0.040
Hexane	1.9	0.038	0.020

Source: Data from Hayes, B. L., *Microwave Synthesis: Chemistry at the Speed of Light*, CEM Publishing, Matthews, NC, 2006.

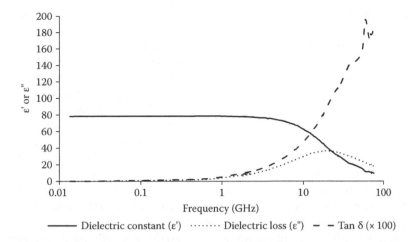

FIGURE 1.3   Dielectric constant (ε'), dielectric loss (ε''), and loss angle (tan δ) are all functions of irradiation frequency. Shown here are the plots for water, which heats most efficiently at approximately 18 GHz. Plot generated from data from Gabriel et al. (1998) and Craig (1995). Tan δ values are scaled ($\times 100$) for clarity.

GHz irradiation source, a small depth of field (1–10 cm), and synthetically relevant temperatures (50–150 °C) (Figure 1.3).

1.2   MICROWAVE EFFECTS

"Microwave heating can enhance the rate of reactions and in many cases improve product yields." This rhetoric typifies that found strewn throughout literature extolling the virtues of utilizing microwave irradiation to "promote" reactions. While that sentence is technically not false, it is every bit as true if one were to remove the word *microwave*, leaving only "Heating can enhance the rate of reactions." That said, microwave heating *can* be different from "conventional," solely convection-based, "stove-top" heating. Numerous attempts have been made to evaluate differences between microwave versus conventional heating, either real or perceived. For the most part, these differences have been divided into two categories: "specific" microwave effects and "nonthermal" microwave effects.

1.2.1   SPECIFIC MICROWAVE EFFECTS

"Specific" microwave effects are conceptually straightforward, grounded in sound theory, and backed up by well-executed experiments. They encompass macroscopic heating events that occur slightly differently under microwave irradiation than when using conventional (convection) heating methods. Additionally, specific microwave effects are often difficult (but not impossible) to reproduce without the use of microwave irradiation. Such examples would include (1) observed heating differences based on microwave absorptivity, (2) inverted temperature gradients, (3) macroscopic superheating, and (4a) selective heating of substances in heterogeneous and potentially in (4b) homogeneous systems.

The first specific microwave has already been addressed: substrates that better convert incident microwave irradiation into heat, heat the bulk faster. Thus, heating 2 mL of water to 100 °C from room temperature will take considerably less time than heating 2 mL of toluene across the same temperature range and utilizing the same applied microwave power at 2.45 GHz. While other attributes certainly impact the rate of heating, because the differences in dielectric loss factors (water: $\varepsilon''= 9.89$; toluene: $\varepsilon'' = 0.096$) are so profound, any variations in heat capacities or heats of vaporization will have negligible impact on the rate of heating. However, it is important to note that there would also be differences in heating rates if heated conventionally, but that any differences would likely show the highest correlation to specific heat capacities. Indeed, it takes a calculated 167.3 J to heat 2 mL water by 80 °C but only 58.7 J to heat the same 2 mL of toluene.

Fortunately, differences in microwave absorptivity generally have little impact: commercial monomode units are able to heat effectively just about any pure solvent. Furthermore, as reactions generally have multiple components such as acid, base, or metal catalysts, and one or more reactants, reaction mixtures will often heat much more efficiently than the solvent alone. Finally, in the extreme cases where microwave units are unable to heat reactions due to poor substrate or solvent absorptivities, additives can be utilized that allow the bench chemist access to any solvent system. Most commonly, ionic liquids[7] or reusable inserts such as silicon carbide[8] or Weflon™[9] have been used when microwave transparent solvents such as toluene or hexane must be employed for a particular reaction.

The next highly touted specific microwave effect is that of inverted temperature gradients when using microwave irradiation. Conventional heating must heat reactions from the outside in, and the walls of the reaction vessel are generally the hottest part of the reaction, especially during the initial ramp to the desired temperature. Microwave heating, on the other hand, can lead to inversion of this gradient as heat is generated across the entire reaction volume, and a larger cross-section of the reaction may reach the ideal reaction temperature sooner than it would have with conventional heating. However, efficient stirring and controlled heating can generally mitigate temperature gradients in both microwave and conventionally heated reactions. Furthermore, it is important to note that the side-by-side thermal images first published in 2003[10]—and reproduced extensively—illustrate *unstirred* reactions that are heated for *only* 60 s either by microwave irradiation or by a conventional oil bath (Figure 1.4). This image should be used as a warning to chemists comparing conventionally heated reactions to those heated under microwave irradiation, especially when comparing reactions carried out at very high temperatures for short reaction times. Indeed, this phenomenon likely has caused more problems than benefits, and led to unfounded speculation.

A third example of specific microwave effects is the phenomenon of macroscopic superheating.[11,12] Solvents will boil only when they are in contact with their own vapor and, if this is not the case, they can be heated to above their normal (atmospheric) boiling point without the onset of boiling.[13] This phenomenon can be appreciated when heating a degassed solvent in a pristine reaction vessel using microwave

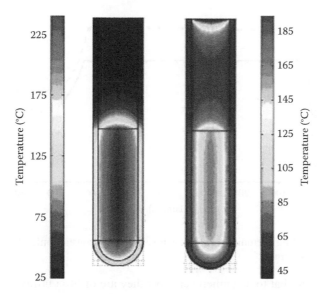

FIGURE 1.4 Infrared thermograph image of temperature gradients across an unstirred reaction heated for 60 s with microwave irradiation (left) and conventionally (right). (Adapted from Schanche, J.-S., *Mol. Diversity* 2003, 7, 293–300. Copyright Springer.)

irradiation. Imperfections in glassware or on boiling stones have areas that cannot be wetted by the solvents, and thus create small pockets of the solvent vapor, termed nucleation sites. Without nucleation sites, solvents are only in contact with their own vapor at the top of the vessel, and thus boiling (and hence release of heat) is limited to this relatively small interface. Using microwave irradiation, solvents have been held well above their boiling points for extended periods of time. For example, acetonitrile has been maintained at over 100 °C (normal b.p. 82 °C) as shown in Figure 1.5. Since the most likely sites for nucleation in the absence of boiling stones are the pits and scratches on glassware walls and, under microwave irradiation, these are likely the coolest part of the system, nucleation events can feasibly be considered less likely. The phenomenon has been further exploited in reaction chemistry. The acid-catalyzed esterification of benzoic acid with hexanol and the solvent-free cyclization of citronellal (ene reaction) were carried out at temperatures well above normal boiling points under open vessel conditions. Presumably this afforded the four diastereomers of isopulegol, though the observed product is not indicated in the published report.[14] In the case of the esterification reaction, temperatures of some 38 °C above the normal boiling point of 1-hexanol were obtained and, in the case of the ene reaction, it was possible to perform the reaction 35 °C above the normal boiling point of citronellal. Accordingly, rate enhancements were observed at these higher temperatures when compared to conventionally heated reactions.

These first three examples of specific microwave effects (observed heating differences based on microwave absorptivity, inverted temperature gradients, and macroscopic superheating) are very real, observable phenomena. While they can occasionally be exploited and have an impact on observed reaction rates, it is

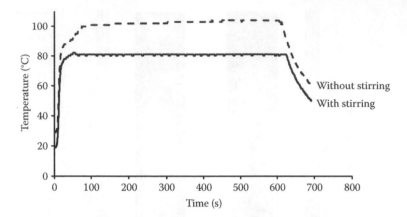

FIGURE 1.5 Heating acetonitrile in an open vessel using constant microwave irradiation with and without stirring.

important to note that to a synthetic chemist they are of little utility and most certainly represent the exception rather than the rule. For example, an inverted temperature gradient is likely manifested only while heating the reaction mixture to the desired temperature. Equilibrium will quickly be reached, and the vessel walls will be only a few degrees cooler than the contents. Furthermore, wall effects as well as the potential for superheating are both virtually *eliminated* with effective stirring. There would be few synthetic chemists willing to give up stirring for a few degrees in reaction temperature, as there are rather few reactions that proceed smoothly without the aid of stirring. Indeed, an esterification where the solvent is a substrate, and neat, unimolecular reactions may represent the majority of such examples. In addition, although reactions may proceed faster under these conditions, there are significant safety concerns. Anyone who has heated a cup of coffee or water in a microwave oven at home and then taken it out and stirred it may well have seen, in some degree, the effects of inducing nucleation. The contents can boil very rapidly and, in some cases, with such vigor as to eject the hot contents onto the person. In the case of a reaction mixture, this can have significant effects including contamination of a considerable area and, at worst, significant personal injury.

The final example of a specific microwave effect is the ability to heat very microwave-absorbent substrates and catalysts selectively under *heterogeneous* reaction conditions. A recent example is in the synthesis of CdSe and CdTe nanomaterials using the nonpolar hydrocarbons heptane, octane, and decane as solvents.[15] It is hypothesized that the precursor substrates are able to absorb the microwave irradiation selectively, this leading to more uniform morphology in the resulting nanomaterials as compared to conventional heating methods. This observation reportedly extends to enzyme-catalyzed transformations. For example, selective heating of green fluorescent protein by microwave irradiation purportedly leads to denaturing of the enzyme and hence an increase in fluorescence that is not consistent with the observed changes in bulk temperature.[16] Similarly, an increase in reactivity in three of four hyperthermophilic enzymes has been observed at bulk temperatures far below their optimal activity window when using microwave irradiation.[17] It is important to note, however, that this phenomenon may be dependent on the particular enzyme, as other studies have found no difference in enzymatic activity whether heated with microwave irradiation or conventionally.[18] Indeed, the microwave-mediated selective heating at the point of reaction seems to be the exception rather than the rule, existing in only very specific instances or highly manipulated protocols.

1.2.2 NONTHERMAL MICROWAVE EFFECTS

Unlike specific microwave effects, venturing into the world of "nonthermal microwave effects" puts the scientist on rather shaky ground.[19] Numerous attempts have been made over the past 20 years to rationalize *perceived* enhancements in reaction rates that could not be explained according to typical models (e.g., the Arrhenius equation). Reactions were often performed side by side, one in a microwave unit and the other in an oil bath, and these reactions were purportedly carried out at identical temperatures, with increased yields or decreased reaction times almost exclusively reported when using microwave as opposed to "conventional" heating. However, when meticulous attention is paid to reaction setup and accurate temperature monitoring, the playing field again becomes level. A number of techniques have been used to examine the impact of microwave energy on reaction rates and also to determine where errors may have previously arisen. For instance, multiple fiber-optic probes placed inside a reaction vessel give a clearer picture of temperature gradients, and hence inaccuracies, in measured and reported microwave reaction conditions.[20,21] Significant variation in reaction temperature has been found, especially under heterogeneous reaction conditions. This effect was most apparent when high initial microwave power was applied, as temperature-monitoring software cannot acquire data at a sufficient rate to be accurate. In these cases, temperature overshoot is common. Additionally, silicon carbide heating inserts[22] and vessels[23] as well as application of simultaneous cooling of vessel walls[24,25] have been used to probe the impact of microwave power on organic reactions at a constant temperature. Similarly, applied power has been reported to have no impact on rates of enzyme-catalyzed reactions, the reaction temperature being the only factor.[19] Raman spectroscopy has been used to investigate the impact of microwave power input on spectroscopic signatures of molecules, and no examples of "localized superheating"[26] have been found. As these results continue to emerge and as previous claims are systematically debunked, one thing becomes ever more clear: *heating is heating.*

1.3 MICROWAVE-ASSISTED SYNTHESIS

The use of microwave irradiation to heat reactions has likely been most widely appreciated and employed by organic chemists, both in academia and industry, and a number of useful books and reviews have been published on this subject.[4,27,28] There are a number of excellent reasons to use the microwave to heat reactions, or to at least have access to a scientific microwave unit. It is a useful tool that exhibits a range of applications that span relatively mundane and routine lab work[29] to affording the bench chemist an opportunity to carry out exciting new chemistry. The use of microwave heating in organic synthesis has been widely adopted since seminal publications in 1986.[30,31] The reported reactions were performed using domestic microwave ovens. The widespread application of microwave irradiation as a tool for heating organic reactions can be appreciated by the increase in total number of publications as well as the increased percentage of publications that cite use of the technology in five major organic chemistry journals from 2002 to 2009 (Table 1.2). Additionally, the use of microwave irradiation in polymer, materials, inorganic and peptide synthesis, as well as other biochemical applications, is seeing a dramatic increase, as highlighted in the chapters of this book

Certainly, the most useful attribute of the scientific microwave is its ability to aid the user when developing new chemistry. Due to the ease of which reactions can be performed under sealed-vessel conditions (autoclave), microwave heating opens access to a range of conditions that are otherwise difficult to attain (though not impossible). For example, an organic chemist generally will select solvents in accordance with boiling point and known or assumed activation energy barriers. For instance, a stubborn reaction may be carried out in refluxing xylenes (b.p. 137–140 °C), 1,2-dichlorobenzene (b.p. 178–180 °C), or possibly N-methyl-2-pyrrolidinone (NMP, b.p. 202 °C). The very reason to choose these solvents (namely, high boiling point) is the same attribute that unfortunately can make them difficult to remove upon workup, especially as scale increases. When using these solvents, the bench chemist is generally relegated to extended evaporation times under reduced pressure or column chromatography in order to isolate the desired compound. However, under sealed-vessel conditions, nearly any solvent the bench chemist selects becomes a viable option, regardless of desired reaction temperature. Ethanol or acetonitrile can replace NMP, ethyl acetate or methyl ethyl ketone (MEK) can serve as an alternative to xylenes, and even dichloromethane (b.p. 40 °C at 1 atm) can be heated to 160 °C within the typical pressure limitations of most commercially available scientific microwave units.[32]

Perhaps the most interesting and underutilized solvent in organic chemistry is water. While there are certainly a number of reactions that do not tolerate the presence of water, for example, alkyl lithium reactions, there are plenty that not only tolerate the presence of water, but in some cases benefit from its addition to the solvent system, or even when it serves as the lone solvent. Furthermore, water is especially suitable for high-temperature organic reactions, and thus is great to pair with microwave heating.[33] The dielectric constant of water changes as a function of temperature and while it is characterized as a very polar solvent at room temperature, at elevated temperatures it becomes quite different. For example, water at 150 °C has a dielectric constant similar to DMSO at room temperature, at 175 °C the dielectric constant becomes similar to DMF at room temperature, water at 200 °C is similar to acetonitrile at 25 °C, and water heated to 300 °C has a dielectric constant on par with room-temperature acetone (Figure 1.6).[34] This attribute is quite useful and certainly can be taken advantage of: water is able to solvate reagents at high temperatures and then, upon cooling, the products become insoluble and facilitate isolation of the newly synthesized compounds.

Freedom in solvent selection not only allows the bench chemist greater flexibility in new methodology development and a reduction in workup time, but also

TABLE 1.2
Percentage of Published Journal Articles for Five Major Organic Chemistry Publications Utilizing Microwave Irradiation (Article Hits for Keyword Search "Microwave" in all Fields/Total Articles Published)

	2002	2003	2004	2005	2006	2007	2008	2009
JOC	36/1465	52/1587	70/1473	98/1633	108/1510	134/1552	118/1524	146/1508
OL	28/1213	43/1305	56/1388	70/1502	66/1565	83/1438	86/1426	101/1470
TET	39/1334	47/1366	62/1480	105/1480	126/1522	167/1569	179/1525	173/1444
TL	61/2503	91/2396	132/2385	188/2207	226/2184	230/2175	213/1981	252/2057
OPRD	7/197	7/198	9/201	12	14/204	19/211	19/202	12/239
Total MW	171	240	329	473	540	633	615	684
% MW	2.55	3.50	4.77	6.75	7.73	9.11	9.24	10.18

Note: JOC—Journal of Organic Chemistry; OL—Organic Letters; TET—Tetrahedron, TL—Tetrahedron Letters; OPRD—Organic Process Research and Development.

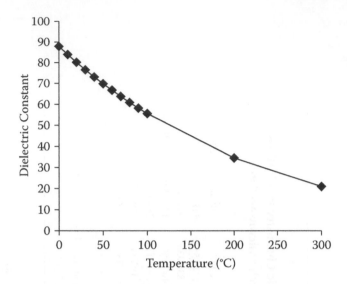

FIGURE 1.6   Plot of the dielectric constant of water as a function of temperature illustrating how water becomes less polar with heating. Points generated from data obtained from *CRC Handbook of Chemistry and Physics*.

affords the potential for "greener" chemistry to be developed. Certainly, ethanol or ethyl acetate could be considered "green" solvent choices as both can be derived from biological sources.[35] Additionally, solvents such as these represent less toxic alternatives and generally require less energy to remove at the end of a synthesis due to their lower boiling points. Indeed, both are found on Pfizer's "Green Solvent List," an in-house solvent selection guide that acts as a reminder to practicing chemists to select more environmentally benign solvents whenever possible.[36] Water can again feature highly. It is easy to extract from, inexpensive, nontoxic, nonflammable, and widely available. However, the true "greenness" of this solvent is very often overstated. Intuitively, something so ubiquitous as water and indeed so essential to life should automatically qualify it as "green," but overlooked is the fact that it cannot be incinerated after use and it takes a considerable amount of energy to distill water in order to purify it. Water purification at treatment plants, too, is a costly and energy-intensive endeavor. Thus, the pros of the use of water as a solvent are balanced by these cons and it is likely no more or less green than solvents such as ethanol, ethyl acetate, or methyl ethyl ketone. Indeed, when evaluated using a full complement of the most essential metrics, it has been reported that, "water is only a truly green solvent if it can be directly discharged to a biological effluent treatment plant."[37] Obviously, dissolved heavy metal catalysts, ionic phase transfer reagents, and trace amounts of newly synthesized organic compounds whose human or aquatic toxicology is likely unknown would render water unfit for this type of disposal. This said, water still represents an attractive solvent and likely a greener choice than most if appropriate predisposal treatments are employed. Furthermore, the ready access to elevated temperatures and the relatively efficient manner with which microwave irradiation heats water make microwave-assisted organic synthesis in water an attractive technique in new methodology development.

1.4 COMMERCIALLY AVAILABLE MICROWAVE EQUIPMENT

As microwaves come into the cavity of a domestic microwave unit, they will move around and bounce off the walls. As they do so, they generate pockets (called modes) of high energy and low energy as moving waves either reinforce or cancel out each other. Domestic microwave ovens are therefore called "multimode." While microwave ovens are useful for heating food, performing chemical reactions in them throws up a number of challenges. The nonuniform microwave field leads to issues with reproducibility; the make and model of the oven, as well as where exactly the reaction vessel is placed inside the cavity are all variables. In addition, domestic microwave ovens are not equipped with a temperature measurement device, nor are they built for safe containment of hot, flammable, organic solvents. To overcome these and other limitations, scientific microwave apparatus has been developed for use in preparative chemistry. As well as being built to withstand explosions of reaction vessels inside the microwave cavity, temperature and pressure monitoring has been introduced, as has the ability to stir reaction mixtures. In addition to larger multimode units, smaller monomode equipment is available. The cavity of a monomode unit is designed for the length of only one wave (mode). By placing the sample in the middle of the cavity, it can be irradiated constantly with microwave energy. It is possible to heat samples of as little as 0.2 mL very effectively. The upper volume limit of a monomode apparatus is determined by the size of the microwave cavity and is in the region of 100 mL. What follows is a brief overview of the commercially available scientific microwave apparatus. For more detailed descriptions, the reader is directed to the Web sites of the major microwave manufacturers: Anton Paar (http://www.anton-paar.com), Biotage (http://www.biotage.com), CEM Corporation (http://www.cem.com), and Milestone (http://www.milestonesrl.com).

1.4.1 SMALL-SCALE EQUIPMENT

The first purpose-built, monomode microwave reactor was introduced in the early 1990s by Prolabo, a French company. Today, the main manufacturers of small-scale, scientific microwave apparatus are Biotage (Initator), CEM (Discover), and Anton-Paar (Monowave). The units are capable of heating reactions, in sealed-vessel format, to temperatures up to 300 °C. Pressure limits of the glass reaction vessels used are 300 psi (~20 bar) for the Initator and Discover units and 435 psi (30 bar) for the Monowave. The CEM Discover can also be used in open-vessel format, accommodating round-bottom flasks of up to 125 mL capacity. The equipment utilizes 300 W (Discover), 400 W (Initiator), and 850 W (Monowave) magnetrons. The waveguide design and intellectual property of the units are quite different, though all are very effective at heating reactions. Temperature is generally monitored via an infrared detector located below or alongside the reaction vessel. Alternatively, a fiber-optic probe can be immersed in the reaction vessel by means of a thermowell. Pressure measurement is generally by means of a load cell. The Anton Paar Monowave and

(a)

(b)

FIGURE 1.7 Two of the small-scale dedicated microwave units for scientific applications. (a) Anton Paar Monowave (Reproduced with permission from Anton Paar.) (b) CEM Discover SF in open-vessel mode. The Biotage Initiator, equipped with an automated vessel handler, is shown in Figure 1.9. (Reproduced with permission from CEM Corp.)

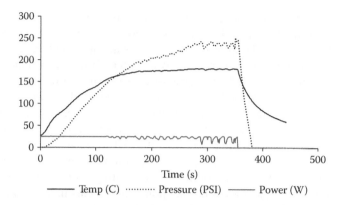

FIGURE 1.8 Typical plot of microwave heating run on the 1 mmol/2 mL scale. Power is modulated by the microwave to maintain the desired reaction temperature of 180 °C.

CEM Discover SF-Class are shown in Figure 1.7 (the Biotage Initiator is shown in Figure 1.9, equipped with an automated vessel handler).

CEM offers three sizes of glass reaction vessel that can be used in conjunction with their monomode microwave line under sealed-vessel conditions: a 10 mL tube (optimal working volume of 2–3 mL), a 35 mL tube (optimal working volume of up to 10 mL), and an 80 mL vessel (optimal working volume of up to 50 mL). Anton Paar and Biotage both also offer the 10 mL reaction vessel together with a 25 mL (Biotage) or 30 mL (Anton Paar) option, both with working volumes of up to 20 mL. Biotage also offers smaller vessels that can be used for volumes as little as 0.2 mL.

All three monomode units can be run either in a stand-alone format or interfaced to a PC. Pressure, temperature, applied microwave power input, and stirring can be monitored in real time (Figure 1.8). Additionally, software either on the unit or PC allows for on-the-fly changes to reaction parameters. Generally, a microwave power should be selected that affords a reasonable ramp time (1–5 °C/s) to the target temperature. Use of too high an initial power in the ramp stage leads to inaccuracies, as the acquisition hardware and software are unable to keep pace with the reaction dynamics. This situation often leads to temperature overshoot, sometimes by 10–20 °C or more. Obviously, this is not desirable, as actual reaction temperatures become nebulous, leading to irreproducible results. It is generally best to provide too little power initially and adjust the applied power, if necessary. Upon completion of the heating phase, all monomode units use pressurized lab air to cool the vessel and its contents back to ambient conditions in short order.

1.4.2   Larger-Scale Equipment

Microwave manufacturers have for the most part developed equipment to meet scale-up needs using three main approaches: (1) continuous-flow, (2) open-vessel batch, and (3) sealed-vessel batch.[38] This topic is discussed in detail in Chapter 4 together with numerous examples of approaches taken to scale up and the equipment available for performing reactions.

Continuous-flow equipment has been touted by microwave chemists as the most likely to be adapted for large-scale synthesis. There are a number of reasons to adopt continuous flow in the scale-up process. Reactions are actually "scaled out" rather than scaled up. Maximum throughput is only a matter of run time and the total number of units operating in parallel. Furthermore, catastrophic loss of a large quantity of valuable substrate can be avoided, as only a small portion of the reaction is subjected to reaction conditions at any given time. Drawbacks to this approach are that reaction mixtures are generally required to be homogeneous before, during, and until out of the microwave apparatus, limiting somewhat the real-world applicability.

An open-vessel approach to scale up has seen limited application, as this eliminates one of the greatest attributes of microwave heating, namely, the ability to heat reactions to well above the normal boiling points of solvents in a safe and effective manner. That said, when removal of a by-product such as water is key to the success of a synthetic transformation, or if a gas is evolved during the course of the reaction, a large-scale open-vessel batch microwave reactor may be an effective tool to carry out the procedure.

A sealed-vessel batch approach represents an attractive choice in the scale up of microwave-promoted reactions. The primary advantage is that most small-scale reactions are developed under sealed-vessel conditions in monomode equipment; thus, scale up is potentially straightforward with little or no reoptimization needed. Disadvantages to this approach are the limits of reaction volume that can be irradiated as well as the safety requirements when working with vessels under pressure.

1.4.3 PERIPHERALS

The field of microwave-assisted synthesis has progressed tremendously over the past two decades. A contributing factor has been the range of peripheral tools available for interface with both monomode and multimode scientific microwave apparatus, ranging from devices to increase throughput to those that allow for novel applications of microwave irradiation, thus opening new avenues for fundamental research.

1.4.3.1 Rotors for Multiple-Vessel Processing

The rotor approach allows the scientist to load from 4 to 192 individual sealed vessels, ranging in volume from 2 to 100 mL (1–70 mL practical working volume) onto a turntable that can then be placed in the microwave reactor.[39] Commercially available rotor-style equipment is offered by Anton-Parr, CEM, and Milestone. While a rotor approach could be applied to the development of synthetic methodology by loading each vessel with a different permutation or combination of catalysts, reagents, and solvents, there are a number of disadvantages to this approach, including nonuniformity of temperature from sample to sample and inefficient stirring of the reaction vessel contents. To address vessel-to-vessel heating homogeneity, Anton-Parr has developed silicon carbide plates in which multiple reaction vessels sit.[40] Since silicon carbide is much more microwave absorbent than the reaction contents, uniform heating is realized. Similarly, with regard to effective agitation of reaction vessels, Milestone has developed a unique unit (MultiSYNTH) that shakes the entire rotor

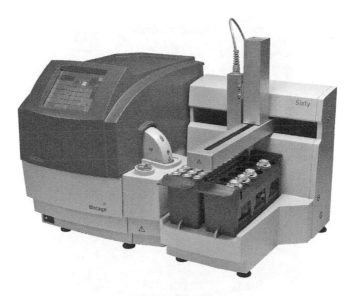

FIGURE 1.9   Biotage Initiator equipped with the sixty-position automated vessel handler. (Reproduced with permission from Biotage.)

while oscillating back and forth through the microwave field in an effort to improve overall mixing efficacy.[41]

1.4.3.2   Automated Sequential Vessel Processing

Both CEM and Biotage have developed automated vessel handlers that can be interfaced with their monomode units. These allow chemists to run a number of reactions sequentially in an automated manner. A robotic arm loads a reaction from a queue into the microwave cavity; the sample is heated for a predetermined time, cooled, returned to the holding area; and the cycle started again with the next sample. Biotage has two commercially available units (Figure 1.9), and CEM has an "Explorer" line with variants that can accommodate 12, 24, 48, 72, or 96 preloaded reaction vials.

1.4.3.3   Stop-Flow Processing

CEM has produced a stop–flow accessory for their Discover platform (Voyager). It interfaces with an 80 mL reaction vessel. The reaction mixture is pumped into and out of the vessel by a peristaltic pump, these functions, as well as running the reaction, being controlled with a PC. This gives a high degree of automation to the process. The reaction mixture can be introduced into the microwave vessel from two separate feed lines. After the reaction is complete, the reaction vessel can be vented to remove an overpressure and then the contents pumped into a collection vessel. If necessary, the reactor can then be cleaned with solvent before the next run. Biotage has developed an automated liquid and solid handler for its larger Advancer microwave unit. The unit allows up to four cycles to be performed sequentially, the reaction vessel being loaded, the contents heated, and then the product mixture ejected into a collection vessel. Application of the stop flow apparatus is discussed in detail in Chapter 5.

FIGURE 1.10   CEM Liberty peptide synthesizer.

1.4.3.4   Peptide Synthesis

Building on its stop flow accessory, CEM has developed an automated peptide synthesizer named the "Liberty" (Figure 1.10). The platform has integrated reservoirs for 20 amino acids as well as 12 additional positions for other reagents, solvents, catalysts, and additives. Additionally, integrated software allows the user to preprogram a multistep peptide sequencing protocol in a couple–deprotect–couple–deprotect strategy that combines the convenience of automation with the ease of microwave heating. Those utilizing the Liberty platform for peptide coupling often report higher yields of the desired peptide with fewer impurities or deletions than when traditional peptide coupling protocols are employed. A comprehensive discussion of this flourishing area is detailed in Chapter 9.

1.4.3.5   Simultaneous Cooling

Simultaneously cooling while heating reactions with the microwave may seem a bit counterintuitive and, indeed, the application of this technique has seen limited practical application in synthesis.[42] While both the CEM and Biotage monomode units use compressed air to cool reaction vessels upon completion, the CEM Discover allows for simultaneous cooling during sample irradiation, dubbed PowerMax by the manufacturers.[43] It seems to display the most utility when the potential exists for selective heating of catalysts or substrates in heterogeneous reaction mixtures. It has been successfully employed when coupling aryl chlorides with phenylboronic acid in a Suzuki reaction using Pd/C as the catalyst (Scheme 1.1).[44] The role of simultaneous cooling was to keep the bulk reaction temperature relatively low while maximizing the applied microwave power, thus heating the Pd/C to the elevated temperatures

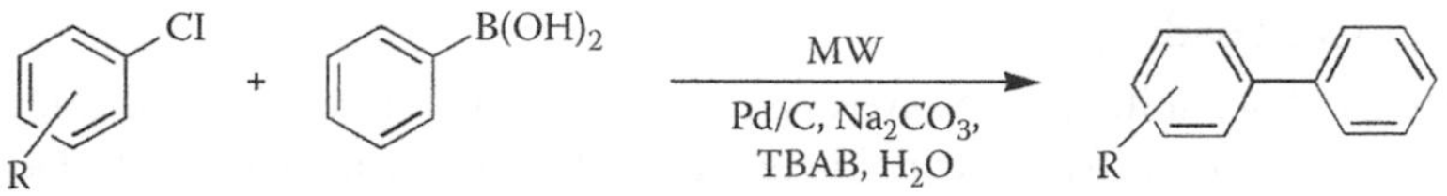

Microwave heating used in conjunction with simultaneous cooling

SCHEME 1.1

needed to affect the cross-coupling. As such, less decomposition of the aryl chloride substrates, and hence higher conversions to the desired cross-coupled products, was observed when using simultaneous cooling as opposed to the same reaction in the absence of cooling. Similarly, when using simultaneous cooling, higher product conversions were obtained in microwave-assisted Suzuki couplings using an encapsulated palladium catalyst, both in batch and flow mode.[45] The technique has also seen use in organocatalysis, where reactions are traditionally performed at ambient temperature or below for extended times. While shorter reaction times have been reported when using microwave irradiation in conjunction with simultaneous cooling,[46] for reasons discussed later, caution needs to be taken in interpreting results due to issues associated with accurate temperature measurement.

In addition to PowerMax cooling, CEM has also developed the CoolMate peripheral that can be fitted to the Discover line of microwave units. This uses a microwave transparent cryogenic fluid that is pumped around a specially designed microwave vessel while a reaction is being irradiated. The fluid can be cooled to temperatures as low as −60 °C, allowing maximum microwave irradiation while keeping the solution relatively cool. There have been only a handful of reports in the scientific literature of the use of the CoolMate or home-made variants in synthetic chemistry.[47] CoolMate has also been used as a tool to probe the existence, or otherwise, of microwave effects.[17,47]

In general, however, it is important to note that cooling reactions while simultaneously heating with microwave irradiation likely have extremely limited practical applications. Obviously, this makes sense as this technique would interfere with one of the greatest attributes of microwave irradiation, namely, rapid heating. Furthermore, it is likely that this technique will open the door to the possibility of data misinterpretation, especially with regard to accurate temperature monitoring, as a significant temperature gradient can be established across the reaction vessel. In addition, if using the built-in infrared sensor on the microwave unit for temperature measurement when applying simultaneous cooling, issues arise around accuracy.[25] Using external temperature measurement, it is possible to, in effect, "trick" the temperature sensor by blowing air over it; the temperature read being significantly lower than the actual bulk temperature in the reaction vessel. Therefore, difference in reactivity could be attributed simply to difference in reaction temperature. As a result, it is important to monitor the internal temperature of a reaction mixture, and this involves the use of a temperature measurement device located inside the reaction vessel.

1.4.3.6   Gas Loading

The design of scientific microwave apparatus to handle elevated temperatures and pressures under sealed-vessel conditions makes this platform ideal to interface with gas-loading systems. As such, there have been recent applications of gas-loading kits to both small- and large-scale commercial microwave apparatus to introduce an atmosphere of reactive gas to the microwave vessel for use as a reagent. In essence, these could be considered to be the modern equivalent of the Parr reactor. Hydrogen gas has been the most widely used gaseous reagent in microwave-assisted reactions, examples including hydrogenation of olefins,[48,49] hydrodechlorination,[50] debenzylation,[48,51] reductive aminations,[49] azide reductions,[48] and dearomitization of pyridine derivatives to form saturated piperidines.[48] In addition to hydrogen, molecular oxygen,[52] carbon monoxide,[53] 1-propyne,[54] and ethylene gas[55] have also seen use as reagents in reactions using microwave heating.

1.4.3.7   In Situ Reaction Monitoring

Reaction temperatures and pressures can be monitored accurately on a second-by-second basis, applied microwave power can be modulated with a precision of ±0.1 W, and most importantly, dedicated scientific microwave apparatus are built with safety in mind. In the event of an "unanticipated pressure release," the unit is designed to automatically cease irradiation and to contain the reaction contents within the cavity. An unavoidable threat to this peace of mind, however, is the inability to monitor the progress of a reaction visually. This leaves the chemist with questions such as, is the reaction stirring adequately? Has a precipitate formed? Has there been a color change? In addition, when optimizing a new protocol or monitoring the progress of reactions, the scientist is generally required to stop it, allow the reaction mixture to cool, and then use standard analysis techniques such as NMR spectroscopy or TLC. As a result, optimization of reaction conditions such as time and temperature is often a matter of trial and error.

The interface of apparatus capable of monitoring the progress of reactions has been an important step forward in terms of increasing the utility of scientific microwave equipment. Perhaps the most beneficial monitoring device in terms of cost to utility is the digital camera.[16] A camera is relatively cheap and allows scientists to monitor a reaction as they would if it were on their bench-top. Indeed, CEM now offers a commercially available digital camera that can be interfaced with any Discover S-Class or SF monomode microwave unit.

In the case of inorganic materials chemistry, neutron and X-ray scattering have been used for in situ reaction monitoring, as discussed in detail in Chapter 8.[57] For organic chemistry, near-IR has successfully been applied in one case.[58] Fluorescence spectroscopy has been used to monitor the emission of green fluorescent protein while under microwave irradiation.[16] Perhaps the most widely examined in situ monitoring technique has been the interface of Raman spectroscopy to microwave units. Recent technological advances such as computing power, diode laser technology, and charge-coupled device (CCD) technology have resulted in a significant drop in cost and increased utility of Raman spectroscopy as a monitoring tool.[59] Furthermore, as borosilicate glass is nearly Raman transparent, Raman spectroscopy lends itself to a "through-the-glass" spectroscopic technique, ideal for the sealed-vessel conditions used in microwave reactions. The first examples of microwave-mediated reactions monitored by Raman spectroscopy were in polymerization[60] and condensation reactions.[61] Recent applications of Raman-interfaced microwave synthesis[62] include qualitative monitoring ligand exchange reactions of organic ligands around a metal,[63] qualitative use of Raman spectroscopy to rapidly optimize organic reactions,[64] investigations to determine the effect of microwave power on rate and outcome of organic reactions,[65] investigations into "nonthermal" microwave effects,[27] and the quantitative use of Raman spectroscopy interfaced to a scientific microwave apparatus to determine physical parameters such as activation energies and/or activation enthalpies, orders of reactions, and rate constants for reactions.[66] Indeed, microwave heating proves a valuable tool for quantitative studies, offering reproducible noncontact heating as well as precise temperature monitoring and data recording.

1.5   CONCLUSIONS

Certainly, the application of microwave heating in preparative chemistry has come a long way since the early reports in the 1980s. Though this tool has seen the greatest application in the field of organic chemistry, other disciplines are catching on to the benefits and the added convenience that a dedicated scientific microwave apparatus can add to the laboratory setting. Accordingly, as these other disciplines tap into the potential of microwave irradiation, new applications will be found, more efficient syntheses will be realized, and the microwave will continue to demonstrate its worth in the laboratory. The remainder of this book will address exciting current applications of microwave irradiation to scientific endeavors.

REFERENCES

1. (a) Horikoshi, S.; Iida, S.; Kajitani, M.; Sato, S.; Serpone, N. *Org. Process Res. Dev.* 2008, *12*, 257–263. (b) Gedye, R. N.; Wei, J. B. *Can. J. Chem.* 1998, *76*, 525–532. (b) Möller, M.; Linn, H. *Key Eng. Mater.* 2004, *264–268*, 735–739. (c) Takizawa, H.; Uheda, K.; Endo, T. *J. Am. Ceram. Soc.* 2000, *83*, 2321–2323. (d) Malinger, A. K.; Ding, Y.-S.; Sithambaram, S.; Espinal, L.; Gomez, S.; Suib, S. L. *J. Catal.* 2006, *239*, 290–298.
2. Hollas, J. M. *Modern Spectroscopy*; Wiley: Chichester, 2004.
3. For an excellent discussion on dielectric heating fundamentals, see: Gabriel, C.; Gabriel, S.; Grant, E. H.; Halstead, B. S. J.; Mingos, D. M. P. *Chem. Soc. Rev.* 1998, *27*, 213–223.
4. Hayes, B. L. *Microwave Synthesis: Chemistry at the Speed of Light*; CEM Publishing: Matthews: NC, 2002.
5. For an excellent discussion of microwave absorptivity and theory from first principles, see: Craig, D. Q. M. *Dielectric Analysis of Pharmaceutical Systems*; Taylor & Francis: Bristol, PA, 1995.
6. Schmink, J. R.; Kormos, C. M.; Devine, W. G.; Leadbeater, N. E. *Org. Process Res. Dev.* *2010*.
7. Leadbeater, N. E.; Torenius, H. M.; Tye, H. *Comb. Chem. High Throughput Screen.* 2004, *7*, 511–528.
8. Kremsner, J. M.; Stadler, A.; Kappe, C. O. *J. Comb. Chem.* 2007, *9*, 285–291.

9. Nüchter, M.; Ondruschka, B.; Tied, A.; Lautenschläger, W.; Borowski, K. J. *American Genomic/Proteomic Technology Magazine* 2001, 34–37.

10. Schanche, J.-S. *Mol. Diversity* 2003, *7*, 293–300.

11. Saillard, R.; Poux, M.; Berlan, J. *Tetrahedron* 1995, *51*, 4033–4042.

12. Baghurst, D. R.; Mingos, D. M. P. *J. Chem. Soc., Chem. Commun.* 1992, 674–677.

13. Lienhard, J. H. IV; Lienhard, J. H. V. *A Heat Transfer Textbook,* 3rd ed.; Phlogiston Press: Cambridge MA, 2008, 457–463.

14. Chemat, F.; Esveld, E. *Chem. Eng. Technol.* 2001, *24*, 735–744.

15. (a) Washington, A. L.; Strouse, G. F. *Chem. Mat.* 2009, *21*, 2770–2776. (b) Washington, A. L.; Strouse, G. F. *J. Am. Chem. Soc.* 2008, *130*, 8916–8922.

16. Copty, A.; Sakran, F; Popov, O.; Ziblat, R.; Danieli, T.; Golosovsky, M.; Davidov, D. *Synth. Met.* 2005, *155*, 422–425.

17. Young, D. D.; Nichols, J.; Kelly, R. M.; Deiters, A. *J. Am. Chem. Soc.*, 2008, *130*, 10048–10049.

18. Leadbeater, N. E.; Stencel, L. M.; Wood, E. C. *Org. Biomol. Chem.* 2007, *5*, 1052–1055.

19. (a) Perreux, L.; Loupy, A. in *Microwaves in Organic Synthesis,* 2nd ed.; Loupy, A. Ed.; Wiley-VCH: Weinheim, 2006, Ch 4, pp 134–218. (b) De La Hoz, A.;Diaz-Ortiz, A.; Moreno, A. *Chem. Soc. Rev.* 2005, *34*, 164–178.

20. For a discussion of temperature measurement in microwave chemistry, see: Nüchter, M.; Ondruschka, B.; Weiss, D.; Beckert, R.; Bonrath, W.; Gum, A. *Chem. Eng. Technol.* 2005, *28*, 871–881.

21. (a) Obermayer, D.; Kappe, C. O, *Org. Biomol. Chem.* 2010, *8*, 114–121. (b) Herrero, M. A.; Kremsner, J. M.; Kappe, C. O. *J. Org. Chem.* 2008, *73*, 36–47.

22. (a) Kremsner, J. M.; Kappe, C. O. *J. Org. Chem.* 2006, *71*, 4651–4658. (b) Razzaq, T.; Kremsner, J. M.; Kappe, C. O. *J. Org. Chem.* 2008, *73*, 6321–6329.

23. Obermayer, D.; Gutmann, B.; Kappe, C. O. *Angew. Chem. Int. Ed.* 2009, *48*, 8321–8324.

24. Leadbeater, N. E.; Pillsbury, S. J.; Shanahan, E.; Williams, V. A. *Tetrahedron* 2005, *61*, 3565–3585.

25. Hosseini, M.; Stiasni, N.; Barbieri, V.; Kappe, C. O. *J. Org. Chem.* 2007, *72*, 1417–1424.

26. Schmink, J. R.; Leadbeater, N. E. *Org. Biomol. Chem.* 2009, *7*, 3842–3846.

27. A number of relevant books reviewing microwave assisted organic synthesis have been published, including: (a) Loupy, A. Ed., *Microwaves in Organic Synthesis,* 2nd ed.; Wiley-VCH: Weinheim, 2006. (b) Kappe, C. O.; Stadler, A. *Microwaves in Organic and Medicinal Chemistry*; Wiley-VCH: Weinheim, 2005. (c) Lidström, P.; Tierney, J. P., Eds., *Microwave-Assisted Organic Synthesis*; Blackwell: Oxford, 2005. (d) van der Eycken, E.; Kappe, C. O. ,Eds. *Microwave-Assisted Synthesis of Heterocycles*; Springer: New York, 2006. (e) Larhed, M., Olofsson, K., Eds. *Topics in Current Chemistry Vol 266: Microwave Methods in Organic Chemistry*; Springer: Berlin, 2006.

28. For recent reviews highlighting the use of microwave heating in organic synthesis, see: (a) Kappe, C. O.; Dallinger, D. *Mol. Divers.* 2009, *13*, 71–193. (b) Caddick, S.; Fitzmaurice R. *Tetrahedron* 2009, *65*, 3325–3355. (c) Kappe, C. O. *Chem. Soc. Rev.* 2008, *37*, 1127–1139. (d) Kappe, C. O. *Angew. Chem. Int. Ed.* 2004, *43*, 6250–6284.

29. Performing a simple, fractional, or vacuum distillation from a scientific microwave allows an exquisite level of control not easy to duplicate in an oil or sand bath. Unpublished results by the authors.

30. Gedye, R.; Smith, K.; Westaway, H. *Tetrahedron Lett.* 1986, *27*, 279–282.

31. Giguere, R. J.; Bray, T. L.; Duncan, S. M.; Majetich, G. *Tetrahedron Lett.* 1986, *27*, 4945–4948.

32. Goodman, J. M.; Kirby, P. D.; Haustedt, L. O. *Tetrahedron Lett.* 2000, *41*, 9879–9882. (The embedded applet described in this article can be found at: http://www-jmg.ch.cam. ac.uk/tools/magnus/boil.html)

33. For recent reviews of water in microwave-assisted synthesis, see: (a) Strauss, C. A. *Aust. J. Chem.* 2009, *62*, 3–15. (b) Polshettiwar, V.; Varma, R. S. *Chem. Soc. Rev.* 2008, *37*, 1546–1557. (c) Dallinger, D.; Kappe, C. O. *Chem. Rev.* 2007, *107*, 2563–2591.

34. Values for organic solvents obtained from: Anslyn, E. V.; Dougherty, D. A. *Modern Physical Organic Chemistry;* University Science Books: Sausalito, CA, 2006.

35. That said, bio-based ethanol and ethyl acetate represent a minute fraction of the volume used. The majority of ethanol utilized in the lab has been synthesized from the hydrolysis of ethylene, which has been distilled from crude oil. Similarly with ethyl acetate, the acetic acid portion was likely produced by the Monsanto (Rhodium) or Cativa (Irridium) acetic acid processes whose feedstock begins with methanol, again originating from crude oil.

36. Cue, B. W. Oral presentation at "2009 American Chemical Society School for Green Chemistry and Sustainable Energy," Golden, CO, July 24 2009.

37. (a) Blackmond, D. G.; Armstrong, A.; Coombe, V.; Wells, A. *Angew. Chem. Int. Ed.* 2007, *46*, 3798–3800. (b) See also: the Water Framework Directive issued by the European Commission. http://www.euwfd.com.

38. For a recent evaluation of a wide range of equipment, see: (a) Moseley, J. D.; Lenden, P.; Lockwood, M.; Ruda, K.; Sherlock, J.-P.; Thomson, A. D.; Gilday, J. P. *Org. Process Res. Dev.* 2008, *12*, 30–40. (b) Bowman, M. D.; Holcomb, J. L.; Kormos, C. M.; Leadbeater, N. E.; Williams, V. A. *Org. Process Res. Dev.* 2008, *12*, 41–57.

39. (a) Kappe, C. O.; Matloobi, M. *Comb. Chem. High Throughput Screening* 2007, *10*, 735–750. (b) Dai, W. M.; Shi, J. Y. *Comb. Chem. High Throughput Screening* 2007, *10*, 837–856. (c) Nüchter, M.; Ondruschka, B. *Mol. Divers.* 2003, *7*, 253–264.

40. (a) Avery, K. B.; Devine, W. G.; Kormos, C. M.; Leadbeater, N. E. *Tetrahedron Lett.* 2009, *50*, 2851–2853. (b) Kremsner, J. M.; Stadler, A.; Kappe, C. O. *J. Comb. Chem.* 2007, *9*, 285–291. (c) Pisani, L.; Prokopcová, H.; Kremsner, J. M.; Kappe, C. O. *J. Comb. Chem.* 2007, *9*, 415–421.

41. Schmink, J. R.; Leadbeater, N. E. *Tetrahedron* 2007, *63*, 6764–6773.

42. (a) Humphrey, C. E.; Easson, M. A. M.; Tierney, J. P.; Turner, N. J. *Org. Lett.* 2003, *5*, 849–852 (b) Chen, J. J.; Deshpande, S. V. *Tetrahedron Lett.* 2003, *44*, 8873–8876. (c) Mathew, F.; Jayaprakash, K. N.; Fraser-Reid, B.; Mathew, J.; Scicinski, J. *Tetrahedron Lett.* 2003, *44*, 9051–9054. (d) Jachuck, R. J. J.; Selvaraj, D. K.; Varma, R. S. *Green Chem.* 2006, *8*, 29–33.

43. Hayes, B. L.; Collins, M. J. Jr., World Patent WO04002617 2004.

44. Arvela, R. K.; Leadbeater, N. E. *Org. Lett.* 2005, *7*, 2101–2104.

45. Baxendale, I. R.; Griffiths-Jones, C. M.; Ley, S. V.; Tranmer, G. T. *Chem. Eur. J.* 2006, *12*, 4407–4416.

46. (a) Rodriguez, B.; Bolm, C. *J. Org. Chem.* 2006, *71*, 2888–2891. (b) Mosse, S.; Alexakis, A. *Org. Lett.* 2006, *8*, 3577–3580.

47. (a) Lange, S. M.; Torok, B. *Catal. Lett.* 2009, *131*, 432–439. (b) Horikoshi, S.; Tsuzuki, J.; Kajitani, M.; Abe, M.; Serpone, N. *New J. Chem* 2008, *32*, 2257–2262. Horikoshi, S.; Ohmori, N.; Kajitani, M.; Serpone, N. *J. Photochem. Photobiol. A.* 2007, *189*, 374–379. (c) Singh, B. K.; Appukkuttan, P.; Claerhout, S.; Parmar, V. S.; Van der Eycken, E. *Org. Lett.* 2006, *8*, 1863–1866.

48. (a) Heller, E.; Lautenschläger, W. Holzgrabe, U. *Tetrahedron Lett.* 2005, *46*, 1247–1249. (b) Vanier, G. S. *Synlett* 2007, 131–135.

49. (a) Heller, E.; Lautenschläger, W. Holzgrabe, U. *Tetrahedron Lett.* 2005, *46*, 1247–1249. (b) Vanier, G. S. *Synlett* 2007, 131–135.

50. (a) Wada, Y.; Yin, H; Kitamura, T.; Yanagida, S. *Chem. Lett.* 2000, 632–633. (b) Pillai, U. R.; Sahle-Demessie, E.; Varma, R. S. *Green Chem.* 2004, *6*, 295–298.
51. Kennedy, D. P.; Kormos, C. M.; Burdette, S. C. *J. Am. Chem. Soc.* 2009, *131*, 8578–8586.
52. (a) Kormos, C.M; Hull, R.M.; Leadbeater, N.E., *Aust. J. Chem.*, 2009, *62*, 51–57. (b) Andappan, M. M. S.; Nilsson, P.; von Schenck, H.; Larhead, M. *J. Org. Chem.* 2004, *69*, 5212–5218.
53. (a) Kormos, C. M.; Leadbeater N. E., *Synlett* 2006, 1663–1666. (b) Kormos, C. M.; Leadbeater N. E. *Org. Biomol. Chem.* 2007, 65–68. (c) Kormos, C. M.; Leadbeater, N. E. *Synlett,* 2007, 2006–2010. (d) Leadbeater, N. E.; Shoemaker, K. M., *Organometallics* 2008, *27*, 1254–1258. (e) Iannelli, M.; Bergamelli, F.; Kormos, C.M.; Paravisi, S; Leadbeater, N.E. *Org. Process Res. Dev.* 2009, *13*, 634–637.
54. Miljanić, O. Š.; Vollhardt, K. P.; Whitener, G. D. *Synlett* 2003, 29–34.
55. (a) Kaval, N. Dehaen, W.; Kappe, C. O.; Van der Eycken, E. *Org. Biomol. Chem.* 2004, *2*, 154–156. (b) Kormos, C. M., Leadbeater, N. E. *J. Org. Chem.* 2008, 3854–3858.
56. (a) Leadbeater, N. E.; Shoemaker, K. M. *Organometallics* 2008, *27*, 1254–1258. (b) Bowman, M. D.; Leadbeater, N. E.; Barnard, T. M., *Tetrahedron Lett.* 2008, *49*, 195–198.
57. Tompsett, G. A.; Conner, W. C.; Yngvesson, K. S. *Chem Phys Chem* 2006, *7*, 296–319.
58. Hocdé, S.; Pledel-Boussard, C.; Le Coq, D.; Fonteneau, G.; Lucas, J. *Proc. SPIE-Int. Soc. Opt. Eng.* 1999, *3849*, 50–59.
59. For a comprehensive overview of Raman spectroscopy fundamentals, theory, and applications, see: (a) McCreery, R. L. in *Chemical Analysis Vol 157*, Winefordner J. D, Ed.; Wiley: New York, 2000. (b) Lewis, I. R.; Edwards, H. G. M. *Handbook of Raman Spectroscopy, From the Research Laboratory to the Process Line;* Marcel Dekker: New York, 2001. (c) Pivonka, D. E.; Chalmers, J. M.; Griffiths, P. R. Eds. *Applications of Vibrational Spectroscopy in Pharmaceutical Research and Development;* Wiley: New York, 2007. (d) Dollish, F. R.; Fateley, W. G.; Bentley, F. F. *Characteristic Raman Frequencies of Organic Compounds;* Wiley: New York, 1974.
60. Stellman, C. M.; Aust, J. F.; Myrick, M. L. *Appl. Spectrosc.* 1995, *3*, 392–394.
61. Pivonka, D. E.; Empfield, J. R. *Appl. Spectrosc.* 2004, *58*, 41–46.
62. For a detailed overview of equipment interface, see: Leadbeater, N. E.; Schmink, J. R. *Nat. Prot.* 2008, *3*, 1–7.
63. Barnard, T. M.; Leadbeater, N. E. *Chem. Commun.* 2006, 3615–3616.
64. Leadbeater, N. E.; Smith, R. J. *Org. Biomol. Chem.* 2007, 2770–2774.
65. Leadbeater, N. E.; Smith, R. J. *Org. Lett.* 2006, *8*, 4589–4591.
66 (a) Schmink, J. R.; Holcomb, J. L.; Leadbeater, N. E. *Chem. Eur. J.* 2008, *14*, 9943–9950. (b) Schmink, J. R.; Holcomb, J. L.; Leadbeater, N. E. *Org. Lett.* 2009, *11*, 365–368.

2 Microwave Heating as a Tool for Organic Synthesis

Robert A. Stockland, Jr.

CONTENTS

2.1 INTRODUCTION

The design and development of clean organic transformations remains a current and challenging goal for the synthetic community. To this end, minimization of decomposition and secondary reactions while simplifying the purification of target compounds has been the subject of intense research. Microwave-assisted reactions have become very popular in the synthetic community due to the drastically reduced reaction times and minimization of secondary reactions.[1] The operational simplicity of microwave apparatus may be one of its greatest attributes as researchers can run dozens of screening reactions in a single day. Another advantage of microwave-assisted chemistry is the precise control over several reaction parameters such as microwave power, time, and temperature. Thus, finding the optimal conditions for a specific reaction has almost been trivialized.

In the early years of developing microwave-assisted organic reactions, the observed differences in reaction times between conventional and microwave heating gave rise to intense debate concerning the existence of nonthermal microwave effects. If these nonthermal microwave effects could be proved and predicted, they would provide the synthetic chemist another way to manipulate the outcome of a specific reaction. To investigate the existence of nonthermal effects, a host of detailed studies have been carried out on a range of transformations.[2–4] In all of these careful studies, no nonthermal microwave effects were observed; however, only a small fraction of reactions have been screened, and it would be premature to conclude from the available data that there will never be an organic transformation that will exhibit/benefit from any microwave effects. As for studies where differences were observed between conventional and microwave heating, the differences have been attributed to several factors, including inaccurate temperature measurement and, in some cases, the inability of conventional heating to reproduce precisely the thermal environment generated in the microwave-assisted reaction.[5] This chapter will not cover the general principles of microwave heating or the current discussions concerning nonthermal microwave effects, but will focus on providing researchers with the fundamentals of how microwave heating can be used to facilitate the clean and sustainable synthesis of organic compounds.

Developing clean organic reactions will be a critical area of research in the coming years due to the growing scarcity of resources and the rising cost of energy, and the viability of running a reaction that results in poor conversion or preparing a compound that is challenging to purify will continue to diminish. Clean organic chemistry can mean different things to different chemists. From the standpoint of the synthetic chemist, a high-yielding transformation with minimal decomposition or secondary products is highly desirable. From a sustainable point of view, reactions that generate high yields of the desired products, but also minimize the use of toxic solvents, catalysts, or additives are highly valued. Arguably, solvent use remains an area where significant progress can be made. In a number of cases, performing a reaction solvent free is possible and also quite attractive; however, there are many cases where the solvent plays an important role in mediating the reaction, and its removal would be detrimental. Not all solvents are created equal, and several of the following sections will describe the benefits and drawbacks in using different solvents and solvent combinations. Additionally, a scientist charged with the preparation of enantiopure materials will have a different view of clean organic chemistry since they require a reaction to be not only high yielding but also selective.

2.2   SOLVENT-FREE REACTIONS

For a synthetic organic chemist, an attractive reaction would be a transformation that generates the desired products from the reagents without the addition of solvents or additives. Ideally, the desired products are of sufficient purity for the envisioned application without extensive purification by crystallization or chromatography. Microwave-assisted chemistry is well suited for this application since the reaction conditions can be readily manipulated through precise control of reaction parameters. Although few reactions can be included in this category, they do offer powerful ways to functionalize small and large molecules. Along these lines, a number of solvent-free microwave-assisted reactions have been reported using solid supports such as silica and various clays as promoters and supports.[6] This section will, however, focus on representative examples of organic and metal-catalyzed transformations performed solvent free without use of a solid support.

2.2.1   HYDROELEMENTATION REACTIONS

The addition of E–H bonds to unsaturated substrates is one of the most useful ways to prepare materials containing carbon-heteroelement bonds. This class of reactions also represents a very efficient use of reagents since every atom in the reagents is incorporated into the products. Although atom efficient in terms of substrate, a number of these processes require the use of additives or catalysts. This section will focus on how those additives and solvents can be minimized or eliminated through microwave-assisted hydroelementation reactions.

The addition of labile P-H bonds to unsaturated substrates will serve as an example of a hydroelementation reaction. This chemistry has been used to prepare a wide range of materials for use in catalysis, medicine, and organic synthesis.[7,8] Traditionally, these reactions required a strong base such as a metal alkoxide to promote the process;[9] however, a recent report described the use of a monomode microwave reactor for the clean addition of secondary phosphine oxides and cyclic hydrogen phosphinates to Michael acceptors without the addition of solvent or additives (Scheme 2.1).[10] The reaction was operationally straightforward and simply adding the reagents to the flask (in air) followed by evacuation and heating afforded the addition products. As is the case with many "neat" reactions, the reagents needed to be thoroughly mixed prior to heating. Crude yields from these reactions were typically excellent (≥90%). Of note is that the initial microwave power setting was a critical parameter in this chemistry, optimum results being obtained using 25–50 W. If the initial power level used was higher in order to reach the desired temperature more rapidly, lower yields of the hydrophosphinylated product were isolated due to decomposition.

In related work, the addition of diphenylphosphine oxide to alkynes was investigated under solvent-free conditions.[11] Similar to the chemistry involving alkenes, the alkynes needed to contain an electron-withdrawing group for a successful addition

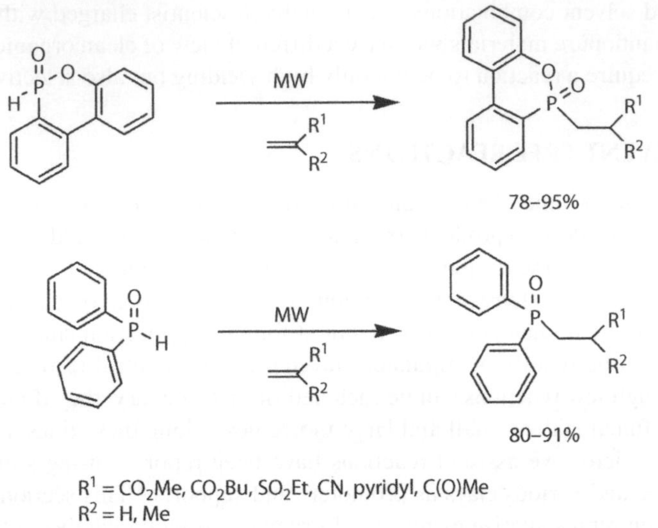

78–95%

80–91%

R^1 = CO$_2$Me, CO$_2$Bu, SO$_2$Et, CN, pyridyl, C(O)Me
R^2 = H, Me

SCHEME 2.1

reaction. Activated alkynes such as dimethylacetylene dicarboxylate (DMAD) were quite reactive under these conditions and cleanly underwent the hydrophosphinylation reaction at 125 °C. While it would have been attractive if the reaction stopped at a single addition, this product was never observed. Instead, the reaction always proceeded to generate a 1:1 mixture of the double-addition product and unreacted starting material. Presumably, the incorporation of the additional electron-withdrawing group (phosphine oxide) increased the reactivity of the single addition product relative to the parent alkyne. Performing the reaction with an additional equivalent of the secondary phosphine resulted in high yields of the double-addition products.

While the addition of secondary phosphine oxides and hydrogen phosphinates to activated alkenes and alkynes proceeded without the addition of a solvent or catalyst, unactivated alkynes were unreactive. To overcome this problem, Rh(I) complexes were employed as catalysts.[11] Using common sources of Rh(I), the addition of diphenylphosphine oxide to unactivated alkynes cleanly generated alkenylphosphine oxides under solvent-free conditions. Similar to the hydrophosphinylation of activated alkenes, the reaction is operationally simple and does not require the use of a glove box. This addition reaction was also used for the functionalization of ethynyl steroids (Scheme 2.2).[12] In addition to the solvent-free conditions, the addition reactions involving secondary phosphine oxides were also successful in a range of solvents, including pure water, tetrahydrofuran, and ethyl lactate. However, hydrogen phosphinates were not as tolerant of the reaction conditions, and high yields were obtained only in tetrahydrofuran.

2.2.2 AMINATION OF HETEROCYCLES

The α-amination of heterocycles is an efficient way to prepare a range of materials that are valuable components for the preparation of biologically active materials.

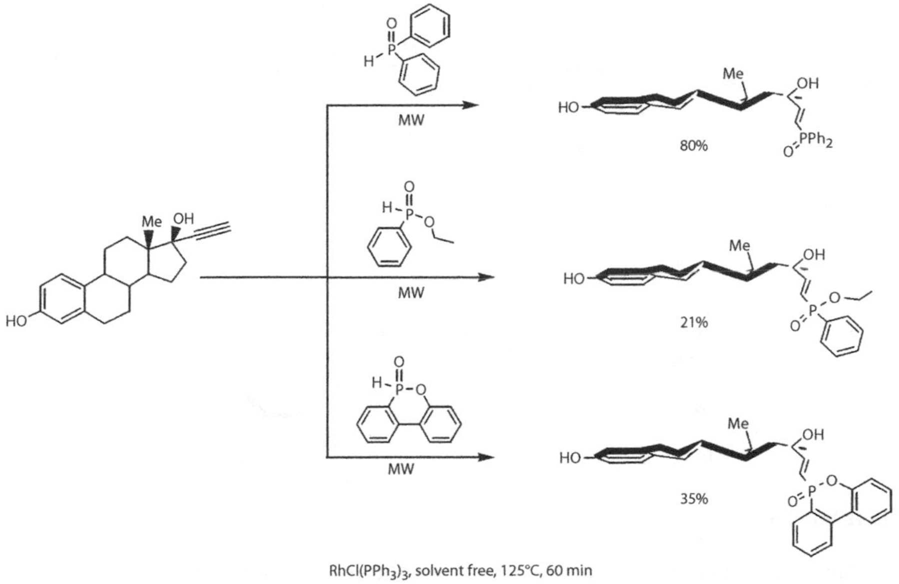

RhCl(PPh$_3$)$_3$, solvent free, 125°C, 60 min

SCHEME 2.2

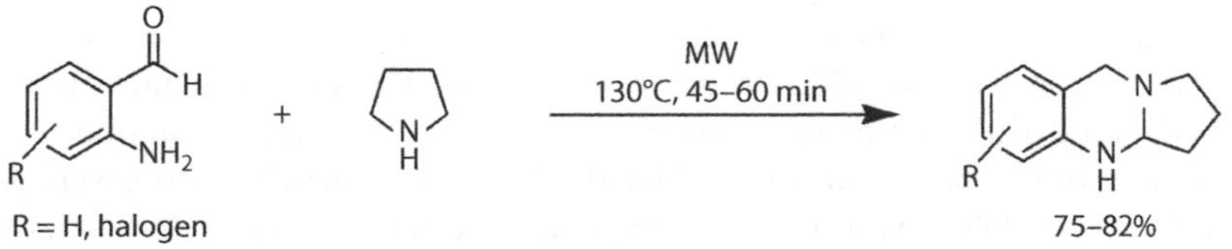

R = H, halogen

MW
130°C, 45–60 min

75–82%

SCHEME 2.3

The synthesis of ring-fused aminals under solvent-free conditions has recently been reported (Scheme 2.3).[13] Simply heating the neat reagents at 130 °C for 45–60 min generated the desired products in moderate-to-high yields. Although a 1:4 ratio of aldehyde to amine was needed for a successful reaction, this one-pot process represents an efficient way to prepare ring-fused aminals.

2.2.3 PYRIMIDINE AND NUCLEOSIDE SYNTHESIS

Over the last few decades, a wide range of methodologies have been developed for the preparation and manipulation of pyrimidine derivatives. Recently, a microwave-assisted approach to the synthesis of pyrazolo[1,5-a]pyrimidines was reported (Scheme 2.4).[14] Equimolar amounts of the pyrazole and chromen-4-one were heated for 2 min without added solvent. This afforded pyrimidine derivatives that were fairly pure directly from the reaction vessel. Recrystallization from ethanol gave pure materials in excellent yield (88–93%). Since the chemistry was performed neat, the only solvent used was a small amount of ethanol in the purification step, this representing a very clean and straightforward way to generate these valuable materials.

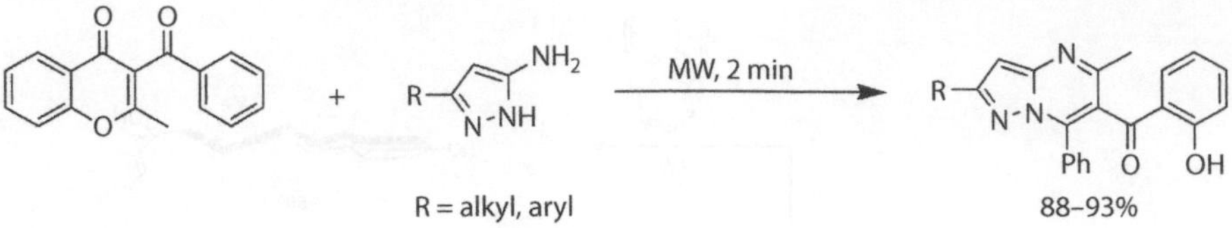

R = alkyl, aryl   88–93%

SCHEME 2.4

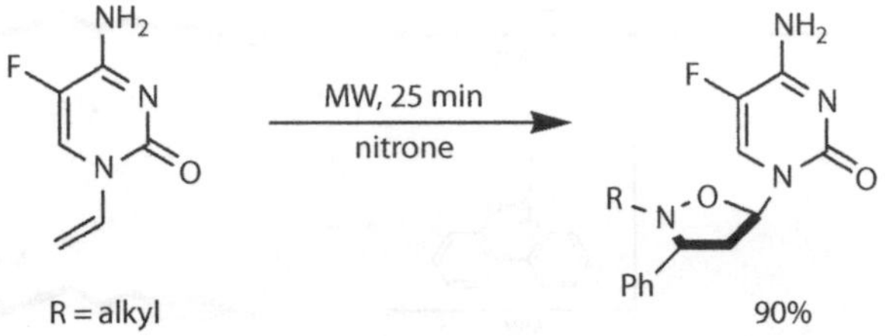

R = alkyl   90%

SCHEME 2.5

The development of new methodology for the preparation of nucleoside derivatives continues to be an area of intense research since many of these derivatives exhibit significant activity toward a range of autoimmune illnesses.[15] To this end, the microwave-assisted 1,3-dipolar cycloaddition between nitrones and vinyl nucleobases has been investigated (Scheme 2.5).[16] The reaction proceeds in a domestic microwave oven with no added solvent or catalyst. The experimental setup was straightforward since the reagents were simply mixed by grinding and heated for the requisite time. The overall isolated yields of the functionalized bases were good to excellent (50–90%), with moderate-to-good selectivity for the endo-isomer. For many synthetic manipulations of nucleobases, extensive protection/deprotection is needed to prevent secondary reactions and decomposition from occurring. The authors found that the vinyl nucleobases could be successfully used without protection; thus, the overall preparation of these materials was simplified.

2.2.4 SONOGASHIRA COUPLING

The metal-catalyzed coupling of terminal acetylenes with aryl halides remains one of the most powerful ways to make aryl-acetylene linkages.[17] While most protocols require the use of an organic solvent for a successful reaction, a recent report outlined a solvent-free palladium-catalyzed protocol using microwave heating (Scheme 2.6).[18] The authors were able to remove the traditional copper cocatalyst from the reaction

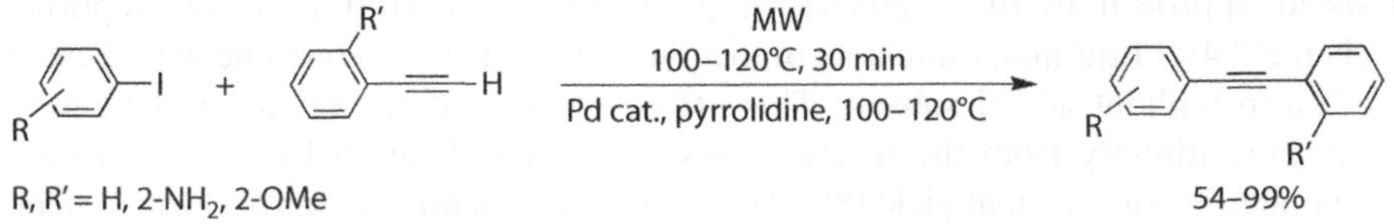

R, R' = H, 2-NH$_2$, 2-OMe   54–99%

SCHEME 2.6

without any effects on product yields. Using a commercially available supported palladium catalyst, the coupling of aryl iodides and substituted phenylacetylenes proceeded in good-to-excellent yields (59–99%) after 30 min at 100–120 °C. Another attractive aspect of this chemistry was the low catalyst loading (0.01 mol%).

2.2.5 ARYLATION OF HETEROCYCLES

The copper-catalyzed arylation of pyrazole and imidazole-based compounds has recently been reported under solvent-free conditions using microwave heating (Scheme 2.7).[19] Using copper iodide as the catalyst and amino acids as the supporting ligands, a range of aryl bromides were coupled with imidazole or pyrazole. Para- and meta-substituted substrates afforded good yields of the coupling product, while ortho substitution of the aryl bromides resulted in lower yields of the coupling product. While both heterocyclic substrates provided moderate-to-good yields of the targets, the authors found that reactions involving imidazole generally afforded higher yields of the arylated compounds.

2.2.6 PROPARGYLAMINE SYNTHESIS

The coupling between an aldehyde, terminal acetylene, and amine is a useful way to prepare propargylamines.[20] A microwave-assisted solvent-free version of this reaction was recently used for the preparation of aza(–)-Steganacin derivatives (Scheme 2.8).[21] (–)-Steganacin analogs have been the subject of significant research due to their potent biological activity.[22,23] In addition to the solvent-free studies, the scope of the coupling reaction was also investigated using a variety of solvents: dioxane/ionic liquid, THF, and DMF. Analyzing the results from the screening reactions revealed that the solvent-free conditions were superior and generated excellent yields (89%) of the reactive intermediates needed for the preparation of the Steganacin derivatives.

R = alkyl, alkoxy, nitro   51–99%

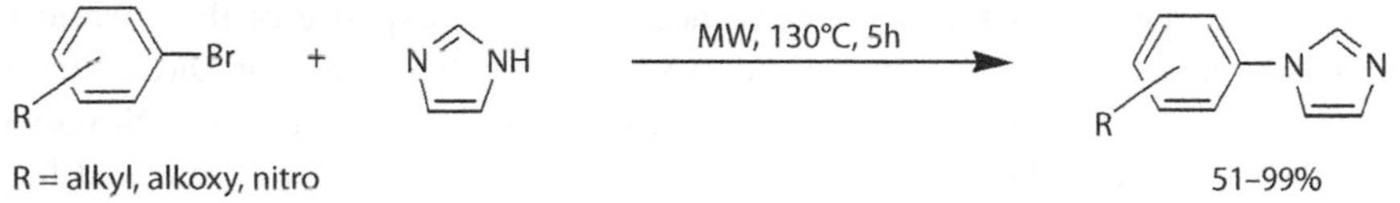

SCHEME 2.7

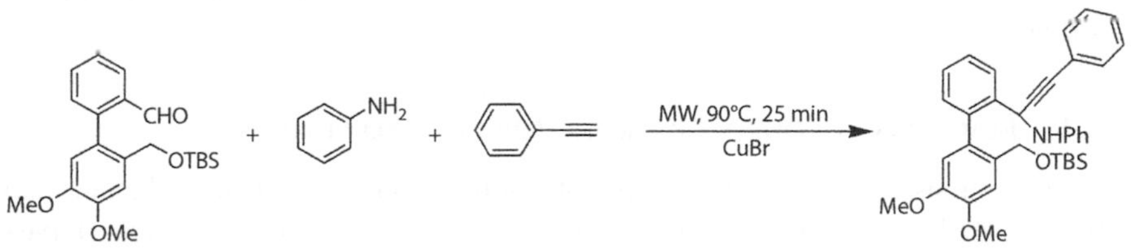

SCHEME 2.8

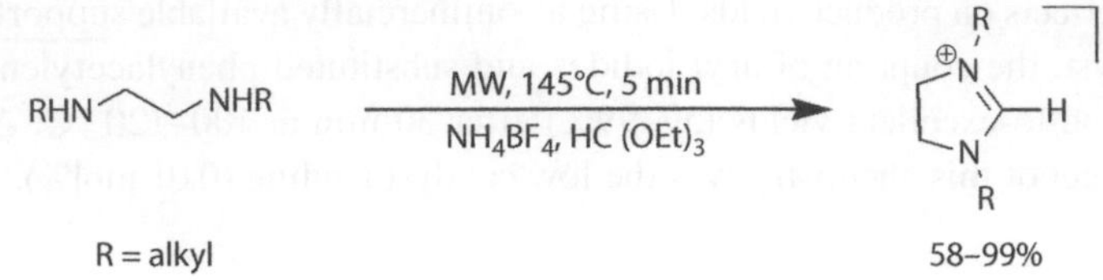

RHN⌒NHR $\xrightarrow[\text{NH}_4\text{BF}_4,\ \text{HC (OEt)}_3]{\text{MW, 145°C, 5 min}}$ [cyclic amidinium structure] $BF_4^{\ominus}$

R = alkyl   58–99%

SCHEME 2.9

2.2.7 CYCLIC AMIDINIUM SALTS

A remarkably clean and operationally simple preparation of cyclic amidinium salts using microwave heating has been reported (Scheme 2.9).[24] Heating a mixture containing a dialkyldiamine, a triethyl orthoester, and an ammonium salt such as ammonium tetrafluoroborate for 5 min at 145 °C resulted in high conversions to the cyclic amidinium salts. The products precipitated during cooling or upon addition of diethyl ether. Thus, isolation was easily carried out by simple filtration and drying. Isolated yields were moderate to excellent (58–99%), with smaller alkyl substituents on nitrogen generally affording higher yields of the amidinium salts.

2.3 REACTIONS USING ORGANIC SOLVENTS

While performing a reaction solvent free is perhaps a desirable option from a clean chemistry perspective, there are many reactions that simply cannot be performed using these conditions. If a solvent is needed, there are many factors to consider during the selection process, including assessment of toxicity, availability, and flammability. In addition, in terms of generating organic compounds cleanly, one of the most important considerations is the removal of the solvent from the product at the end of the reaction. This is typically accomplished through evaporation under reduced pressure or precipitation/crystallization of the product from the solution. One of the most important characteristics to consider when selecting a solvent for a reaction using microwave heating is the response of the solvent to microwave irradiation.[25] In general, polar solvents such as DMF or DMSO interact well with microwave energy and thus can be heated very rapidly. Nonpolar solvents such as toluene and hexane are almost microwave transparent and heat very slowly via indirect heating from the reaction vessel walls. This section will acquaint the reader with the fundamentals required to approach the design of a microwave-promoted reaction using a solvent through a number of representative examples.

2.3.1 REACTIONS IN POORLY ABSORBING ORGANIC SOLVENTS

Microwave-assisted reactions using nonpolar solvents such as toluene or hexane can still be heated to high temperatures using additives such as ionic liquids or passive heating elements. Ionic liquids heat extremely rapidly upon microwave irradiation.[26,27] For example, heating 2 mL of pure hexane at 200 W in a monomode microwave unit resulted in a temperature rise of only 20 °C after 10 s. Adding a small amount of the ionic liquid (10–50 mg) resulted in a temperature of 217 °C under the same conditions. In addition to ionic liquids, disks made of silicon carbide can act as passive heating elements.[28] In a control study, heating neat hexane (150 W, 77 s) resulted in a final temperature of 42 °C. Adding a silicon carbide disk to the hexane solution and performing the same study resulted in a temperature of 158 °C. Both approaches are attractive from a clean chemistry perspective, studies of several reactions using solvent/ionic liquid and solvent/SiC protocols showing no contamination of the organic phase following the reaction.

The use of small amounts of ionic liquids to heat weakly absorbing solvents has been studied in a number of reactions, including *aza*-Michael additions and Diels–Alder cycloadditions.[27] Using methyl acrylate and imidazole as prototypical substrates along with triethylamine as the base, very small amounts of the ionic liquid (~50 mg) successfully heated solvents such as hexane and toluene far above the temperatures reached using the solvent alone and rapidly formed the Michael adduct. Removal of the organic layer followed by analysis using ¹H NMR spectroscopy revealed high conversion into the desired product and no contamination from the ionic liquid. It should be noted that not all ionic liquids are completely nonmiscible with nonpolar organic solvents; however, judicious selection of the ionic liquid and solvent will provide the researcher a very effective way of promoting microwave-assisted transformations in poorly absorbing solvents.

The use of silicon carbide inserts has proved valuable in the uncatalyzed *aza*-Michael addition between piperazine and methyl acrylate.[28] The reaction was low yielding using pure toluene as the solvent since the maximum temperature that could be attained was 170 °C. Adding the passive heating elements to the reaction mixture allowed for temperatures of 200 °C to be reached and, with this, a dramatic increase in the yield for the reaction was observed (Scheme 2.10). The product isolation procedure was trivial and consisted of physical removal of the heating element using a pair of tweezers followed by removal of the solvent.

The thermal Claisen rearrangement of allyl phenyl ethers is one of the most well-known organic transformations and has been the subject of many discussions in undergraduate classrooms. The reaction using microwave heating was very slow using only toluene as the solvent since the temperatures attained were too low for the thermal rearrangement to occur. Adding an ionic liquid to the reaction resulted in an increase in the temperature of the reaction (250 °C) and promoted clean conversion into the allyl phenol. Analogous experiments using silicon carbide heating elements produced similar results.[28] Heating a toluene solution of the allyl phenyl ether in the presence of a passive heating element allowed temperatures of 250 °C to be reached,

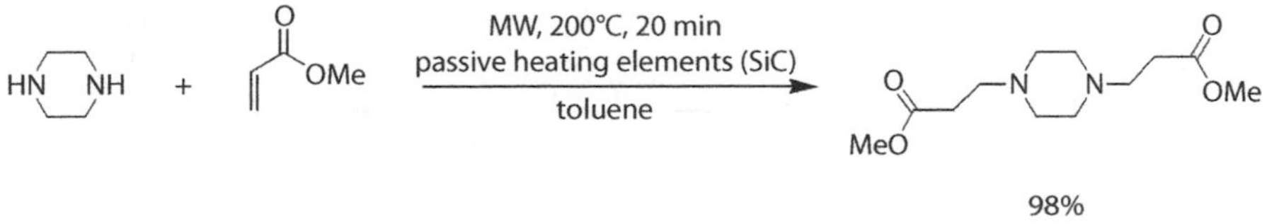

SCHEME 2.10

and clean conversion to the allyl phenol (99%) was possible. Analysis of the reaction mixture revealed only the pure rearrangement product with no decomposition or secondary products. Thus, the purification of the product was trivial and consisted of physical removal of the heating element followed by removal of the solvent.

2.3.2 TANDEM METATHESIS/MICHAEL ADDITION

A tandem cross-metathesis/intramolecular *aza*-Michael addition protocol has recently been described using microwave heating (Scheme 2.11).[29] The study used a range of vinyl ketones as model acceptors and protected amines bearing a pendant alkene as model donors. The Hoveyda–Grubbs catalyst promoted the cross-metathesis step in the reaction. The key to the tandem procedure was finding an additive that would promote the *aza*-Michael reaction and not interfere with the cross-metathesis. This requirement was met by $BF_3 \cdot Et_2O$, and high yields of the desired products were obtained. When substituted amines were employed, the authors found a reversal in the diastereoselectivity for the microwave-heated reactions when compared to the conventionally heated examples.

2.3.3 METAL-CATALYZED CARBONYLATION

Palladium-catalyzed carbonylation reactions have grown in popularity over the past few decades due to the wide range of compounds that can be prepared using this approach.[30] A number of catalyst systems have been developed, with the majority of approaches requiring a supporting ligand for the metal center such as a phosphine. Variants of these protocols using microwave heating have appeared in the literature and often report the clean formation of the desired materials with minimal product purification required. A number of these reports use metal carbonyl complexes as sources of carbon monoxide.[31–34] Recent reports of hydroxy- and alkoxy-carbonylation of aryl iodides have appeared in the literature. They use palladium acetate as the catalyst and gaseous carbon monoxide as the source of CO (Scheme 2.12).[35–38] This approach is attractive since there are no toxic metal sources of CO and no phosphine

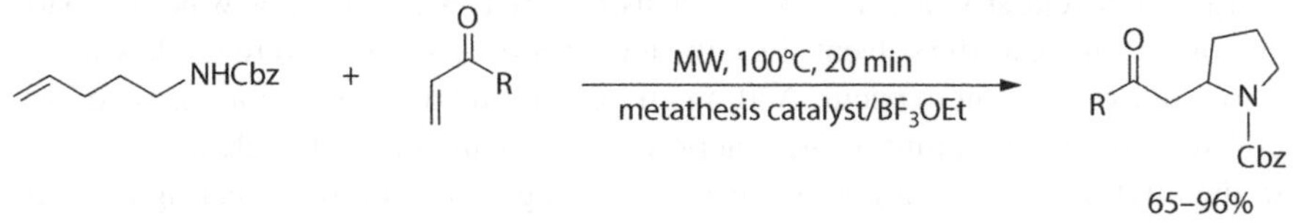

MW, 100°C, 20 min
metathesis catalyst/BF₃OEt

65–96%

SCHEME 2.11

MW, 125°C, 20 min
Pd(OAc)₂, DBU

R = alkyl, alkoxy, F, NO₂

89–99%

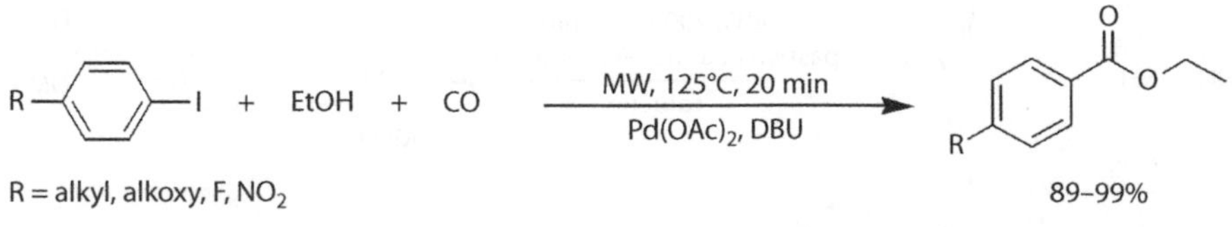

SCHEME 2.12

ligands to remove at the end of the reaction. The authors found that low catalyst load-ings (0.1 mol%) were successful in promoting the carbonylation chemistry. Good-to-excellent yields of the esters were obtained in the alkoxycarbonylation protocol for primary and secondary alcohols (76–99%). The chemistry was also moderately tolerant of steric bulk on either the aryl halide or the alcohol; the addition of a single ortho substituent on the aryl halide still resulted in high yields of the ester. Moving from secondary to tertiary alcohols had a significantly deleterious effect on the yield of the ester products. The scope of the chemistry was also quite broad from the perspective of substituent electronic effects, with a range of aryl halides bearing electron-withdrawing and electron-donating groups being used successfully.

2.3.4 CYCLOADDITION REACTIONS

The copper-catalyzed 1,3-dipolar cycloaddition of organic azides and alkynes (click chemistry) has been the subject of intense research due to wide functional group tolerance, operational ease, and clean formation of the 1,2,3-triazoles.[39] These reac-tions are particularly amenable to microwave heating, and a host of new compounds and materials have been created using this methodology.[40–43]

One of the earliest applications of microwave heating to click chemistry was in the development of a one-pot procedure using an alkyl halide, terminal alkyne, and sodium azide as the substrates (Scheme 2.13).[44] A mixture of copper turnings and copper sulfate served as the precatalyst in these reactions, and the solvent was a 1:1 mixture of water and ⁱBuOH. Heating the reaction mixture for 10 min at 125 °C afforded high conversions (84–93%) with excellent regioselectivity (100%). Isolation of clean product was trivial since it precipitated from the aqueous solution and was pure after washing. No chromatography was needed.

A significant amount of research has been focused on developing adaptations of click chemistry that do not require a copper catalyst. Some advances toward this goal have been made with most of the successful copper-free reactions using activated or strained alkynes. An interesting example of a metal-free version of this reaction employing microwave heating involved the use of a benzyne species to react with the organic azide (Scheme 2.14).[45] The authors used a two-step approach, generating the aryl azide prior to the addition of a benzyne precursor. Heating this reaction mixture for 15 min at 150 °C afforded moderate-to-high yields of the benzotriazoles. Higher yields were observed for azides incorporating electron-donating substituents. The absence of a copper catalyst remains a noteworthy achievement in this chemistry.

If the substrate contains both azide and alkyne functional groups, intramo-lecular and intermolecular cycloaddition reactions are possible. While the former

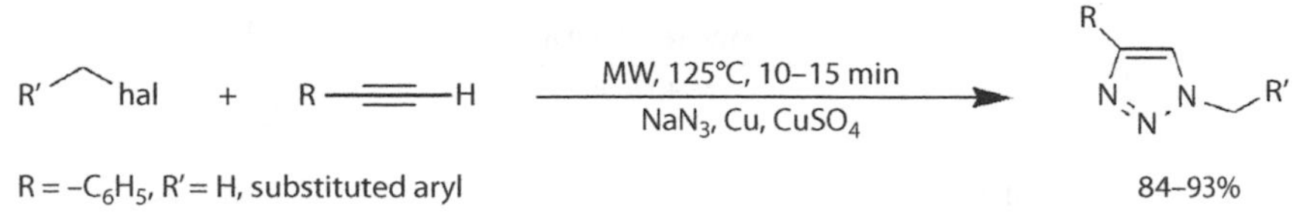

R = –C₆H₅, R′ = H, substituted aryl

MW, 125°C, 10–15 min
NaN₃, Cu, CuSO₄

84–93%

SCHEME 2.13

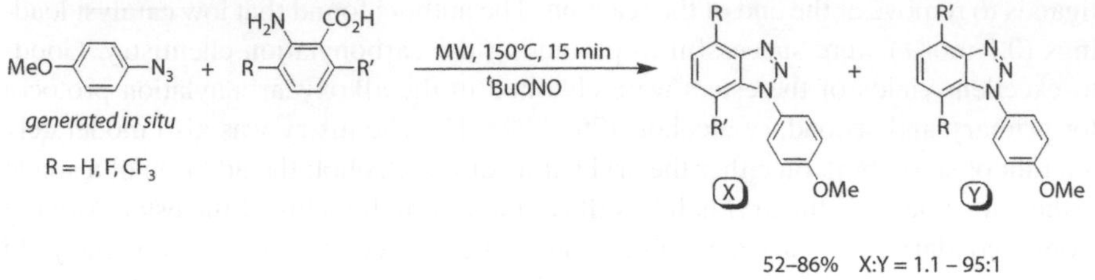

52–86%   X:Y = 1.1 – 95:1

SCHEME 2.14

would generate an internal triazole, the latter would generate oligomers/polymers held together by triazole linkers. An investigation into the control of these potentially competing processes was recently undertaken using amino-acid-based model complexes derived from phenylalanine as substrates (Scheme 2.15).[46] The model complexes were prepared using classic organic methodology and then subjected to elevated temperatures (160–170 °C) using microwave heating with the objective of performing a copper-free intramolecular cycloaddition. For substrates containing an ester linkage, significant oligomerization occurred, and none of the monomeric product was observed. Analogous reactions carried out on an amide-containing derivative gave exclusive formation of the intramolecular cyclization product. In an effort to understand these results, the authors carried out computational studies on the two substrates and proposed a two-stage mechanism in which a conformational rearrangement was needed prior to the cycloaddition step. The amide-containing derivative in the *cis* orientation had a lower energy than the corresponding isomer in the ester-containing species and facilitated the intramolecular cycloaddition reaction.

2.3.5 Amination Reactions

The Buchwald–Hartwig amination of aryl halides using amines as substrates has had a dramatic effect on how synthetic chemists prepare compounds containing carbon–nitrogen bonds.[47] This palladium-catalyzed process typically entails the use of aryl iodides or bromides as substrates and a catalyst system comprising a palladium precursor (often Pd(OAc)$_2$ or Pd$_2$dba$_3$), a large bite-angle diphosphine, and a strong

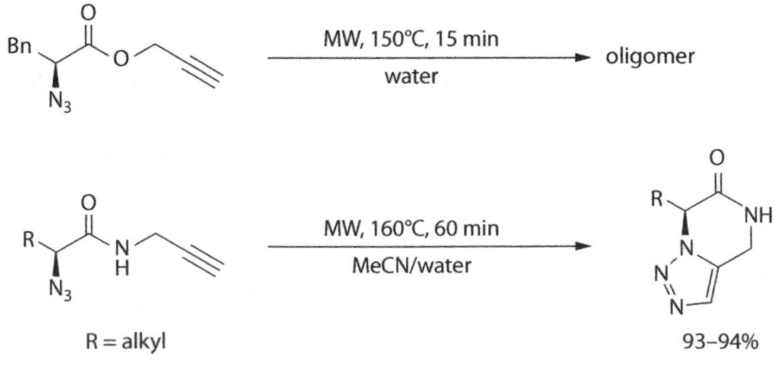

SCHEME 2.15

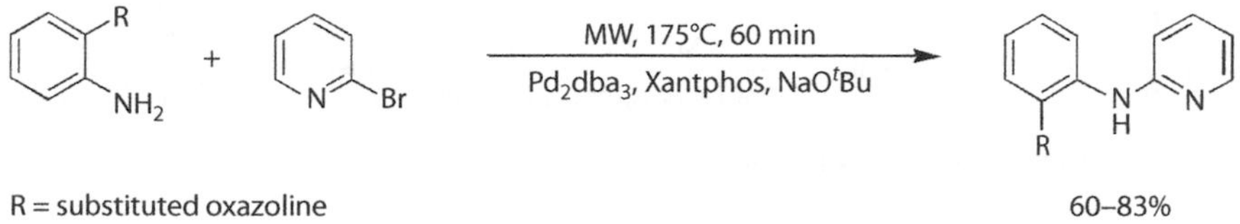

R = substituted oxazoline   60–83%

SCHEME 2.16

base such as KO^{t}Bu. While a host of common aryl halides generate high yields of amination products using conventional heating, aryl halides bearing heteroatoms have been less studied. Recently, the microwave-assisted coupling of halogenated pyridine derivatives with oxazoline-containing anilines has been reported using toluene as the solvent (Scheme 2.16).[48] Performing the reaction for 1 h at 175 °C afforded the amination products in moderate-to-excellent yields (60–83%) without significant contamination due to palladium–pyridyl complexes. The polar reagents and base interacted strongly with the microwave irradiation such that the reaction mixtures could be effectively heated despite the fact that a relatively microwave-transparent solvent was used.

2.3.6 C–H ACTIVATION

The selective activation of carbon–hydrogen bonds remains one of the most challenging synthetic transformations, although it is one of the most direct and efficient ways to construct carbon–carbon and carbon–heteroelement bonds. Metal-catalyzed arylation reactions have blossomed in the past several decades and are now a viable method for the construction of complex organic frameworks.[49,50] The intramolecular arylation of pyridazin-3(2H)-ones illustrates how microwave heating can benefit these transformations (Scheme 2.17).[51] Using conventional heating, the arylation reactions required extended reaction times and >20 mol % PdCl$_2$(PPh$_3$)$_2$ as the catalyst to afford modest yields of the target compound (46%). Due to the high catalyst loading, product isolation was problematic. The protocol using microwave heating generated 90% of the target compound after 10 min at 180 °C but using a catalyst loading of only 0.2–1 mol%. Thus, this procedure doubled the yield of the desired compound while drastically reducing the amount of the catalyst and reaction time.

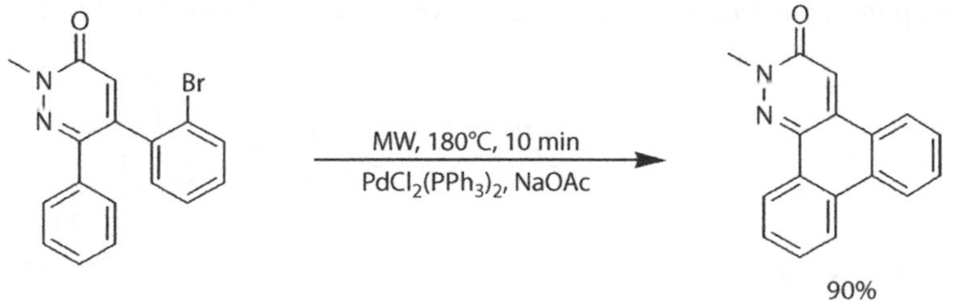

90%

SCHEME 2.17

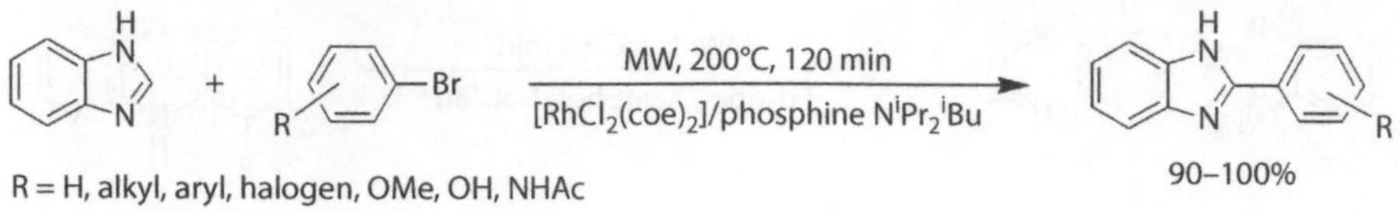

R = H, alkyl, aryl, halogen, OMe, OH, NHAc

MW, 200°C, 120 min

[RhCl$_2$(coe)$_2$]/phosphine N^{i}Pr$_2$^{i}Bu

90–100%

SCHEME 2.18

Microwave-assisted rhodium-catalyzed arylation reactions involving benzimidazole and related heterocycles were recently reported by Bergman (Scheme 2.18).[52] The critical aspect of this intermolecular arylation chemistry was the use of an intriguing phosphine ligand bearing a pendant alkene moiety along with the highly active rhodium precursor [RhCl(coe)$_2$]$_2$ (coe = cyclooctene). Excellent yields (90–100%) of the arylation products were obtained after heating for 2 h at 200 °C using THF as the solvent. The rhodium catalyst and ligand used in this chemistry are air sensitive and must be used in a glove box. The more robust rhodium precursor [Rh(cod)Cl]$_2$ (cod = cyclooctadiene) historically exhibited low reactivity; however, using microwave heating, moderate-to-high yields (55–99%) of the functionalized heterocycles were obtained. Using [Rh(cod)Cl]$_2$ in conjunction with the air-stable HBF$_4$ salt of the phosphine as the catalyst precursor had the advantage that no glove box was required.

Typically, arylation reactions involving substrates containing free –NH$_2$ groups are problematic due to arylation at nitrogen instead of carbon. Protection/deprotection strategies are often required when these compounds are functionalized. These extra synthetic steps can be cumbersome and often reduce the yield of the overall transformation. The use of microwave heating to facilitate the arylation of adenine derivatives without protection of the free –NH$_2$ group was recently reported (Scheme 2.19).[53] This palladium-catalyzed process also required additional additives such as copper iodide and an added base such as piperidine or Cs$_2$CO$_3$. Various palladium sources were screened for activity, with Pearlmans catalyst (Pd(OH)$_2$/C) providing the highest conversion into the C-arylation product without competing N-arylation. A noteworthy point of this chemistry is that no additional ligands for the palladium were needed. The authors found that the chemistry was very substrate dependent, and incorporating electron-withdrawing groups on the aryl halide and ortho substituents typically resulted in lower yields of C-arylation product, although addition of a –CF$_3$ group in the meta position of an aryl bromide still gave high yields (82%).

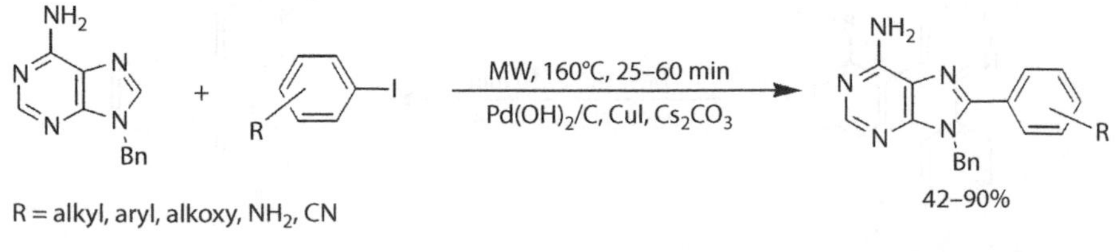

R = alkyl, aryl, alkoxy, NH$_2$, CN

MW, 160°C, 25–60 min

Pd(OH)$_2$/C, CuI, Cs$_2$CO$_3$

42–90%

SCHEME 2.19

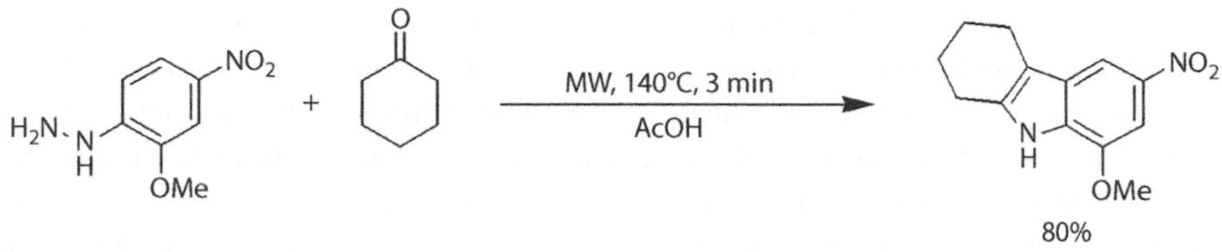

80%

SCHEME 2.20

2.3.7　FISCHER INDOLE SYNTHESIS

The preparation of functionalized nitrogen heterocycles continues to be the subject of intense research due to the biological activity of the resulting compounds. The Fischer indole synthesis is an effective way to generate these valuable compounds. The use of microwave heating for Fischer indole synthesis cleanly generates a range of heterocycles with minimal decomposition or formation of side-products. The preparation of tetrahydrocarbazole species illustrates the benefits of this approach (Scheme 2.20).[54] Using conventional heating, low-to-moderate yields of the indoles were obtained (36–62%; 4–10 h; 100 °C) but, when using microwave heating, the desired indoles could be isolated in high yields (78–95%) after 3 min at 140 °C. The procedure was clean enough that the pure indoles could be isolated after a simple quench with sodium carbonate and extraction with ether followed by drying. No chromatography was needed.

2.3.8　LIEBESKIND–SROGL CROSS-COUPLING

One of the more recent strategies for functionalizing heteroaromatic substrates entails a coupling of a thioether with a boronic acid using common palladium catalysts such as $Pd(PPh_3)_4$ along with cocatalysts such as copper thiophene-2-carboxylate; this is known as the Liebeskind–Srogl cross-coupling (Scheme 2.21).[55,56] The reaction often proceeds under mild conditions and is attractive for the preparation of compound libraries due to the wide range of boronic acids that are readily available. When heating is required, microwave heating can be used, an example being the coupling of 3,4-dihydropyrimidine-2-thiones and boronic acids to generate 2-aryl-1,4-dihydro-pyrimidines.[57] In a more recent case, the functionalization of resin-bound pyrazinones was investigated.[58] Using conventional heating, the conversion into the biaryl species stopped at moderate conversion (63–77%) after 1 day at 65 °C. Analogous reactions using microwave irradiation in conjunction with simultaneous (cryogenic)

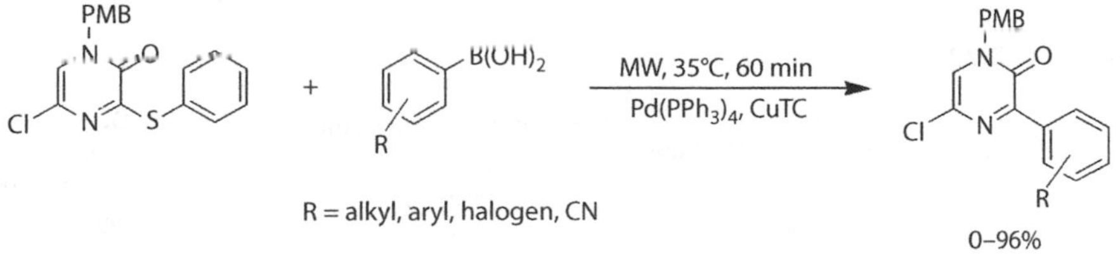

R = alkyl, aryl, halogen, CN

0–96%

SCHEME 2.21

cooling afforded the coupling products in yields as high as 96%. However, this result may be a little misleading, given that accurate temperature measurement is very hard when using cryogenic cooling of reaction vessel walls while heating the contents with microwave irradiation. As a result, comparing conventional and microwave heating protocols performed at the same temperature is difficult to undertake reliably. A number of reactions where a rate acceleration over conventional heating has been reported when using microwave heating with simultaneous cooling at the same temperature have been reassessed.[59,60] The results lead to the general conclusion that the real bulk temperature of the reaction mixture can in fact be very different between conventional and microwave heating, the latter being underestimated. When the reactions are performed under strictly isothermal conditions, similar outcomes are observed.

2.3.9 ULLMANN COUPLINGS

The generation of diaryl and alkyl–aryl ethers through a copper-catalyzed Ullmann coupling remains one of the most powerful ways to construct carbon–oxygen bonds.[61] From a sustainability standpoint, there are several common drawbacks to the original procedure, including the use of stoichiometric quantities of copper and incompatibility with a range of functional groups. One compound that has remained challenging to employ in an Ullmann-type coupling is tetrahydrofurfuryl alcohol. A solution to this problem was recently reported and involves the use of an organo-soluble copper cluster as a reservoir of catalytically active Cu(I) instead of copper powder (Scheme 2.22).[62] Using this catalyst, high yields of alkyl–aryl ethers can be obtained. The reaction proceeds rapidly at 125 °C using microwave heating. Added advantages of the procedure include low catalyst loadings and no need for solvent. Many of the products can be obtained in high yield after a simple filtration through a short plug of silica and removal of the volatiles (no chromatography).

In an effort to eliminate heavy metals (Pd, Cu) from the preparation of diaryl ethers, a procedure using a sterically hindered proazaphosphatrane promoter has recently been reported using microwave heating (Scheme 2.23).[63] Using TBDMS-protected phenols, aryl halides, 1–10% of the promoter and toluene as a solvent, this

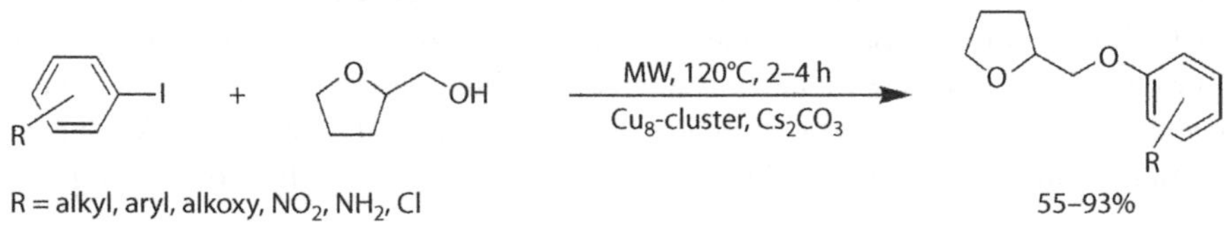

R = alkyl, aryl, alkoxy, NO$_2$, NH$_2$, Cl   55–93%

SCHEME 2.22

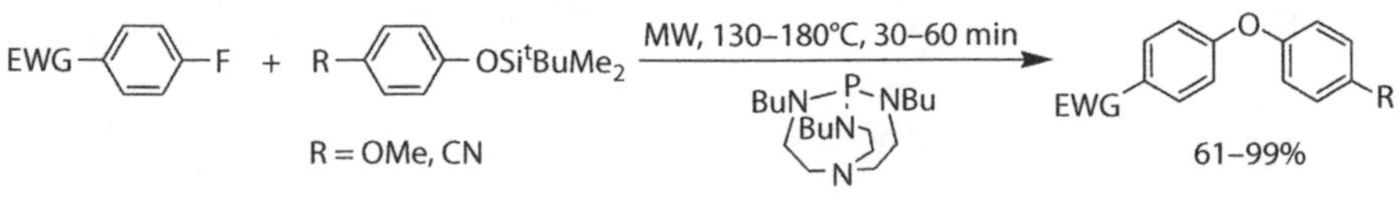

R = OMe, CN   61–99%

SCHEME 2.23

chemistry afforded moderate-to-excellent yields (61–99%) of the diaryl ethers upon heating at 130–180 °C for 30 min to 5 h, depending on the substrate. The chemistry is quite tolerant of both electronic and steric variation of the protected phenol, but only aryl halides bearing electron-withdrawing groups gave high yields of the desired diaryl ether products. A noteworthy achievement in this chemistry was the use of aryl fluorides as the aryl halide coupling partner. Aryl fluorides are often challenging to employ in metal-catalyzed transformations.

2.4 REACTIONS IN WATER

Water is the most abundant and benign solvent on the planet. Despite this, its use as a solvent in organic reactions is overshadowed by the myriad reactions using organic solvents. Over the past decade, however, there has been a surge of interest in the use of water as a medium for organic reactions. For many metal complexes (and organometallic intermediates), rapid reaction with water prevents the use of water as a solvent. However, a number of late transition metal complexes can be used for performing organic transformations in water with great success. Microwave heating has been used extensively as a tool for performing reactions using water as a solvent, and a number of reviews highlighting different aspects of this chemistry have appeared.[64–67] This section will cover the development of several representative reactions.

2.4.1 SUZUKI COUPLINGS USING ULTRA-LOW CATALYST LOADING

One of the most challenging aspects of a metal-catalyzed reaction is the removal of the metal and supporting ligands from the products at the end of the reaction. However, if the concentration of metal catalyst used is extremely low, separation is often simplified and might even be unnecessary. Recently, it has been demonstrated that remarkably low levels (50 ppb–2.5 ppm) of simple palladium salts are sufficient to catalyze the Suzuki coupling of aryl bromides and iodides with aryl boronic acids (Scheme 2.24).[68–70] In addition to the excellent yields (93–99%), the use of water as the solvent for this reaction facilitated the separation of the biaryl product since it typically precipitated from the aqueous solution and could be collected by filtration. Alternatively, ethyl acetate could be added to extract the biaryl product from the reaction mixture. Simply using a stock solution of elemental palladium was sufficient as the catalyst; a palladium complex comprising an exotic supporting ligand is not necessary. The clean reaction chemistry combined with the availability of the metal catalyst, the use of water as the solvent, and the fact that the reaction can be scaled from the milligram to the multigram level makes this process an attractive way to prepare compounds with biaryl motifs.[71,72]

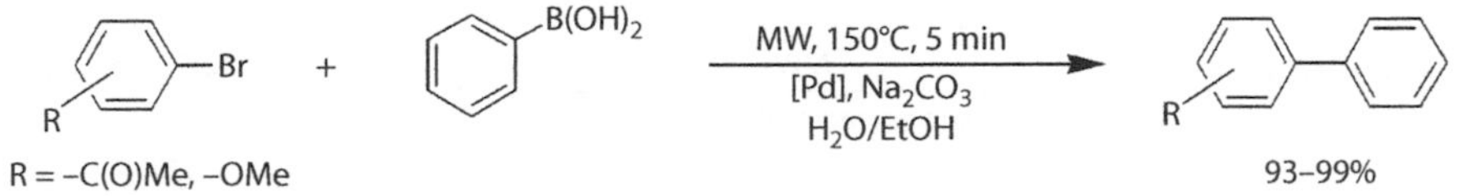

R = –C(O)Me, –OMe

MW, 150°C, 5 min
[Pd], Na$_2$CO$_3$
H$_2$O/EtOH

93–99%

SCHEME 2.24

2.4.2 HECK REACTIONS USING ULTRA-LOW CATALYST LOADING

The palladium-catalyzed coupling of alkenes with aryl halides (Heck reaction) is one of the most popular ways to prepare functionalized internal alkenes. This reaction is typically carried out in a polar solvent such as DMF or DMSO with a range of palladium catalysts. Not surprisingly, a number of protocols using microwave heating have been reported. A particularly interesting example focused on developing a Heck protocol that used ultra-low catalyst loadings, analogous to the Suzuki coupling chemistry already discussed (Scheme 2.25).[73] Using 4-bromoanisole and styrene as prototypical substrates along with water as the solvent, palladium loadings of 1–2 ppm were successful in generating high yields of the 4-methoxystilbene product that could be isolated by simple filtration or extraction with a solvent. The palladium source was again a commercially available stock solution of elemental palladium. The operational ease of the chemistry coupled with the ready availability of the catalyst makes this a successful and convenient way to carryout Heck coupling reactions.

2.4.3 ORGANOCATALYSIS

Despite the popularity of organocatalysis, the problematic separation of the catalyst from the reaction mixture still persists in many cases. This is typically accomplished by crystallization or chromatography, with the latter arguably being the most popular method. To provide a solution to this problem, a new approach has recently been described. Using magnetic nanoparticles as a solid support, glutathione was attached.[74] The resulting supported organocatalyst was used to promote a number of organic transformations such as Paal–Knorr reactions, *aza*-Michael additions, and pyrazole synthesis using microwave heating (Scheme 2.26). After optimization, the Paal–Knorr reactions proceeded in good-to-excellent yield (72–92%) using pure water as the solvent after heating at 140 °C for 20 min. At the end of the reactions, the magnetic nanoparticles attached to the stirring bar inside the reaction vessel and could be removed easily and reused. The organic products phase-separated and could be isolated by simple decantation. The addition of benzyl amine to methyl

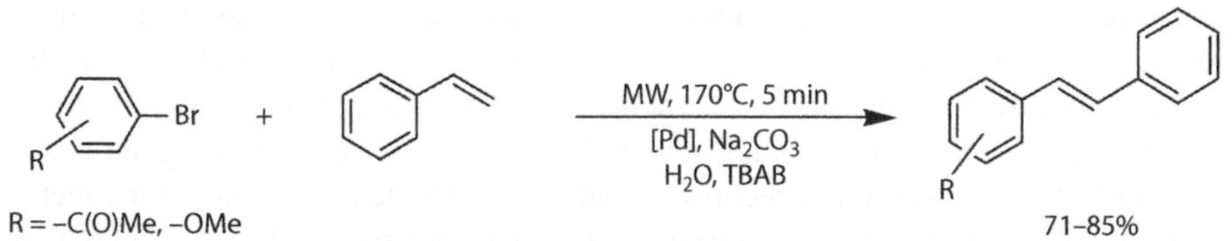

R = –C(O)Me, –OMe   71–85%

SCHEME 2.25

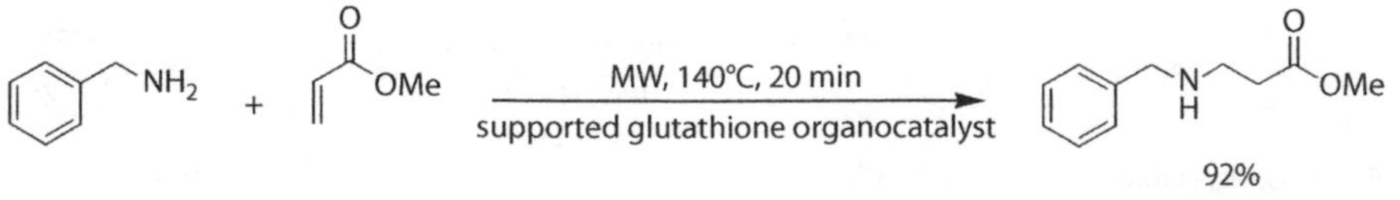

92%

SCHEME 2.26

acrylate (*aza*-Michael addition) also proceeded in excellent yield (90%); however, a solvent extraction step was needed due to the solubility of the Michael adducts in the water. Similarly, the pyrazole synthesis proceeded in excellent yield (96%) and could be separated by simple decantation. The supported catalyst retained its activity for several cycles. For example, the yields of isolated pyrazole for the first and second reactions using recycled catalyst were 97% and 96%, respectively. Due to the ease of catalyst separation and subsequent recycling, the use of pure water as the solvent, and clean synthesis of several different products, this supported organocatalyst could be an attractive choice for other transformations.

2.4.4 REACTIONS IN HIGH-TEMPERATURE AND NEAR-CRITICAL WATER

In the temperature window between 250 °C and 350 °C, water exhibits an intriguing set of properties.[75,76] For example, at 300 °C, the dielectric constant of water is roughly equivalent to that of acetone. The practical result of this observation is that a range of organic compounds are quite miscible in high-temperature water. This observation is attractive to the synthetic chemist as the need for a phase-transfer catalyst could be eliminated. Near-critical water (NCW) also has great potential for acid-catalyzed reactions, as the ionization constant for water is several orders of magnitude higher at 250 °C.

Although the properties of NCW have been known for some time, it has rarely been used in chemical transformations. Arguably, this is due to the myriad reactions that were developed in the 20th century using organic solvents. In the drive to develop sustainable chemical reactions, NCW is a medium that is attracting attention.[77,78] As a result, a number of chemical transformations have been reported using high-temperature as well as NCW.[79–83] In addition to conventionally heated approaches, microwave heating has been used for performing reactions in high-temperature water.[84,85] However, accessing the NCW regime is challenging because the ability of water to absorb microwave energy decreases dramatically with increasing temperature. To overcome this issue, silicon carbide heating elements can be used.[86] Given the potential applications and the scarcity of reports, the use of microwave heating as a tool for performing chemistry in NCW is an attractive area for future research.

2.5 ASYMMETRIC TRANSFORMATIONS

To a synthetic chemist charged with the preparation of enantiopure compounds and materials, "clean organic synthesis" is relative. Performing a reaction that proceeds to 99% conversion might be very clean in terms of remaining unreacted starting materials and unwanted secondary products. However, if the process generates a racemic mixture of compounds, it is disingenuous to describe the reaction as "clean." Although the number of reports relating to microwave-assisted asymmetric synthesis is currently small, they provide proof of concept that selected enantioselective reactions can be performed using microwave heating. Some reactions benefit from significantly elevated temperatures, but most are performed at moderate temperatures in order to maintain high enantioselectivity. This often requires the use of simultaneous cooling in conjunction with controlled microwave heating. A note of caution again is that, when directly comparing conventional and microwave heating protocols, accurate temperature measurement is essential. Rate accelerations or enhancements in asymmetric induction observed could be due to differences in the bulk temperature as opposed to the heating method used.[3,87] Indeed, the results obtained either with microwave irradiation or with microwave irradiation with simultaneous cooling can generally be reproduced by conventional heating at the same reaction temperature and time in an oil bath. This does not decrease the applicability of microwave heating as a tool for performing asymmetric transformations. It is a very versatile tool for optimization of reaction conditions because it offers reproducible, reliable, noncontact heating as well as precise temperature monitoring and data recording.

2.5.1 HECK COUPLING REACTIONS

Chiral architectures abound in nature and are often the key features of a compound that gives rise to the potency of a drug toward a specific target. One such interesting compound is the *Strychnos* alkaloid, minfiensine, and it proves to be a challenging architecture to construct or mimic. An elegant synthesis has recently been reported by Overman in which one of the key steps is an asymmetric intramolecular Heck coupling in which the use of microwave heating is leveraged (Scheme 2.27).[88] The transformation utilized palladium acetate as the metal source and a chiral (phosphinoaryl)oxazoline as the ligand. The coupling reactions generated good-to-excellent yields of the intramolecular coupling product with outstanding enantioselectivity (99%) upon heating at 170 °C for 30–45 min using toluene as the solvent. Another noteworthy aspect of this chemistry was the lack of alkene isomerization. Changing the solvent to acetonitrile still gave high enantioselectivity (96% ee); however, a significant amount of the alkene rearrangement product (ca. 10%) was observed.

The microwave-assisted asymmetric Heck reaction has also been used with success for intermolecular bond formation (Scheme 2.28).[89] Using 2,3-dihydrofuran and phenyl triflate as model substrates, the most active catalyst screened was a combination of Pd$_2$dba$_3$ and a phosphine–thiazole supporting ligand. Heating a THF/DIPEA

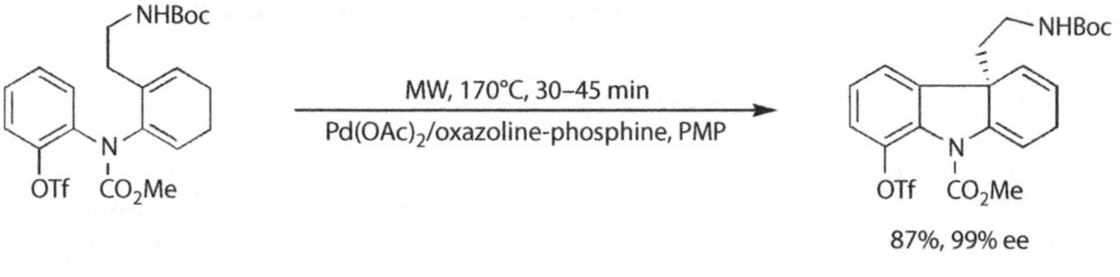

$$\text{MW, 170°C, 30–45 min} \atop \text{Pd(OAc)}_2/\text{oxazoline-phosphine, PMP}$$

87%, 99% ee

SCHEME 2.27

R = alkyl, aryl, alkoxy

$$\text{MW, 120°C, 1–4 h} \atop \text{Pd}_2\text{dba}_3/\text{chiral phosphine, DIPEA}$$

98%, 94–98% ee

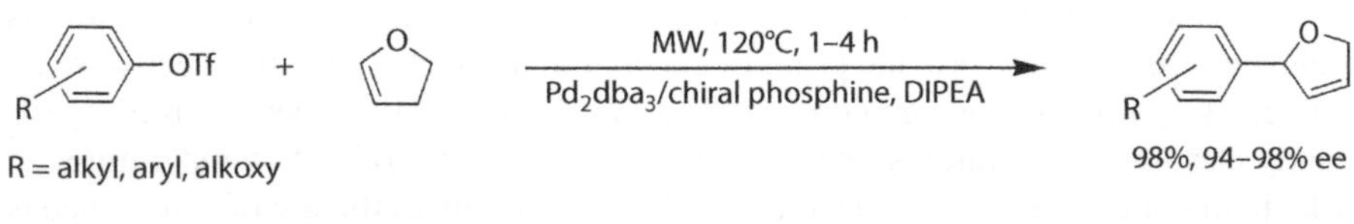

SCHEME 2.28

solution of the model system at 120 °C for 1–4 h cleanly furnished the coupling product in high yield (98%) with excellent enantioselectivity (96% ee). A range of amine bases could be used in the coupling. The authors found that incorporating electron-donating groups into the aryl triflate resulted in slightly higher enantiose-lectivities (relative to phenyl triflate). Additionally, the use of the most sterically crowded phosphine–thiazole ligand consistently resulted in the highest yields and enantioselectivities for the couplings.

2.5.2 PREPARATION OF AMINO ALCOHOLS

The enantioselective preparation of amino alcohols is of great interest to both syn-thetic and medicinal chemists due to their popularity as synthetic intermediates and high level of biological activity.[90,91] As a result, a range of synthetic procedures have been developed for the preparation of these compounds. While many of these approaches are successful, they often require stoichiometric quantities of additives. In an effort to generate enantiopure internal β-amino alcohols cleanly, microwave heating has been used in conjunction with a catalytic quantity of trifluoroacetic anhy-dride (Scheme 2.29).[92] After screening a range of conditions, the authors found that THF was the optimal solvent together with 20 mol% of the anhydride as the catalyst. Heating the reaction mixture at 100–180 °C for 2 h resulted in the desired β-amino alcohols in moderate-to-high yields (74–99%) and excellent enantioselectivities.

2.5.3 ANOMERIZATION REACTIONS

An organocatalytic approach to the generation of β-C-glycosylmethyl aldehydes has been reported using microwave heating.[93] These nonnatural compounds are valu-able building blocks for complex organic structures. The known approaches to the generation of the β-anomers are cumbersome. To provide a solution to this synthetic challenge, L-proline has been used as an organocatalyst to promote the anomer-ization of the known α-anomers into the corresponding β-anomers (Scheme 2.30). Using methanol as the solvent, the reaction mixture was heated to 50 °C and held at

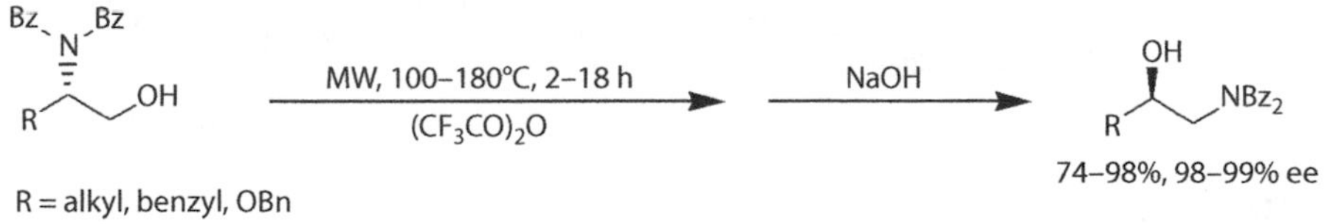

R = alkyl, benzyl, OBn

74–98%, 98–99% ee

SCHEME 2.29

95% β:α = 19:1

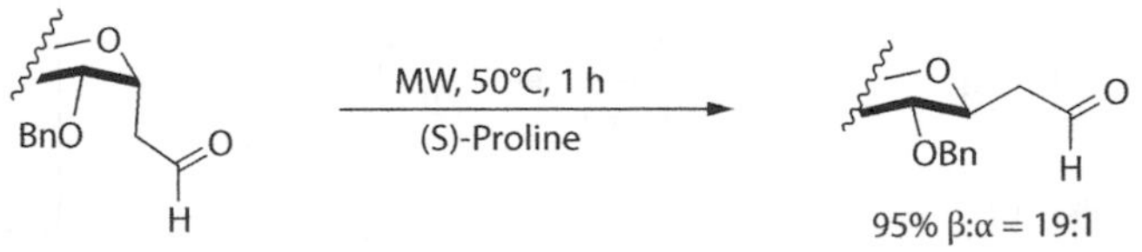

SCHEME 2.30

this temperature for 1 h using a low microwave power. The reaction was remarkably clean, affording 95% recovery of the β-C-glycosylmethyl aldehydes with a β:α ratio of 19:1.

2.5.4 MANNICH REACTIONS

The organocatalyzed enantioselective Mannich reaction is growing in popularity and typically employs catalysts such as proline.[94] The effects of microwave heating on the conversion and enantioselectivity of this valuable transformation have been probed using a standard three-component Mannich reaction as a model system (Scheme 2.31).[95] Cyclohexanone, formaldehyde, and substituted anilines were selected as the substrates along with DMSO as the solvent. Using (S)-proline as the catalyst, even low loadings (0.5 mol%) were effective in promoting the Mannich reaction at 46°C (96% with 98% ee). Since DMSO is a strongly microwave-absorbing solvent, only a low microwave power input was required. For comparison, the authors also carried out analogous reactions using conventional heating and found that increasing the catalyst loading to 1 mol% successfully promoted the reaction at 75–77 °C without detrimentally affecting the enantioselectivity (98% ee).

2.5.5 MICHAEL ADDITIONS

Asymmetric conjugate addition is one of the most effective ways to add nucleophiles to alkenes in a highly selective manner. To probe the effectiveness of microwave heating as a tool for this reaction, the addition of aldehydes to a common Michael acceptor (β-nitrostyrene) has been investigated (Scheme 2.32).[96] The study used proline and a bipyrrolidine derivative as organocatalysts for the reaction. Employing simultaneous cooling in conjunction with constant low microwave power gave good-to-excellent yields and enantioselectivities of the Michael adducts. The simultaneous cooling maintained the recorded temperature of the reaction at ~27 °C. However, it should be noted that an external IR (infrared) sensor was used for temperature

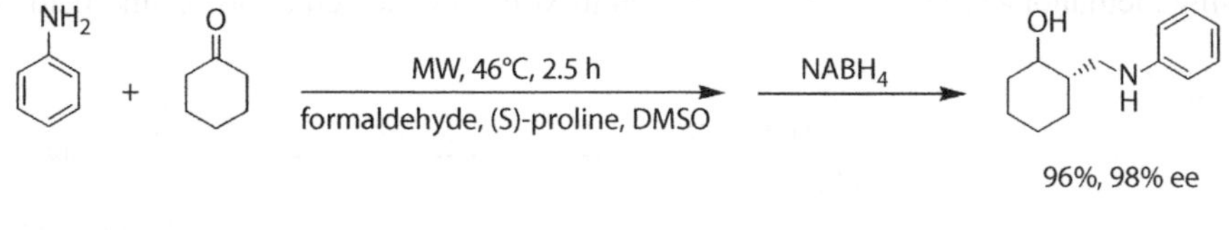

96%, 98% ee

SCHEME 2.31

R = alkyl, H; R′ = alkyl, OH

79–99% 73–98% ee

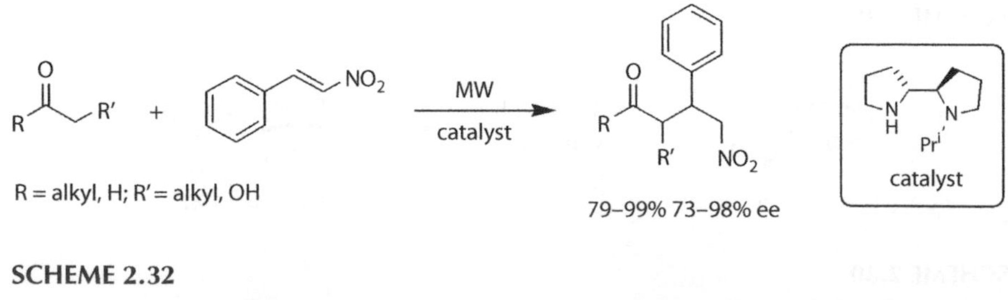

SCHEME 2.32

measurement and, as a result, the real bulk temperature was probably significantly higher. The use of functionalized prolinamide organocatalysts has also been studied for activity toward the conjugate addition of nucleophiles to Michael acceptors.[97] Using N-methyl pyrrolidinone as the solvent, the addition reaction between 3-pentanone and β-nitrostyrene proceeded in excellent yield (99%) with good enantioselectivity (72% ee) upon heating at 48 °C for 2.5 h. Again, simultaneous cooling was applied during the course of the reaction and, again, an external IR sensor was used for temperature measurement, indicating that real bulk temperature was probably significantly higher than that recorded.

2.5.6 Reduction of Ketones

The asymmetric reduction of ketones remains one of the most popular methods for the generation of chiral secondary alcohols. Microwave heating has proved valuable for the preparation of a precatalyst used in a version of this reaction (Scheme 2.33).[98] The reduction reaction was catalyzed by a diazaborolidine species prepared from BH_3 etherate and a chiral diamine derived from (S)-proline. Heating a toluene solution of the chiral diamine and borane (120 °C for 15 min) rapidly and cleanly generated the chiral catalyst as evidenced from the changes in the [11]B chemical shift. The catalyst prepared was pure enough to be used directly. This was attractive since it removed an isolation step from the overall synthetic procedure. The asymmetric reduction of prochiral ketones could then be accomplished in a "one-pot" reaction. The overall reaction consisted of catalyst generation followed by cooling to −78 °C and addition of the ketone. Yields and enantioselectivities of the desired chiral secondary alcohols were excellent. It is noteworthy that no epimerization of the chiral diamine occurred during the high-temperature catalyst preparation step.

2.5.7 Chromanone Synthesis

Enantiopure chromanones have been prepared using microwave heating in a procedure employing a resin-bound organocatalyst derived from proline (Scheme 2.34).[99] Adding the 2-hydroxylaryl ketone and a cyclic ketone to the catalyst in methanol followed by irradiation (100 °C for 11 min) generated the chromanones in good-to-excellent yields (71–94%) with 100% diastereoselectivity and excellent enantioselectivity (90–99% ee). Both electron-withdrawing and electron-donating groups were well tolerated using this methodology. The single-pot reaction and clean generation of highly enantiopure material make this procedure an attractive way to generate these valuable natural products.

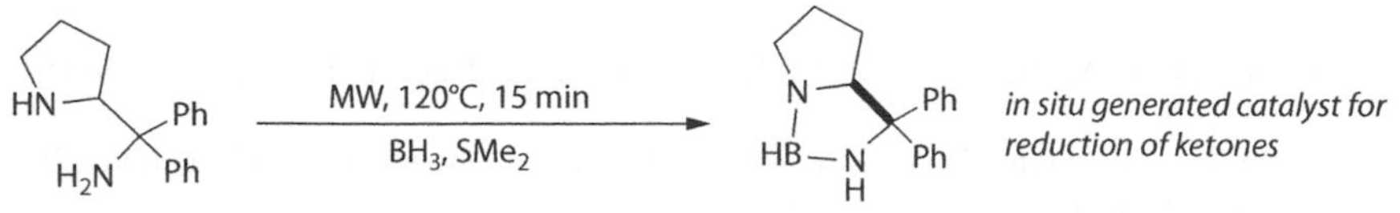

SCHEME 2.33

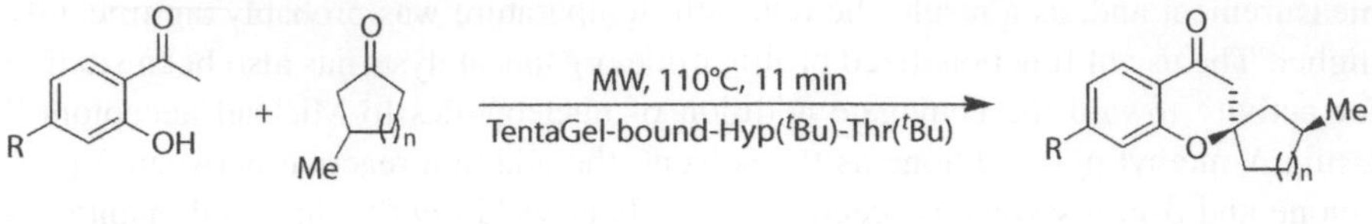

R = OEt, OMe, F   71–94%, 90–97% ee

SCHEME 2.34

2.6 CONCLUDING REMARKS

The use of microwave heating has dramatically affected the way that scientists approach the synthesis of new compounds, primarily due to the short reaction times that are possible and the high product yields and purity that can be obtained. A wide range of organic transformations encompassing classic organic, metal-catalyzed, and organocatalyzed processes can benefit from microwave heating. One of the most notable advances has been in the area of metal-catalyzed reactions, where the catalyst loading can often be decreased to remarkably low levels. The combination of this approach with a nontoxic solvent such as water makes this an attractive approach for the sustainable synthesis of organic compounds. Even nonpolar solvents such as toluene or hexane can be employed in microwave-assisted reactions by adding small amounts of ionic-liquids or passive heating elements. Furthermore, a number of asymmetric reactions have been found to benefit from heating to moderate temperatures with no loss of selectivity. In sum, microwave heating has changed in a fundamental way the approach synthetic organic chemists take to the design and development of reactions and will continue to be the source of intriguing discoveries in the years ahead.

REFERENCES

1. Kappe, C. O.; Dallinger, D. *Mol. Divers.* 2009, *13*, 71–193.
2. Razzaq, T.; Kremsner, J. M.; Kappe, C. O. *J. Org. Chem.* 2008, *73*, 6321–6329.
3. Hosseini, M.; Stiasni, N.; Barbieri, V.; Kappe, C. O. *J. Org. Chem.* 2007, *72*, 1417–1424.
4. Lombard, C. K.; Myers, K. L.; Platt, Z. H.; Holland, A. W. *Organometallics* 2009, *28*, 3303–3306.
5. Herrero, M. A.; Kremsner, J. M.; Kappe, C. O. *J. Org. Chem.* 2008, *73*, 36–47.
6. Varma, R. S. *Green Chem.* 1999, *1*, 43–55.
7. Montchamp, J. L. *J. Organomet. Chem.* 2005, *690*, 2388–2406.
8. Ribiere, P.; Bravo-Altamirano, K.; Antczak, M. I.; Hawkins, J. D.; Montchamp, J. L. *J. Org. Chem.* 2005, *70*, 4064–4072.
9. Enders, D.; Saint-Dizier, A.; Lannou, M. I.; Lenzen, A. *Eur. J. Org. Chem.* 2005, 29–49.
10. Stockland, R. A., Jr; Taylor, R. I.; Thompson, L. E.; Patel, P. B. *Org. Lett.* 2005, *7*, 851–853.
11. Stone, J. J.; Stockland, R. A.; Reyes, J. M.; Kovach, J.; Goodman, C. C.; Tillman, E. S. *J. Mol. Cat. A—Chem.* 2005, *226*, 11–21.
12. Stockland, R. A. Jr.; Lipman, A. J.; Bawiec, J. A., III; Morrison, P. E.; Guzei, I. A.; Findeis, P. M.; Tamblin, J. F. *J. Organomet. Chem.* 2006, *691*, 4042–4053.

13. Polshettiwar, V.; Varma, R. S. *Tetrahedron Lett.* 2008, *49*, 7165–7167.
14. Quiroga, J.; Portilla, J.; Abonía, R.; Insuasty, B.; Nogueras, M.; Cobo, J. *Tetrahedron Lett.* 2008, *49*, 6254–6256.
15. Huryn, D. M.; Okabe, M. *Chem. Rev.* 1992, *92*, 1745–1768.
16. Bortolini, O.; D'Agostino, M.; De Nino, A.; Maiuolo, L.; Nardi, M.; Sindona, G. *Tetrahedron* 2008, *64*, 8078–8081.
17. Chinchilla, R.; Najera, C. *Chem. Rev.* 2007, *107*, 874–922.
18. Carpita, A.; Ribecai, A. *Tetrahedron Lett.* 2009, *50*, 204–207.
19. Chow, W. S.; Chan, T. H. *Tetrahedron Lett.* 2009, *50*, 1286–1289.
20. Shi, L.; Tu, Y. Q.; Wang, M.; Zhang, F. M.; Fan, C. A. *Org. Lett.* 2004, *6*, 1001–1003.
21. Mont, N.; Pravinchandra Mehta, V.; Appukkuttan, P.; Beryozkina, T.; Toppet, S.; Van Hecke, K.; Van Meervelt, L.; Voet, A.; DeMaeyer, M.; Van der Eycken, E. *J. Org. Chem.* 2008, *73*, 7509–7516.
22. Tomioka, K.; Ishiguro, T.; Mizuguchi, H.; Komeshima, N.; Koga, K.; Tsukagoshi, S.; Tsuruo, T.; Tashiro, T.; Tanida, S.; Kishi, T. *J. Med. Chem.* 1991, *34*, 54–57.
23. Zavala, F.; Guenard, D.; Robin, J. P.; Brown, E. *J. Med. Chem.* 1980, *23*, 546–549.
24. Aidouni, A.; Bendahou, S.; Demonceau, A.; Delaude, L. *J. Comb. Chem.* 2008, *10*, 886–892.
25. Hayes, B. L. *Microwave Synthesis: Chemistry at the Speed of Light*, CEM Publishing: Matthews, NC, 2002.
26. Hoffmann, J.; Nuchter, M.; Ondruschka, B.; Wasserscheid, P. *Green Chem.* 2003, *5*, 296–299.
27. Leadbeater, N. E.; Torenius, H. M. *J. Org. Chem.* 2002, *67*, 3145–3148.
28. Kremsner, J. M.; Kappe, C. O. *J. Org. Chem.* 2006, *71*, 4651–4658.
29. Fustero, S.; Jimenez, D.; Sanchez-Rosello, M.; del Pozo, C. *J. Am. Chem. Soc.* 2007, *129*, 6700–6701.
30. Barnard, C. F. J. *Organometallics* 2008, *27*, 5402–5422.
31. Odell, L. R.; Savmarker, J.; Larhed, M. *Tetrahedron Lett.* 2008, *49*, 6115–6118.
32. Wannberg, J.; Larhed, M. *J. Org. Chem.* 2003, *68*, 5750–5753.
33. Wu, X. Y.; Ronn, R.; Gossas, T.; Larhed, M. *J. Org. Chem.* 2005, *70*, 3094–3098.
34. Appukkuttan, P.; Axelsson, L.; Van der Eycken, E.; Larhed, M. *Tetrahedron Lett.* 2008, *49*, 5625–5628.
35. Kormos, C. M.; Leadbeater, N. E. *Synlett* 2007, 2006–2010.
36. Kormos, C. M.; Leadbeater, N. E. *Synlett* 2006, 1663–1666.
37. Kormos, C. M.; Leadbeater, N. E. *Org. Biomol. Chem.* 2007, *5*, 65–68.
38. Iannelli, M.; Bergamelli, F.; Kormos, C. M.; Paravisi, S.; Leadbeater, N. E. *Org. Process Res. Dev.* 2009, *13*, 634–637.
39. Meldal, M.; Tornøe, C. W. *Chem. Rev.* 2008, *108*, 2952–3015.
40. Morales-Sanfrutos, J.; Ortega-Muñoz, M.; Lopez-Jaramillo, J.; Hernandez-Mateo, F.; Santoyo-Gonzalez, F. *J. Org. Chem.* 2008, *73*, 7772–7774.
41. Song, Y.; Kohlmeir, E. K.; Meade, T. J. *J. Am. Chem. Soc.* 2008, *130*, 6662–6663.
42. Zhang, X.; Xiao, Y.; Qian, X. *Org. Lett.* 2008, *10*, 29–32.
43. Pokrovskaya, V.; Belakhov, V.; Hainrichson, M.; Yaron, S.; Baasov, T. *J. Med. Chem.* 2009, *52*, 2243–2254.
44. Appukkuttan, P.; Dehaen, W.; Fokin, V. V.; Van der Eycken, E. *Org. Lett.* 2004, *6*, 4223–4225.
45. Zhang, F.; Moses, J. E. *Org. Lett.* 2009, *11*, 1587–1590.
46. Balducci, E.; Bellucci, L.; Petricci, E.; Taddei, M.; Tafi, A. *J. Org. Chem.* 2009, *74*, 1314–1321.
47. Hartwig, J. F. *Acc. Chem. Res.* 2008, *41*, 1534–1544.
48. Coeffard, V.; Muller-Bunz, H.; Guiry, P. J. *Org. Biomol. Chem.* 2009, *7*, 1723–1734.
49. Lewis, J. C.; Bergman, R. G.; Ellman, J. A. *Acc. Chem. Res.* 2008, *41*, 1013–1025.

50. Alberico, D.; Scott, M. E.; Lautens, M. *Chem. Rev.* 2007, *107*, 174–238.
51. Franck, P.; Hostyn, S.; Dajka-Halász, B.; Polonka-Bálint, Á.; Monsieurs, K.; Mátyus, P.; Maes, B. U. W. *Tetrahedron* 2008, *64*, 6030–6037.
52. Lewis, J. C.; Berman, A. M.; Bergman, R. G.; Ellman, J. A. *J. Am. Chem. Soc.* 2008, *130*, 2493–2500.
53. Sahnoun, S.; Messaoudi, S.; Peyrat, J.; Brion, J.; Alami, M. *Tetrahedron Lett.* 2008, *49*, 7279–7283.
54. Barbieri, V.; Ferlin, M. G. *Tetrahedron Lett.* 2006, *47*, 8289–8292.
55. Liebeskind, L. S.; Srogl, J. *J. Am. Chem. Soc.* 2000, *122*, 11260–11261.
56. Prokopcová, H.; Kappe, C. O. *Angew. Chem. Int. Ed.* 2009, *48*, 2276–2286.
57. Lengar, A.; Kappe, C. O. *Org. Lett.* 2004, *6*, 771–774.
58. Singh, B. K.; Mehta, V. P.; Parmar, V. S.; Van der Eycken, E. *Org. Biomol. Chem.* 2007, *5*, 2962–2965.
59. Leadbeater, N. E.; Pillsbury, S. J.; Shanahan, E.; Williams, V. A. *Tetrahedron* 2005, *61* 3565–3585.
60. Herrero, M. A.; Kremsner, J. M.; Kappe, C. O. *J. Org. Chem.* 2008, *73*, 36–47
61. Hassan, J.; Sevignon, M.; Gozzi, C.; Schulz, E.; Lemaire, M. *Chem. Rev.* 2002, *102*, 1359–1470.
62. Manbeck, G. F.; Lipman, A. J.; Stockland, R. A., Jr.; Freidl, A. L.; Hasler, A. F.; Stone, J. J.; Guzei, I. A. *J. Org. Chem.* 2005, *70*, 244–250.
63. Raders, S. M.; Verkade, J. G. *Tetrahedron Lett.* 2008, *49*, 3507–3511.
64. Dallinger, D.; Kappe, C. O. *Chem. Rev.* 2007, *107*, 2563–2591.
65. Polshettiwar, V.; Varma, R. S. *Acc. Chem. Res.* 2008, *41*, 629–639.
66. Polshettiwar, V.; Varma, R. S. *Chem. Soc. Rev.* 2008, *37*, 1546–1557.
67. Roberts, B. A.; Strauss, C. R. *Acc. Chem. Res.* 2005, *38*, 638–661.
68. Arvela, R. K.; Leadbeater, N. E.; Sangi, M. S.; Williams, V. A.; Granados, P.; Singer, R. D. *J. Org. Chem.* 2005, *70*, 161–168
69. Leadbeater, N. E. *Chem. Commun.* 2005, 2881–2902.
70. Arvela, R. K.; Leadbeater, N. E.; Collins, M. J. *Tetrahedron*, 2005, *61*, 9349–9355.
71. Bowman, M. D.; Holcomb, J. L.; Kormos, C. M.; Leadbeater, N. E.; Williams, V. A. *Org. Process Res. Dev.* 2008, *12*, 41–57.
72. Bowman, M. D.; Schmink, J. R.; McGowan, C. M.; Kormos, C. M.; Leadbeater, N. E. *Org. Process Res. Dev.* 2008, *12*, 1078–1088.
73. Arvela, R. K.; Leadbeater, N. E. *J. Org. Chem.* 2005, *70*, 1786–1790.
74. Polshettiwar, V.; Baruwati, B.; Varma, R. S. *Chem. Commun.* 2009, 1837–1839.
75. Siskin, M.; Katritzky, A. R. *Chem. Rev.* 2001, *101*, 825–836.
76. Katritzky, A. R.; Nichols, D. A.; Siskin, M.; Murugan, R.; Balasubramanian, M. *Chem. Rev.* 2001, *101*, 837–892.
77. Clark, J. H.; Tavener, S. J. *Org. Process Res. Dev.* 2007, *11*, 149–155.
78. Reichardt, C. *Org. Process Res. Dev.* 2007, *11*, 105–113.
79. Leikoski, T.; Kaunisto, J.; Alkio, M.; Aaltonen, O.; Yli-Kauhaluoma, J. *Org. Process Res. Dev.* 2005, *9*, 629–633.
80. Eckert, C. A.; Liotta, C. L.; Bush, D.; Brown, J. S.; Hallett, J. P. *J. Physical Chem. B* 2004, *108*, 18108–18118.
81. Nolen, S. A.; Liotta, C. L.; Eckert, C. A.; Glaser, R. *Green Chem.* 2003, *5*, 663–669.
82. Hunter, S. E.; Savage, P. E. *Ind. Eng. Chem. Res.* 2003, *42*, 290–294.
83. Dunn, J. B.; Savage, P. E. *Green Chem.* 2003, *5*, 649–655.
84. An, J.; Bagnell, L.; Cablewski, T.; Strauss, C. R.; Trainor, R. W. *J. Org. Chem.* 1997, *62*, 2505–2511.
85. Kormos, C. M.; Leadbeater, N. E. *Tetrahedron* 2006, *62*, 4728–4732.
86. Kremsner, J. M.; Kappe, C. O. *Eur. J. Org. Chem.* 2005, *2005*, 3672–3679.

87. Hoogenboom, R.; Wilms, T. F. A.; Erdmenger, T.; Schubert, U.S. *Aust. J. Chem.* 2009, *62*, 236–243
88. Dounay, A. B.; Humphreys, P. G.; Overman, L. E.; Wrobleski, A. D. *J. Am. Chem. Soc.* 2008, *130*, 5368–5377.
89. Kaukoranta, P.; Källström, K.; Andersson, P. G. *Adv. Synth. Cat.* 2007, *349*, 2595–2602.
90. Bergmeier, S. C. *Tetrahedron* 2000, *56*, 2561–2576.
91. Ager, D. J.; Prakash, I.; Schaad, D. R. *Chem. Rev.* 1996, *96*, 835–876.
92. Metro, T.; Pardo, D. G.; Cossy, J. *J. Org. Chem.* 2007, *72*, 6556–6561.
93. Massi, A.; Nuzzi, A.; Dondoni, A. *J. Org. Chem.* 2007, *72*, 10279–10282.
94. List, B.; Pojarliev, P.; Biller, W. T.; Martin, H. J. *J. Am. Chem. Soc.* 2002, *124*, 827–833.
95. Rodriguez, B.; Bolm, C. *J. Org. Chem.* 2006, *71*, 2888–2891.
96. Mosse, S.; Alexakis, A. *Org. Lett.* 2006, *8*, 3577–3580.
97. Almasi, D.; Alonso, D. A.; Gomez-Bengoa, E.; Nagel, Y.; Najera, C. *Eur. J. Org. Chem.* 2007, *2007*, 2328–2343.
98. Luis Olivares-Romero, J.; Juaristi, E. *Tetrahedron* 2008, *64*, 9992–9998.
99. Carpenter, R. D.; Fettinger, J. C.; Lam, K. S.; Kurth, M. J. *Angew. Chem. Int. Ed.* 2008, *47*, 6407–6410.

3 Microwave Heating as a Tool for Sustainable Polymer Chemistry

Mauro Iannelli

CONTENTS

3.1 INTRODUCTION

Polymer chemistry covers the broad field of the preparation, characterization, and manipulation of substances having molecular weights ranging from thousands to millions.[1] Polymers or macromolecules have properties that are quite different from those of low-molecular-weight compounds. Descriptors such as degree of polymerization, molar mass distribution, tacticity, degree of branching, and crystallinity, together with thermal properties such as glass transition temperature and melting temperature, are used to fully characterize them. Structural polymers are defined by their mechanical strength together with their thermal and chemical properties and are mainly used as construction materials. Functional polymers can have special electrical, optical, or biological properties and find their main application in bio-related domains or for the manufacture of microelectronic devices. Polymers can be obtained from natural sources, many naturally occurring substances being macromolecules, for example, starch, cellulose, proteins, nucleic acids, and natural rubber. They can also be prepared through synthetic procedures, either starting from opportune polymerizable monomers or through modification of natural or synthetic polymers.

The impact of polymer chemistry on modern society can be easily evaluated considering the large number of materials used in almost any application of daily life. Their application spans adhesives, coatings, foams, packaging materials to textile and industrial fibers, and electronic, biomedical, and optical devices. Plastic containers of all shapes and sizes, clothing, floor coverings, automobile parts, windshields, pipes, tanks, and playground equipment are all made of polymeric materials. The total polymer production in the United States for 2007 has been estimated to be more than 50 million tons.[2] Worldwide, plastics and polymer consumption will have an average growth rate of 5%, and it will touch a mammoth 227 million tons by 2015.[3] Considering that a large majority of these materials are made from nonrenewable resources and the large amount of greenhouse gases associated with their production, it is clear that new innovative approaches toward a more sustainable future are needed.

Sustainable chemistry can be broadly defined as the design, manufacture, and use of efficient, effective, safe, and more environmentally benign chemical products and processes.[4] It aims to maximize resource efficiency through activities such as energy and nonrenewable resource conservation, pollution prevention, minimization of waste at all stages of a product lifecycle, and the development of durable products that can be reused and recycled. When considering the polymer industry, concepts such as clean efficient synthesis, ready degradability, and recycling or reuse of plastics are clear driving forces toward sustainability.

The advent of microwave heating in preparative chemistry represents a major step forward toward the use of nonconventional energy sources to promote chemical reactions. Since the first reports in the late 1980s,[4,5] the number of papers and the scientific interest in microwave-assisted synthesis has grown dramatically and touches almost every field of the chemical sciences.[6] Advantages of microwave heating include the ability to access elevated temperatures easily and safely, short reaction times and, often, high product yields. Microwave irradiation offers an efficient noncontact heating method without the waste of energy heating the surroundings; this is very different from the case of an electrically heated oven, mantle, or autoclave. The conversion of electricity into microwave energy, however, is not a very efficient process.[7]

The tenets of sustainable chemistry need to be applied to the whole lifecycle of a polymer, starting from its conception at the research scale, through its production on industrial scales and, finally, to its reuse or recycling. There is increasing interest in trying to extend the advantages of microwave chemistry from lab scale to production scale. To this end, microwave heating has been used for synthesis on scales of up to few kilograms.[8–10] This can be considered satisfactory for a number of highly valued intermediates or specialty chemicals, but may not comply with the scale-up requirements for production of many polymers. With a few exceptions,[11,12] reports on microwave-assisted polymer synthesis present work performed at the research scale.[12–15] While this can appear as a limitation, with developments in equipment, scale-up solutions can be expected in the foreseeable future. From a chemistry perspective, research scale discovery can be seen as the key starting point to future innovation.

The analysis of reports from the polymer chemistry literature relating to the use of microwave heating is the focus of this chapter, with a particular emphasis on the development of sustainable approaches. It is now well established that microwave-assisted reactions should be performed using dedicated scientific instruments on the grounds of safety and ease of reproducing results. Such apparatus allows for control of a wide range of reaction parameters, including temperature, pressure, microwave power, and stirring. For this reason, attention will be focused on reports using scientific microwave equipment. However, much of the early work in the field was performed using domestic (household) microwave ovens, and even today a significant number of research groups still report reactions performed using these units, albeit somewhat adapted for synthetic chemistry. While reproducibility cannot be guaranteed, some of the results reported do merit discussion and so are included here.

3.2 USE OF ALTERNATIVE SOLVENTS

The choice of reaction solvent is a critical factor for development of sustainable synthetic pathways to any chemical product in general, and for polymers in particular. To this end, it is important to try to avoid volatile organic solvents, chlorinated solvents, and solvents that can damage the environment (e.g., fluorinated hydrocarbons). The most widely used green solvents for polymer synthesis are water, ionic liquids, and supercritical CO_2.[16] In addition, polymerizations can often be performed solvent free.

3.2.1 MICROWAVE-ASSISTED POLYMER SYNTHESIS IN WATER

Water as a reaction medium for organic synthesis has always attracted great interest because it is inexpensive, nontoxic, and easy to recycle.[17] In addition, it often displays unique properties for performing organic chemistry. In the field of polymer chemistry, aqueous media are largely used for the production of particles and latex for applications such as paints and coatings. Water is an excellent solvent for use in conjunction with microwave heating. Although it has a dielectric loss factor that puts it into the category of a medium microwave absorber (tan $\delta = 0.123$), it can be efficiently heated under microwave irradiation (2.45 GHz) up to and above its normal boiling point in open and sealed vessels, respectively.[18]

Emulsion and miniemulsion polymerizations are widely used techniques in the polymer industry since they have a number of inherent operational advantages, including efficient heat transfer and ease of mixing as well as having environmental benefits. Nitroxide-mediated radical polymerization (NMRP) has the drawback of requiring long reaction times for most monomers.[19] This can be overcome by using microwave heating, an example being the synthesis of stable and well-defined polystyrene latexes using a scientific microwave unit.[20] Working at a temperature of 135 °C in a sealed reaction vessel, monomer conversion was approximately 62% after 4 h when using microwave heating as opposed to only 46% conventionally. Probing the reaction in more detail showed that the induction time was reduced from 1 h to a few minutes when using microwave heating. This was mainly attributed to a faster decomposition rate of the initiator (potassium persulfate).

The emulsion polymerization of methyl methacrylate (MMA) and butyl acrylate (BuA) using microwave heating has been studied and the results compared with those obtained using conventional heating.[21] Again, potassium persulfate was used as an initiator. In the case of MMA, both continuous and pulsed microwave irradiation (scientific apparatus) resulted in increased polymerization rates and smaller particle sizes. Polymerization of BuA seemed not to be affected by the heating source, showing comparable kinetics and particle dimensions in both cases. These findings were attributed to different water solubilities and dielectric parameters of the two monomers that could play a role in the mechanism of heat transfer: direct coupling with microwave energy for MMA and conduction through the hot water for BuA.

The atom transfer radical emulsion polymerization (ATREP) of styrene using PEG-Cl as a macroinitiator was studied using a modified domestic oven.[22] Microwave heating was found to have effects on the morphology, size, and size distribution of the resultant nanoparticles. In general, they were smaller (30–50 nm) compared with those obtained using conventional heating. In addition, stable dispersions were obtained after only 1 h of microwave heating compared to 20 h for the conventional procedure, working at 75 °C in both cases.

An interesting example of microwave-assisted polymerization in aqueous media is the preparation of magnetic molecularly imprinted polymer beads (mag-MIP) (Figure 3.1).[23] Using a scientific microwave unit, suspension polymerization of a self-assembled mixture of triazine (template molecule) and methacrylic acid (MAA) in toluene using surface PEGylated iron oxide particles as magnetic cores and divinyl benzene as a crosslinker was efficiently accomplished in less than a tenth of the time required under conventional heating (2 h compared to 24 h). In addition, using microwave heating, mag-MIP beads were obtained with a narrower size distribution (100–200 µm) and revealed a higher imprinting efficiency (4.7) compared to those

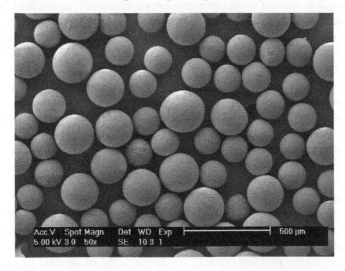

FIGURE 3.1 SEM micrograph of mag-MIP obtained by microwave irradiation. (Reproduced with permission from Zhang, Y.; Liu, R.; Hu, Y.; Li, G. *Anal. Chem.* 2009, *81*, 967–976. Copyright American Chemical Society.)

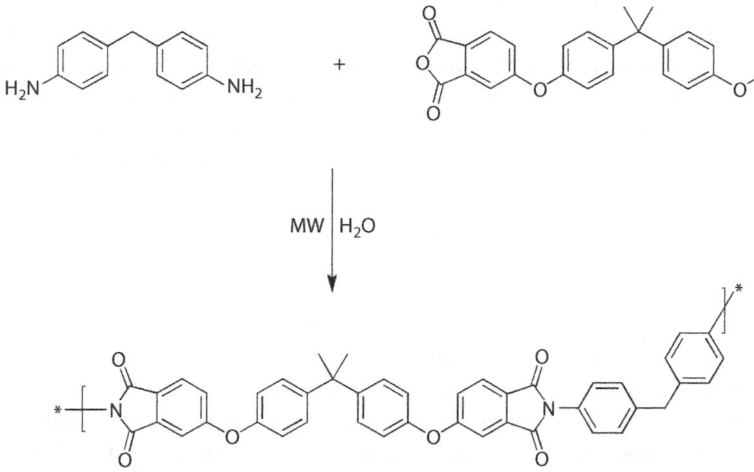

SCHEME 3.1

prepared by conventional heating (100–1200 μm, 2.7). These materials were successfully used for trace analysis and selective absorption of triazines from complex analytical mixtures.

Titanium oxide/polystyrene core-shell nanospheres have been prepared using a combined sol-gel/microwave-assisted emulsion polymerization procedure using a modified domestic oven.[24] The sol-gel process gave stable TiO_2 colloids that were used as seeds for the subsequent polymerization step accomplished in 1.5 h at 70 °C.

A limited number of reports have appeared in the literature showing the use of water as a solvent for the microwave-promoted synthesis of thermoplastics. An example is the synthesis of water-borne polyimides using the standard polycondensation reaction of a dianhydride with a diamine (Scheme 3.1).[25] In a scientific microwave unit, polymers with high molecular weights (M_n up to 35.460 g/mol) were obtained within 22 min of heating using a one-pot two-step procedure. The dianhydride was first hydrolyzed in water to obtain the corresponding tetracarboxylic acid. This was then condensed with the diamine. The obtained polymers were completely comparable in their chemical and thermal properties to those obtained by conventional polymerization in *m*-cresol as solvent.

3.2.2 MICROWAVE-ASSISTED POLYMER SYNTHESIS IN IONIC LIQUIDS

Ionic liquids (ILs) by definition are liquids composed predominantly of ions and ion pairs at some given temperature. More recently, the terminology has come to be used primarily for salts whose melting point is relatively low (below 100 °C).[26] Through their modular design, the physicochemical properties of ILs can be varied, and a very wide range is commercially available.[27] Characteristics of ILs such as thermal and chemical stability, nonflammability, and their ability to act as catalysts make them very good candidates to substitute volatile organic solvents in synthetic chemistry applications. The use of ILs in polymer synthesis has been explored for free[28] and controlled radical polymerizations,[29] living cationic polymerizations,[30] and for the synthesis of block[31] and statistical copolymers.[32] Due to their ionic nature, ILs can be used very effectively in conjunction with microwave heating.[33] As an example, the cationic ring-opening polymerization (CROP) of 2-ethyl-2-oxazoline (EtOx) in ILs has been reported using a scientific microwave unit.[34] Experiments were conducted using 1-butyl-3-methyl-imidazolium hexafluorophosphate, [bmim][PF$_6$] as a solvent and methyl tosylate as an initiator. Kinetic comparisons with the same polymerizations performed in pure acetonitrile and in a mixture of [bmim][PF$_6$] and acetonitrile (50/50 w/w) under identical experimental conditions showed increased reaction rates when using the pure IL. Activation energies derived from the kinetic data were found to be in the order [bmim][PF$_6$] (63.0 kJ/mol) < acetonitrile/[bmim] [PF$_6$] (67.4 kJ/mol) < acetonitrile (73.4 kJ/mol). Moreover, the use of the pure IL facilitated polymer isolation and solvent recycling due to the higher solubility of poly(2-ethyl-2-oxazoline) in water than in the hydrophobic [bmim][PF$_6$]. A simple extraction with water was suitable for isolating the polymer and recycling the solvent. The synthesis of hydrophobic polymers has also been studied in water soluble ILs.[35] The free radical polymerization of MMA as well as the CROP of 2-(m-difluorophenyl)-2-oxazoline (F$_2$Ox) and 2-phenyl-2-oxazoline (PhOx) was performed in a scientific microwave unit using 1-butyl-3-methylimidazolium trifluoromethanesulfonate, [bmim][CF$_3$SO$_3$], and 1-butyl-3-methylimidazolium tetrafluoroborate, [bmim][BF$_4$], as reaction media (Scheme 3.2).[35] Both are water soluble and miscible with the polymer products. Comparison with polymerizations performed in the absence of IL revealed a progressive reduction in yield as the concentration of IL in the starting reaction mixture was increased. However, the use of the hydrophilic ILs as solvents resulted in an easier recovery and purification of the final polymers. The ILs obtained after precipitation of the polymers could be purified for reuse by microwave flash distillation of any residual

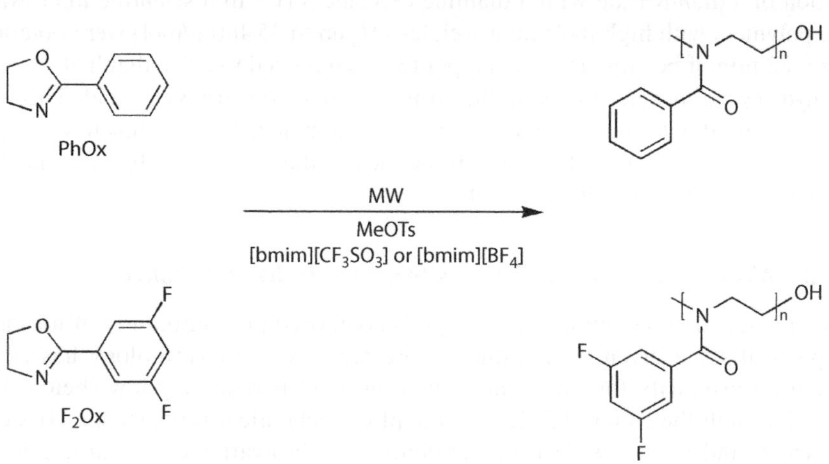

SCHEME 3.2

water and organics, taking advantage of the very fast heating rates achievable when irradiating ILs with microwave energy. The purified ILs could be reused many times, the polymers obtained being identical to those prepared using fresh ILs.

Poly(trimethylene carbonate) is among the most investigated biodegradable poly(alkyl carbonate)s due to its biocompatibility, biodegradability, and good mechanical properties.[36] The ring-opening polymerization of trimethylene carbonate in the presence of [bmim][BF$_4$] has been studied using a scientific microwave unit (Scheme 3.3).[37] The work was performed by heating the reaction mixture at a constant microwave power for a set time. In the presence of 5 wt% [bmim][BF$_4$], an 84% conversion to poly(trimethylene carbonate) was observed over the period of 1 h when using an appropriate microwave power to hold the reaction mixture at 111 °C. The resultant polymer had a number-average molar mass (M_n) of 36,400 g/mol. Performing the reaction at this temperature using conventional heating (oil bath) and the same amount of IL, only a 5% conversion to polymer was obtained, and the M_n was also significantly lower (1,800 g/mol).

The rapid heating of ILs when using microwave energy has been leveraged for the synthesis of foamed polymeric materials.[38] A number of ILs based on 1-vinyl-3-tert-butoxycarbonylmethylene imidazolium salts with different counterions were polymerized using conventional heating and then foamed using microwave promoted pyrolysis (scientific unit) of the tert-butyl ester groups and elimination of isobutene (Scheme 3.4). The foaming process was accomplished in a few seconds due to the ionic character of the poly(IL)s. In addition, the presence of the electron-withdrawing imidazolium group allowed isobutene elimination at temperatures ranging from 150 to 170 °C (in function of the specific counterion) while the same reaction with somewhat more typical polymers occurs only at around 300 °C.[39]

SCHEME 3.3

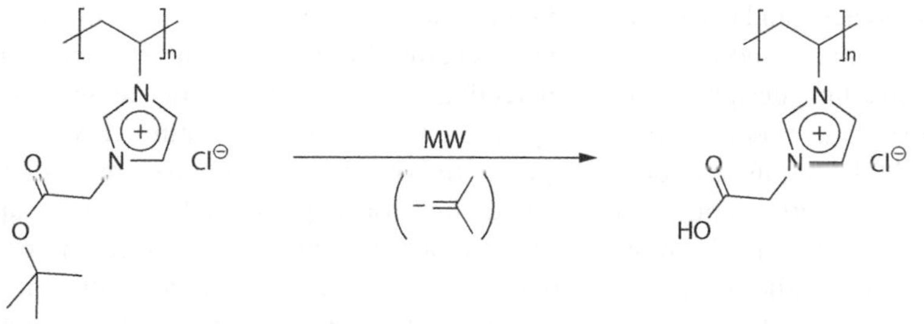

SCHEME 3.4

3.2.3 MICROWAVE-ASSISTED POLYMER SYNTHESIS USING SUPERCRITICAL CO_2

The use of supercritical CO_2 (scCO_2) as a replacement for volatile organic solvents in polymer synthesis and processing has attracted much attention during the last decade.[40] The critical conditions of 31.1 °C and 73.8 bar are relatively easy to achieve, allowing scCO_2 to be used for large-scale applications. As a solvent, it has been used for the polymerization of styrene and methyl methacrylate.[41-44] Due to its dielectric properties, it readily dissolves fluorinated monomers. As a result, it is a good medium for the preparation of fluoropolymers that normally have to be synthesized in chlorofluorocarbons (CFCs).[45] The use of scCO_2 for the production of drug delivery systems and in the formation of highly porous tissue engineering scaffolds has also been reported.[46]

Despite the general interest in the topic, there have been no reports to date on the use of scCO_2 as a polymerization medium (solvent or comonomer) in conjunction with microwave heating. The lack of suitable dedicated microwave reactors that can be fed with gases and withstand the pressures needed has been a limiting factor. However, palladium-catalyzed alcoxycarbonylation reactions performed under microwave irradiation have recently been reported using gaseous CO as a source for the carbonyl functionality.[47] The syntheses were performed on the 1 mol scale using a new microwave reactor dedicated to high-pressure gas applications (up to 200 bar).[48] This instrument could be considered as well for performing polymer chemistry using scCO_2. Another issue is that scCO_2 interacts very poorly with microwave energy. However, as with the use of nonpolar solvents such as hexane or toluene in microwave-promoted chemistry, it may be possible to use passive heating elements[49] or to explore differential heating conditions with solvent mixtures where one component is scCO_2.[50] This area certainly offers an interesting avenue for future research.

3.2.4 MICROWAVE-ASSISTED POLYMER SYNTHESIS USING SOLVENT-FREE CONDITIONS

The possibility to work efficiently without solvent is very attractive from a cleaner-chemistry perspective. In such cases, in addition to eradicating the need for the solvent, it is often possible to increase productivity and simplify purification procedures. However, solvents in polymerization reactions are not only required to solvate monomers (and eventually polymers) but also play an important role as thermal sinks, absorbing and dissipating heat generated during polymerization. Scientific microwave unit design is based on feedback algorithms linking temperature and input microwave power. Thus, it is possible to control temperature very accurately, using just the requisite microwave power to hold the reaction mixture isothermally. Also, since energy input to the sample starts or stops immediately upon application or cessation of microwave irradiation, it is possible to reproduce an optimized temperature profile for a specific reaction very accurately. As a result, microwave heating has proved a useful technique for controlled solvent-free polymer synthesis. The first example dates back to 1983 when the polymerization of 2-hydroxyethyl methacrylate (HEMA) without any radical initiator was performed using a domestic microwave oven.[51] The bulk copolymerization of HEMA with MMA could also be performed solvent free and resulted in copolymers with molecular weight twice as high and of narrower molecular weight distribution compared to those obtained under conventional conditions.[52]

Preparation of poly(ε-caprolactone) (PCL), a biodegradable polymer that finds applications in the biomedical field for prostheses, bandages, or controlled drug release matrix for active substances, was successfully accomplished by heating the monomer in the presence of nontoxic, biologically acceptable lanthanide halides (SmBr$_3$, SmCl$_3$, YbBr$_3$) as initiators in a scientific microwave unit (Scheme 3.5).[53] PCL with a M_n of 14,100 g/mol was obtained after heating at 200 °C for 45 min in the absence of solvent.

The ring-opening polymerization (ROP) of p-dioxanone using triethylaluminum or tin powder as catalysts has also been performed solvent free using a scientific microwave unit (Scheme 3.6).[54] Increased-molecular-weight products were obtained in shorter reaction times as compared with conventional heating.

Optically active poly meth(acryl) amides were synthesized under solvent-free conditions by microwave heating of a mixture of meth(acrylic) acid and (R)-1-phenylethylamine in the presence of 2,2′-azoisobutyronitrile (AIBN) as a free radical initiator (Scheme 3.7).[55,56] Characterization of the prepared materials revealed the presence of imide moieties, presumably due to the high reaction temperatures achieved and, as a result, facile intramolecular dehydration. Interestingly, in the case of the acrylic acid derivatives, the molecular weight of the obtained polymers was found to be inversely proportional to the applied microwave power.

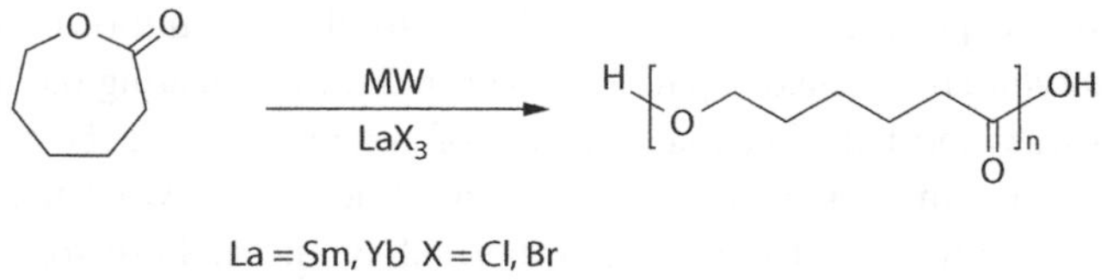

La = Sm, Yb X = Cl, Br

SCHEME 3.5

SCHEME 3.6

R = H, Me

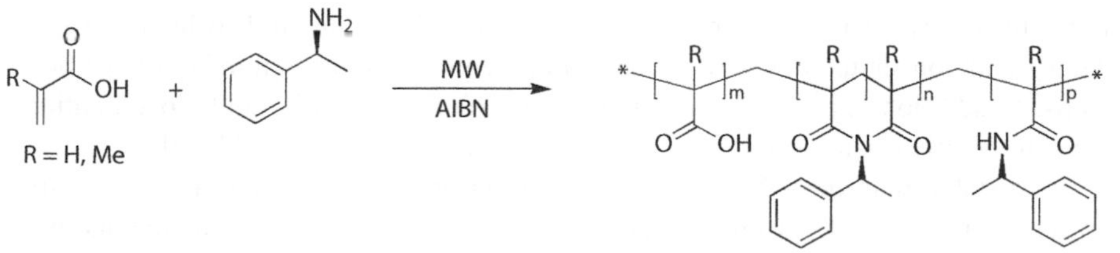

SCHEME 3.7

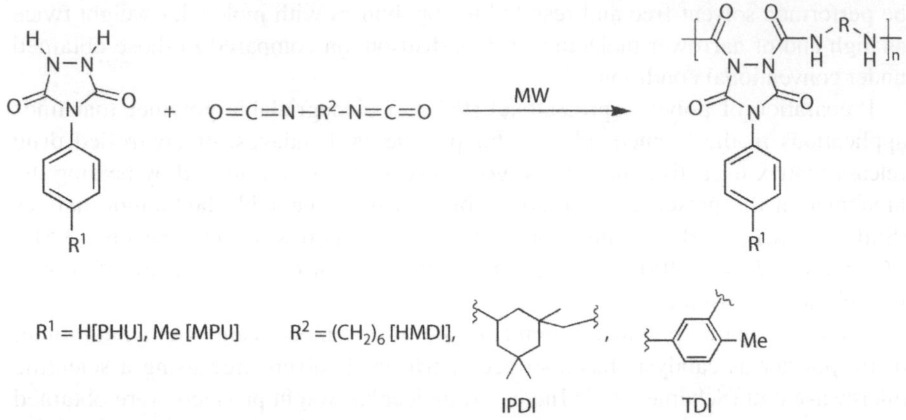

SCHEME 3.8

Microwave heating has been used effectively as a tool for solid-state polymerization (SSP), this process being a postmelt polycondensation largely used to manufacture high-molecular-weight polymers such as poly(ethylene terephthalate), nylon, and bisphenol-A-based polycarbonates.[57] As an example, 4-phenylurazole (PHU) and 4-(4-methoxyphenyl)urazole (MPU) have been condensed with hexamethylene diisocyanate (HMDI), isophorone diisocyanate (IPDI), and tolulene-2,4-diisocyanate (TDI) under solvent-free conditions using a domestic microwave oven (Scheme 3.8).[58] Optimized experimental conditions allowed for the preparation of high-molecular-weight poly(urea)s in yields up to 92% within 5 min of heating (no temperature measurement was reported). The materials are of potential use for the encapsulation of pharmaceuticals, inks, and dyes or for the modification of wool fibers by interfacial grafting techniques.[59] Conventional thermal synthesis, both solvent free and also in dimethylacetamide solution, afforded polymers with comparable properties but only after 12 h at 120 °C.

3.3　USE OF ALTERNATIVE FEEDSTOCKS

The use of renewable feedstocks for the production of energy, chemicals, and materials is one of the key factors for sustainable living.[60] Besides the advantages related to a diminished dependence on fossil fuels, a number of additional benefits can be obtained from using bioderived resources, perhaps the most important being that they inherently have a higher degree of functionality compared to hydrocarbons. Thus, their conversion to useful products may require a reduced number of synthetic steps. In addition, they may also exhibit better biodegradability and biocompatibility. On the other hand, even if the cost of carbohydrates or vegetable oil derivatives is more stable than that of crude oil, the market price of bio-derived products still suffers from a lack of optimized technologies available for their extraction and processing. Following success using conventional methods for polymerization of bio-derived feedstocks,[61,62] a number of studies using microwave heating and aimed at chemical modification of carbohydrates that lead to materials for the production of straining, emulsifying, and softening agents have been reported.[63]

Bulk CROP of a soy-based 2-oxazoline (SoyOx) monomer has been performed using a scientific microwave unit, with methyl tosylate (MeOTs) serving as an initiator (Scheme 3.9).[64] Full conversion was achieved within 8 min, leaving the double bonds of the fatty acids side chains unaffected and thus available for further cross-linking by UV irradiation. Statistical copolymers of SoyOx with 2-ethyl-2-oxazoline were also prepared. Rubbery cross-linked materials showing shape-persistent swelling upon absorption of water from the air could be prepared with as little as 13 wt% (5 mol%) of SoyOx.[65]

Poly(lactic acid) (PLA) is among the most promising biodegradable polymers due to its transparency, high mechanical properties, and good biocompatibility. Single-step direct polycondensation of lactic acid (derived from bio-based resources such as corn or sugarcane) to produce high-molecular-weight PLA has been explored using a scientific microwave unit.[66] Using $SnCl_2$ and p-TsOH as a catalyst pair, PLA with M_w of up to 16,000 g/mol was obtained in good yields (54%) after 30 min of irradiation at 200 °C and 30 mmHg. Comparable molecular weights can be achieved only under thermal heating after multistep reactions, each taking from a few hours to several days.

PLA was also blended with poly(glycolic acid) through microwave heating of a physical mixture of the two polymers, the former solubilized in and the latter suspended in chloroform. Thermal characterization of the materials obtained after removal of the solvent supported the presence of a polymeric blend with chemical linkages between the two polymers instead of a mere physical blend.[67]

Poly(ether-ester)s have been prepared by condensation of adipoyl or terephthaloyl chloride with an isosorbide-based ether-diol (Scheme 3.10).[68] The parent isosorbide is obtained from renewable resources, hydrogenation, and subsequent dehydration of D-glucose. The polymerizations were accelerated about five times under microwave heating using a scientific microwave unit as compared to conventional methods, and a 95% yield of the desired polymers was obtained within 5 min at 180 °C.

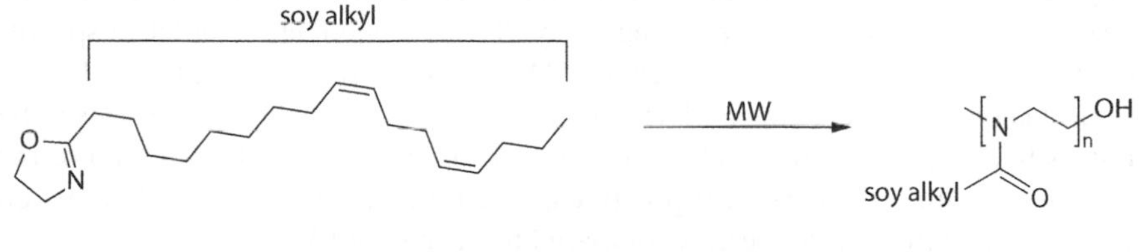

SCHEME 3.9

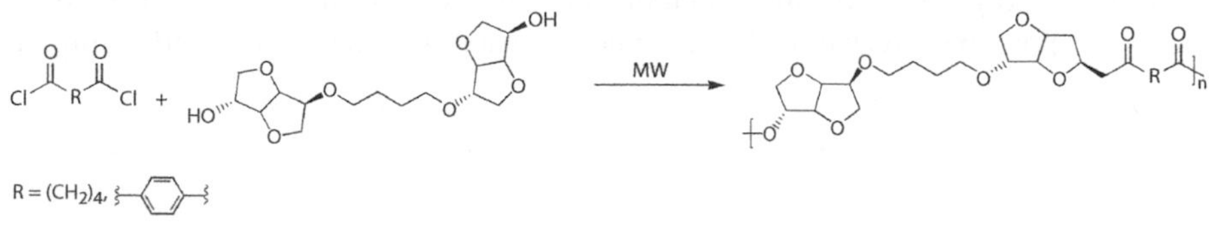

SCHEME 3.10

The solvent-free synthesis of poly(β-pinene), poly(α-pinene), poly(α-pinene-*co*-styrene), and poly(β-pinene-*co*-styrene) has been accomplished by ring-opening homopolymerization and free radical copolymerization using $AlCl_3$ and AIBN as iniators, respectively.[69] Microwave heating using a scientific apparatus proved to be the only method to achieve incorporation of significant amounts of β-pinene in the copolymer with styrene. Reaction times were in the range of 5 h as opposed to the 7 days previously reported for the conventional process.[70] In the case of α-pinene, the copolymers obtained had higher molecular weight than those prepared by conventional heating but contained mainly styrene units.

3.4 DESIGN OF DEGRADABLE POLYMERIC MATERIALS

Even though there is increased drive toward reusing and recycling, not all products at the end of their lifecycle can be recycled and, in some cases, the process may not be economically sustainable. Thus, to avoid the increasing accumulation of wasted plastics and other polymeric materials, there is interest in the design of biodegradable alternatives. These materials need to be easily degraded by microorganisms present in soil and water without producing byproducts that may themselves cause environmental concerns. Aliphatic polyesters are one of the most widely utilized classes of biodegradable polymers. They find use in fibers, packaging materials, injection molded products, bottles, and for medical and biomedical applications. Common synthetic pathways for this class of compounds are direct polycondensation using equimolar amounts of diacids and diols in the presence of a suitable catalyst or ROP of lactones. Again, microwave chemistry has proved useful in driving the field forward. High-molecular-weight ($M_w = 23,500$ g/mol) poly(butylene succinate) has been prepared via the condensation of butane-1,4-diol and succinic acid in the presence of a tin catalyst (2 mol%). The reaction reached 80% yield after heating in a scientific microwave instrument for 20 min at 220 °C (Scheme 3.11).[71]

Poly(vinyl alcohol) (PVA), a well-known water-soluble and biodegradable polymer, has been used as an initiator for the microwave-assisted bulk ROP of ε-caprolactone in a domestic microwave oven.[72] The graft procedure proved to be an excellent method for functionalizing presynthesized polymers in order to specifically tailor their properties. In contrast to PVA, poly(ε-caprolactone) (PCL) is hydrophobic and degrades very slowly. The combination of the two polymers proves to be an attractive way to control biodegradability of the final material. In addition, the resultant poly(vinyl alcohol)-*graft*-poly(ε-caprolactone) (PVA-*g*-PCL) had improved mechanical and thermal properties compared to the parent PVA.

Poly(alkylene hydrogen phosphonate)s are a relatively new family of biodegradable polymers that find domain in the biomedical field.[73,74] The synthesis of poly(oxyethylene hydrogen phosphonate) has been reported using a

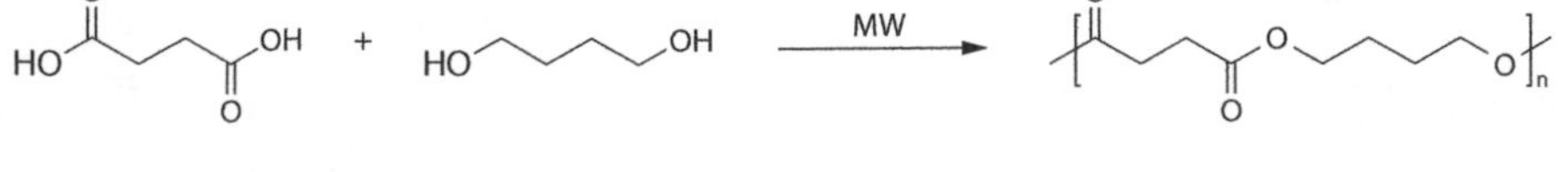

SCHEME 3.11

scientific microwave unit. Transesterification of dimethyl hydrogen phosphonate with poly(ethylene glycol) (PEG 400) led to the desired polymer with an Mn of 3100 g/mol within 55 min (Scheme 3.12).[75] Under conventional conditions, an M_n of 4900 g/mol was obtained only after 9.5 h of heating.

Peptide-based polymers can be used to design biomaterials that can be specifically cleaved by endogenous proteases.[76] Repeating units typical of natural polymers such as silk, elastin, and collagens can be used for a wide range of applications, including drug delivery systems and scaffolds for tissue engineering. The classical peptide chemistry approach suffers from time-consuming protection–deprotection steps and requires the use of preactivated building blocks such as N-carboxy anhydrides, diphenylphosphorylazide, or carbodiimides. A useful method to connect unprotected functional peptides to yield longer polymeric chains of well-defined structure is the copper(I)-catalyzed cycloaddition reaction between terminal alkynes and organic azides that yields the corresponding 1,4-disubstituted-1,2,3-triazoles (click reaction).[77,78] Using this approach, azido-phenylalanyl-alanyl-lysyl-propargyl amide (N3-Phe-Ala-Lys-PA) and azidophenylalanyl-alanyl-glycolyl-lysyl-propargyl amide (N3-Phe-Ala-Glyc-Lys-PA) have been polymerized using microwave heating (Scheme 3.13).[79] Concentrations of up to 400–500 mg/mL in DMF could be quantitatively polymerized obtaining high-molecular-weight polymers (M_w up to 25,600 g/mol) after heating at 100 °C for 30 min in a scientific microwave unit. The prepared materials contain recognition sites for the model proteases trypsin and chymotrypsin for effective biodegradation.

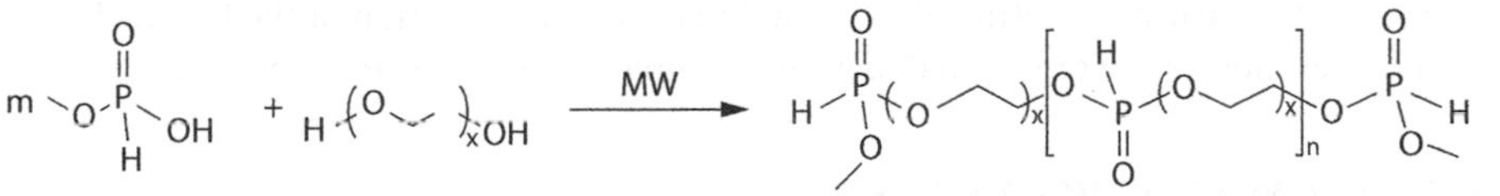

SCHEME 3.12

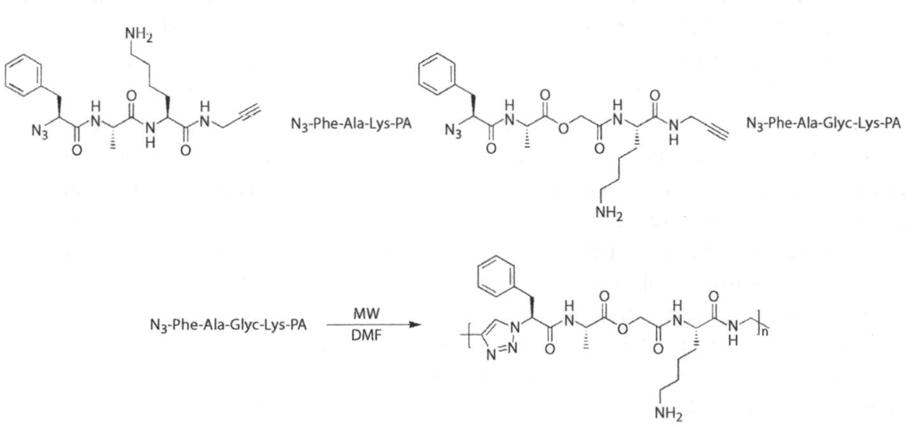

SCHEME 3.13

A valuable approach to reducing the volume of plastic waste is to use partially degradable materials obtained by blending conventional plastics with biodegradable fillers.[80] Starch, a renewable and biodegradable hydrophilic polymer, can be employed for this purpose. In order for the final blend to have suitable mechanical properties, starch has to be chemically modified to reduce the high interfacial tension when blending with hydrophobic polymers such as polyethylene and polypropylene. Modification with alkenyl succinic anhydrides has been found to be a useful method to increase the hydrophobic character of starch.[81] Microwave heating has been employed for the chemical modification of cassava starch with octenyl succinic anhydride (OSA).[82] Heating a starch/OSA (10 wt%) mixture using a domestic microwave oven in the absence of solvent yielded, in 7 min, a grafted starch (S-g-OSA) with a degree of substitution about 30% higher than that obtained by conventional heating for starches with similar amylose contents. The S-g-OSA was used as a compatibilizer for linear low-density polyethylene (LLDPE)/starch blends. An 80/20 (w/w) LLDPE/S-g-OSA blend afforded a material with mechanical properties completely comparable to pure LLDPE.

Hemicelluloses are among the most abundant renewable biopolymers, and their fatty acid ester derivatives are potential candidates for incorporation into biodegradable films for food packaging. Esterification of wheat straw hemicelluloses with a number of acid chlorides (among them lauroyl, palmitoyl, stearoyl, and oleoyl chloride) has been achieved using microwave heating in a domestic oven.[83] The reactions were performed in a DMF/LiCl solution using N-bromosuccinimide as a catalyst. Heating at 78 °C for 5 min was sufficient to prepare hemicelluloses with degrees of substitution (DS) up to 1.34 (DS = 2 for full substitution) in all cases. Of note, however, is that microwave heating also promoted partial degradation of the hemicelluloses that resulted in lower thermal stability compared to the original material.

3.5   RECYCLING POLYMERS

Recycling has often been the first approach used to overcome the environmental issues associated with the accumulation of plastic waste. Different strategies are available to achieve the goal of extending the lifecycle of plastics before final disposal. These include (1) reuse after cleaning or mechanical reprocessing to generally lower-grade materials, (2) chemical depolymerization into a series of still-valuable products or monomers that can be later transformed into new polymeric products, (3) cracking and conversion to general feedstocks such as naphtha (via pyrolysis) or synthesis gas (via gasification), and (4) incineration to use the high calorific content for energy production. Of these, recycling through depolymerization has been widely investigated using microwave heating. Polyethylene terephthalate (PET) can be defined as an intrinsically green polymer. Its industrial manufacture involves low emission, it is nontoxic, and it is very versatile due to its high inertness and easy moldability into various products, particularly for the beverage industry; but it also finds applications as film, chips, and coatings. However, it is nonbiodegradable. PET depolymerization has been performed via hydrolysis, aminolysis, methanolysis, and glycolysis.[84] Glycolysis using ethylene glycol, propylene glycol, or diethylene and dipropylene glycols results in the formation of oligomers with hydroxyl-functionalized terminal groups. Using methanol and propylene glycol as reagents and solvent and zinc acetate as a catalyst, complete PET degradation has been achieved within 6 min working in a pressurized vessel using a scientific microwave unit.[85] The products obtained were identified as a glycol-terephtalic acid adducts together with different oligomers (when using propylene glycol) and crystalline dimethylterephtalate (when using methanol). Other diols, including 1,4-butanediol, 1,5-pentanediol, tetraethylene glycol, and neopentyl glycol, have also been used for the reaction.[86] A series of different base catalysts (K_2CO_3, $NaOCH_3$, $NaHCO_3$, CaO, KH_2PO_4, and $Na_2C_2O_4$) have been screened for the process, complete depolymerization being found only when using K_2CO_3 and $NaOCH_3$.

In an open reaction vessel, potassium hydroxide has been used for PET depolymerization with success. Complete degradation was accomplished in 8 min in 1-butanol, in 4 min in 1-pentanol, and in 7 min in 1-hexanol working at a constant microwave power of 500 W in a scientific apparatus.[87] This method yielded as products (after acidification) terephthalic acid and ethylene glycol of sufficient purity to be reused as monomers for the synthesis of virgin PET (Scheme 3.14). In most cases, similar reactions under conventional heating required up to 8 h at reflux.[88]

Poly(vinyl chloride) (PVC) is the third most used thermoplastic polymer after polyethylene and polypropylene. A large amount of the global PVC production, estimated to be 33.46 million tons per annum in 2002, is used for the construction market because it is cheap and highly durable.[89] To reuse PVC waste, thermal dehydrochlorination processes operating at temperatures higher than 280 °C have been developed.[90] Due to its chemical nature, PVC has a higher capacity to absorb microwaves as compared to other plastic materials. This has been leveraged for the dehydrochlorination of commercially available PVC resins using a domestic microwave.[91] Since the PVC used also contained a relatively large amount of other polymers and low-molecular-weight compounds such as stabilizers and plasticizers, the high temperatures required to initiate the reaction were achieved with the addition of activated carbon and Zn-Mn ferrite as heating aids. Decomposition of the PVC occurred mainly through dehydrochlorination; however, in addition to HCl gas, about 15 wt% of other decomposition products were released. A process operating at lower temperatures has recently been reported using a dedicated microwave reactor and a closed reaction vessel.[92] A flexible PVC resin was treated with a NaOH/ethylene glycol (EG) mixture and then heated to 160 °C for 10 min, yielding residues constituted only of hydrocarbons. An equivalent procedure using conventional heating, also in a closed vessel, required 1 h at 190 °C to achieve comparable results. The differences have been attributed to the fact that EG is itself a strong microwave absorber. In addition, PVC is easily swollen in hot EG, which, once diffused in the bulk of the resin, should promote extraction of the additives and dichlorination from the inside.

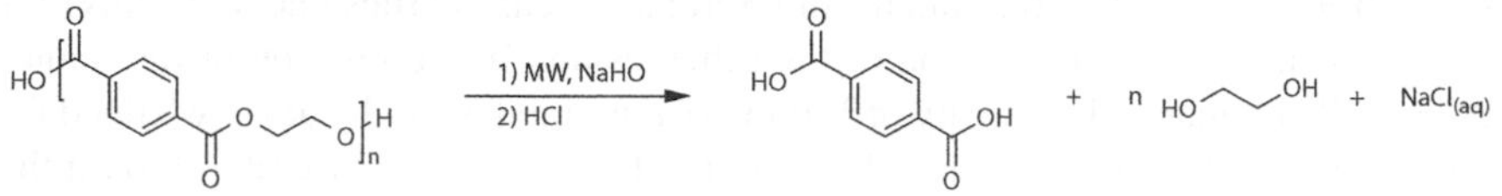

SCHEME 3.14

Chemical recycling of poly(urethane)-based waste (PUs) has been carried out in many different ways, including aminolysis,[93] glycolysis,[94] hydroglycolysis,[95] and hydrolysis.[96] The glycolysis of a flexible cold cure PU foam has been performed using microwave heating.[97] The foam, based on a propylene oxide–ethylene oxide copolyether containing polyol and diphenylmethane diisocyanate, was heated using glycerine as a breaking agent and NaOH or KOH as a catalyst (1 wt% relative to glycerine). Complete dissolution of the foam occurred after heating at 160 °C for 93 and 110 s in the case of the NaOH- and KOH-catalyzed processes, respectively. By increasing the temperature to 220 °C, the same process could be accomplished in 65 and 67 s, respectively, yielding a low-density fraction having a composition very similar to the starting polyol and a higher-density one containing different by-products obtained from the transesterification reactions of urea, urethane, and allophanate derivatives.

The same PU foam was also treated with a mixture of glycerine and water (up to 40%) to investigate the effective decomposition through the combined mechanisms of hydrolysis and glycolysis.[98] The presence of water slowed down the process to 10 min at 160 °C, while working at higher temperatures resulted in undesired transesterification and pyrolysis products. However, this hydroglycolysis approach did dramatically reduce the amount of solvent required. In addition, due to the presence of water, the resultant polyols formed could be easily used in new formulations of water-blown PU foams.

3.6 CONCLUSIONS

In this chapter, the application of microwave heating to the field of polymer synthesis and processing has been discussed from the standpoint of sustainable chemistry front and center. Microwave heating has proved to be a valuable technique for performing a wide range of polymerization reactions, either solvent free or in solvents such as water or ionic liquids. The use of vegetable-oil- and carbohydrate-derived materials as feedstocks for microwave-assisted polymerizations has been found to be an excellent route for producing new or improved polymeric materials. Biodegradable polymers can be synthesized in high yields, and plastic waste efficiently recycled via microwave-promoted chemical depolymerization. As the drive toward sustainability—which inspired this work—continues to gather momentum, we hope that the examples reported here will inspire readers to use microwave heating as an efficient, versatile tool for further innovative polymer science.

ACKNOWLEDGMENTS

We wish to thank Fabio Bergamelli and Serena Lorenzi (Milestone s.r.l, Italy) for their assistance in the preparation of the schemes and in the revision of the manuscript. Laila Fiannaca (University of Milan, Milan, Italy) is kindly acknowledged for assistance with literature searches. Last but not least, we are very grateful to all the authors mentioned in this work (and to the many unmentioned). Their reports have contributed to the foundations for a more sustainable future.

REFERENCES

1. Odian, G. *Principles of Polymerization*; Hoboken: Wiley, 2004.
2. Plastics Industry Producers' Statistics Group, as compiled by Veris Consulting, LLC, http://www.verisconsulting.com.
3. CIPET—Plastics Industry—Statistics, http://www.cipet.gov.in.
4. Gedye, R.; Smith, F.; Westaway, K.; Ali, H.; Baldisera, L.; Laberge, L.; Rousell, J., *Tetrahedron Lett.* 1986, *27*, 279–282.
5. Giguere, R. J.; Bray, T. L.; Duncan, S. M.; Majetich, G. *Tetrahedron Lett.* 1986, *27*, 4945–4958.
6. For a list of review articles and key references in microwave-assisted organic synthesis see: Kappe, C. O.; Dallinger, D. *Mol. Divers.* 2009, *13*, 71–193.
7. Nüchter, M.; Ondruschka, B.; Bonrath, W.; Gum, A. *Green Chem.* 2004, *6*, 128–141.
8. Bowman, M. D.; Holcomb, J. L.; Kormos, C. M.; Leadbeater, N. E.; Williams, V. A. *Org. Process Res. Dev.* 2008, *12*, 41–57.
9. Moseley, J. D.; Lenden, P.; Lockwood, M.; Ruda, K.; Sherlock, J.; Thomson, A. D.; Gilday, J. P. *Org. Process Res. Dev.* 2008, *12*, 30–40.
10. Moseley, J. D.; Lawton, S. J. *Chim. Oggi* 2007, *25*, 16–19.
11. For a recent example see: Paulus, R. M.; Erdmenger, T.; Becer, C. R.; Hoogenboom, R.; Schubert, U. S. *Macromol. Rapid Commun.* 2007, *28*, 484–491.
12. Bogdal, D.; Penczek, P.; Pielichowski, J.; Prociak, A. *Adv. Polym. Sci.* 2003, *163*, 93–263.
13. Wiesbrock, F.; Hoogenboom, R.; Schubert, U. S. *Macromol. Rapid Commun.* 2004, *25*, 1739–1764.
14. Hoogenboom, R.; Schubert, U. S. *Macromol. Rapid Commun.* 2007, *28*, 368–386.
15. Sinnwell, S.; Ritter H. *Aust. J. Chem.* 2007, *60*, 729–743.
16. Li, C.-J. *Green Chem.* 2008, *10*, 151–152.
17. Li, C.-J.; Chan, T.-H. *Organic Reactions in Aqueous Media*; New York: Wiley, 1997.
18. Kremsner, J. M.; Kappe, C. O. *Eur. J. Org. Chem.* 2005, 3672–3679.
19. Hawker, C. J.; Bosman, A. W.; Harth, E. *Chem. Rev.* 2001, *101*, 3661–3688.
20. Li, J.; Zhu, X.; Zhu, J.; Cheng, Z. *Radiat. Phys. Chem.* 2007, *76*, 23–26.
21. Costa, C.; Santos, A.; Fortuny, M.; Araújo, P.; Sayer, C. *Mater. Sci. Eng., C* 2009, *29*, 415–419.
22. Xu, Z.; Hu, X.; Li, X.; Yi, C. *J. Polym. Sci., Part A: Polym. Chem.* 2008, *46*, 481–488.
23. Zhang, Y.; Liu, R.; Hu, Y.; Li, G. *Anal. Chem.* 2009, *81*, 967–976.
24. Luo, H.; Sheng, J.; Wan, Y. *Mater. Lett.* 2008, *62*, 37–40.
25. Brunel, R.; Marestin, C.; Martin, V.; Mercier, R. *High Perform. Polym.* 2010, *22*, 82–94.
26. Rogers, R. D and Seddon, K. R. Eds. *Ionic Liquids as Green Solvents: Progress and Prospects* Washington, DC: American Chemical Society, 2003.
27. Boros, E.; Seddon, K. R.; Strauss, C. R. *Chim. Oggi,* 2008, *26(6)*, 28–30.
28. Harrison, S.; MacKenzie, S. R.; Haddleton, D. M. *Macromolecules* 2003, *36*, 5072–5075.
29. Carmichael, A. J.; Haddleton, D. M.; Bon, S. A. F.; Seddon, K. R. *Chem. Commun.* 2002, 1237–1238.
30. Basko, M.; Biedron, T.; Kubisa, P. *Macromol. Symp.* 2006, *240*, 107–113.
31. Zhang, H.; Hong, K.; Mays, J. W. *Macromolecules,* 2002, *35*, 5738–5741.
32. Zhang, H.; Hong, K.; Jablonsky, M.; Mays, J. W. *Chem. Commun.*, 2003, 1356–1357.
33. Martínez-Palou, R. *J. Mex. Chem. Soc.* 2007, *51(4)*, 252–264.
34. Guerrero-Sanchez, C.; Hoogenboom, R.; Schubert, U. S. *Chem. Commun.* 2006, 3797–3799.

35. Guerrero-Sanchez, C.; Lobert, M.; Hoogenboom, R.; Schubert, U. S. *Macromol. Rapid Commun.* 2007, *28*, 456–464.
36. Dobrzynski, P.; Kasperczyk, K. *J. Polym. Sci. Part A: Polym. Chem.* 2006, *44*, 3184–3201.
37. Liao, L.; Zhang, C.; Gong. S. *J. Polym. Sci., Part A: Polym. Chem.* 2007, *45*, 5857–5863.
38. Amajjahe, S.; Ritter, H. *Macromol. Rapid Commun.* 2009, *30*, 94–98.
39. Taylor, R. *J. Chem. Soc., Perkin Trans.* 2 1975, 1025–1029.
40. Sameer, P. N.; Picchioni, F.; Janssen, L.P.B.M. *Prog. Polym. Sci.* 2006, *31*, 19–43.
41. van Herk, A. M. ; Manders, B. G.; Canelas, D. A.;. Quadir, M. A.; DeSimone, J. M. *Macromolecules* 1997, *30*, 4780–4782.
42. Beuermann, S.; Buback, V.; Schmaltz, C.; Kuchta, F. D. *Macromol. Chem. Phys.* 1998, *199*, 1209–1216.
43. Beuermann, S.; Buback, M.; Isemer, C.; Wahl, A. *Macromol. Rapid Commun.* 1999, *20*, 26–32.
44. Mang, S. A.; Dokolas, P.; Holmes, A. B. *Org. Lett.* 1999, *1*, 125–127.
45. DeYoung, J. P.; Romack, T. J.; DeSimone, J. M. In *Fluoropolymers 1: Synthesis*. Hougham, G. G.; Cassidy, P. E.; Johns, K.; Davidson, T. Eds.; Springer, New York, 1999.
46. Davies, O. R.; Lewis, A. L.; Whitaker, M. J.; Tai, H.; Shakesheff, K. M.; Howdle, S. M. *Adv. Drug Delivery Rev.* 2008, *60*, 373–387.
47. Iannelli, M.; Bergamelli, F.; Kormos, C. M.; Paravisi, S.; Leadbeater, N. E. *Org. Process Res. Dev.* 2009, *13*, 634–637.
48. http://www.milestonesrl.com/analytical/products-microwave-digestion-ultraclave.html.
49. (a) Kremsner, J. M.: Kappe, C. O. *J. Org. Chem.* 2006, *71*, 4651–4658. (b) Leadbeater, N. E.; Torenius, H. M. *J. Org. Chem.* 2002, *67*, 3145–3148.
50. Strauss, C. R.; Trainor, R. W. *Aust. J. Chem.* 1995, *48*, 1665–1692.
51. Teffal, M.; Gourdenne, A. *Eur. Polym. J.* 1983, *19*, 543–555.
52. Palacios, J.; Sierra, M.; Rodriguez, P. *New Polym. Mat.* 1992, *3*, 273–281.
53. Barbier-Baudry, D.; Brachais, L.; Cretu, A.; Gattin, R.; Loupy, A.; Stuerga, D. *Environ. Chem. Lett.* 2003, *1*, 19–23.
54. Chen, Y.-Y.; Wu, G.; Qiu, Z.-C.; Wang, X.-L.; Zhang, Y.; Lu, F.; Wang, Y.-Z. *J. Polym. Sci., Part A: Polym. Chem.* 2008, *46*, 3207–3213.
55. Iannelli, M.; Alupei, V.; Ritter, H. *Tetrahedron* 2005, *61*, 1509–1515.
56. Iannelli, M.; Ritter, H. *Macromol. Chem. Phys.* 2005, *206*, 349–353.
57. Ma, Y.; Agarwal, U. S.; Sikkema, D. J.; Lemstra, P. J. *Polymer* 2003, *44*, 4085–4096.
58. Mallakpour, S.; Dinari, M. *High Performance Polymers* 2009, 1–14.
59. Chantarasiri, N.; Choprayoon, C.; Thussanee, M. P.; Nongnuj, M. N. *Polym. Degrad. Stab.* 2004, *86*, 505–513.
60. Gallezot, P. *Green Chem.* 2007, *9(4)*, 295–302.
61. Meier, M. A. R.; Metzger, J. O.; Schubert, U. S. *Chem. Soc. Rev.* 2007, *36*, 1788–1802.
62. Günera, F. S.; Yagĉi, Y.; Erciyes, A. T. *Prog. Polym. Sci.* 2006, *31*, 633–670.
63. Corsaro, A.; Chiacchio, U.; Pistarà, V.; Romeo, G. 2006. Microwave-assisted chemistry of carbohydrates. In *Microwaves in organic synthesis* 2nd, ed. A. Loupy, Ed.; Wiley-VCH: Weinheim, Germany, pp. 579–614.
64. Hoogenboom, R.; Schubert, U. S. *Green Chem.* 2006, *8*, 895–899.
65. Hoogenboom, R.; Thijs, H. M. L.; Fijten, M. W. M.; Schubert, U. S. *J. Polym. Sci., Part A: Polym. Chem.* 2007, *45*, 5371–5379.
66. Nagahata, R.; Sano, D.; Suzuki, H.; Takeuchi, K. *Macromol. Rapid Commun.* 2007, *28*, 437–442.
67. Pandey, A.; Pandey, G. C.; Aswath, P. B. *J. Mech. Behav. Biomed. Mater.* 2008, *1*, 227–233.

68. Chatti, S.; Bortolussi M.; Bogdal, D.; Blais, J.C.; Loupy, A. *Eur. Polym. J.* 2006, *42*, 410–424.
69. Barros, M. T.; Petrova, K. T.; Ramos, A. M. *Eur. J. Org. Chem.* 2007, *8*, 1357–1363.
70. Ramos, A. M.; Lobo, S. L.; Bordado, J. M. *Macromol. Symp.* 1998, *127*, 43–50.
71. Velmathi, S.; Nagahata, R.; Sugiyama, J.; Takeuchi, K. *Macromol. Rapid Commun.* 2005, *26*, 1163–1167.
72. Yu, Z.;. Liu, L. *J. Appl. Polym. Sci.* 2007, *104*, 3973–3979.
73. Wang, J.; Mao, H.-Q.; Leong, K.W. *J. Am. Chem. Soc.* 2001, *123*, 9480–9481.
74. Georgieva, R.; Tsevi, R.; Kossev, K.; Kusheva, R.; Baljjiska, M.; Petrova, R.; Tenchova, V.; Gitsov, I.; Troev, K. *J. Med. Chem.* 2002, *45*, 5797–5801.
75. Bezdushna, E.; Ritter, H.; Troev, K. *Macromol. Rapid Commun.* 2005, *26*, 471–476.
76. Chiu, H. C.; Kopeckova, P.; Deshmane, S. S.; Kopecek, J. *J. Biomed. Mater. Res.* 1997, *34*, 381–392.
77. Le Droumaguet, B.; Velonia, K. *Macromol. Rapid Commun.* 2008, *29*, 1073–1089.
78. van Dijk, M.; Mustafa, K.; Dechesne, A. C.; van Nostrum, C. F.; Hennink, W. E.; Rijkers, D. T. S.; Liskamp, R. M. J. *Biomacromolecules* 2007, *8*, 327–330.
79. van Dijk, M.; Nollet, M. L.; Weijers, P.; Dechesne, A. C.; van Nostrum, C. F.; Hennink, W. E.; Rijkers, D. T. S.; Liskamp, R. M. J. *Biomacromolecules* 2008, *9*, 2834–2843.
80. Walker, A. M.; Tao, Y.; Torkelson, J. M. *Polymer* 2007, *48*, 1066–1074.
81. Thirathumthavorn, D.; Charoenrein, S. *Carbohydr. Polym.* 2006, *66*, 258–265.
82. Rivero, E. I.; Balsamo. V.; Müller, A. J. *Carbohydr. Polym.* 2009, *75*, 343–350.
83. Xu, F.; Jiang, J.; Sun, R.; She, D.; Peng, B.; Sun, J.; Kennedy, J. F. *Carbohydr. Polym.* 2008, *73*, 612–620.
84. Lorenzetti. C.; Maaresi, P.; Berti, C.; Barbiroli, G.; *J. Poly. Environ.* 2006, *14*, 89–101.
85. Kržan, A. *J. Appl. Polym. Sci.* 1998, *80*, 1115–118.
86. Kržan, A. *Polym. Adv. Technol.* 1999, *10*, 603–606.
87. Nikje, M. M. A.; Nazari, F. *Adv. Polym. Tech.* 2006, *25*, 242–246.
88. Shukla, S.; Harad, A. M.; Jawale, L. S. *Waste Manage.* 2008, *28*, 51–56.
89. Data from "Market Research Report of China Polyvinyl Chloride" http://www.wanfang-data.com.
90. Braun, D. *Prog. Polym. Sci.* 2002, *27*, 2127–2195.
91. Moriwaki, S.; Machida, M.; Tatsumoto, H.; Otsubo, Y.; Aikawa, M.; Ogura, T. *Appl. Thermal Eng.* 2006, *26*, 745–750.
92. Osada, F.; Yoshioka, T. *Journal of Material Cycles and Waste Management* 2009, *11*, 19–22.
93. Kanaya, K.; Takahashi, S. *J. Appl. Polym. Sci.* 1994, *51*, 675–682.
94. Molero, C.; Lucas, A.; Rodriguez, J. F. *Polym. Degrad Stab.* 2006, *91*, 221–228.
95. Simioni, F.; Bisello, S.; Tavan, M. *Cell. Polym.* 1983, *2*, 281–283.
96. Mahoney L. R. Hydrolysis of polyurethane foams. U.S. Patent 4,196,148, 1980.
97. Nikje, M. M. A.; Nikrah, M.; Haghshenas, M. *Polym. Bull.* 2007, *59*, 91–104.

4 Microwave Heating as a Tool for Drug Discovery

Ping Cao and Nicholas E. Leadbeater

CONTENTS

4.1 INTRODUCTION

In early drug discovery, the major focus of medicinal chemists is the design and preparation of new compounds, followed by evaluation of their structure–activity relationship (SAR) based on the results from biological tests.[1] Traditional medicinal chemistry approaches adopted during the 1970s and 1980s were focused primarily on making analogs of endogenous ligands and industry leads. Recent advances in genomics and proteomics are leading to a very large number of possible drug targets. The development of high-throughput technologies removes the hurdles associated with screening these, but creates a greater demand for their rapid preparation. Additional drive for the generation of many new compounds comes from the fact that innovative drug discovery is challenged by high failure rates in clinical trials. This is mainly due to issues associated with poor pharmacokinetics and efficacy together with high toxicity. In order to overcome these problems and avoid failure at a late stage of the development process, new approaches have to be taken in the drug discovery industry.[2] An integral part of this is the development and use of new synthetic chemistry methodology since this enables medicinal chemists to explore the wide chemical space available for screening.[2]

Of all the recent technical innovations for synthetic chemistry, the development of scientific microwave apparatus ranks very high in its impact on drug discovery.[3] The use of microwave heating for synthesis of potential leads offers significant advantages over conventional approaches, including reduced reaction time, improved yield, and increased productivity. The efficiency of microwave technology is further enhanced by the combination with other novel approaches such as combinatorial chemistry, solid-phase and solution-phase parallel synthesis, and continuous-flow processing. Integration of the synthesis platform with automated purification, evaporation, high-throughput screening, and cheminformatics modules further facilitates the drug discovery process.[4]

Alongside advances in technology, a number of developments in synthetic methodology have transformed the drug discovery process, such as click chemistry,[5] transition-metal-catalyzed cross-coupling reactions,[6] DNA-templated organic synthesis,[7] and olefin metathesis.[8] Combining microwave heating with these facilitates the synthesis of new classes of compounds with complex carbon frameworks.

The aim of this chapter is to show the significant contribution made by microwave-promoted synthesis in the drug discovery processes. We will start by illustrating how the technology has been used for generation of discovery chemistry libraries before transitioning to show how it serves as a tool for medicinal chemists as they try to develop drug candidates. There have been reviews published on the use of microwave heating in drug discovery covering material up to 2007.[9] In this chapter, attention will focus on reports published from 2007 to 2009.

4.2 MICROWAVE-BASED SYNTHETIC METHODOLOGIES FOR GENERATION OF DISCOVERY CHEMISTRY LIBRARIES

Small molecule discovery chemical libraries (MW < 1500 Da) play an essential role in chemical biology and drug discovery. Specifically, libraries of 10–100 members derived from a core scaffold by adding various appendages can offer a very effective approach to lead generation and lead optimization.[10] There are two main classes of chemical library, namely, those derived from natural products and those that are purely synthetic. The value of a discovery library is determined by its ability to explore different areas of chemical space. Generally, libraries derived from natural products display greater structural diversity and complexity than purely synthetic ones. However, the preparation, purification, and characterization of these complex molecules are much less straightforward. By contrast, members of chemical libraries that are synthetically derived can often be more easily purified and characterized.

Microwave heating can be used as a tool for the preparation of both classes of library. Attention here will be focused on examples of libraries generated either using monomode microwave apparatus combined with automated vial-handling systems or in parallel employing suitable reaction vessels using multimode microwave equipment. Since the first introduction of the concept of automated sequential microwave-assisted library synthesis in 2001, this method has become increasingly useful to medicinal chemists interested in preparing small focused libraries of 10–100 members.[11] Taking advantage of the benefits of microwave heating and parallel processing, chemists have become interested in performing reactions in well plates.[12] Although an attractive proposition, the use of well plates in a microwave unit has some significant problems. The most important issue to overcome is uneven heating across the plate. Many of the early reports using polypropylene, Teflon, or high-temperature polyethylene plates showed that wells located on the periphery were at a significantly lower temperature than those on the inside due to radiative heat loss as well as lower microwave coupling. In an attempt to overcome this problem, plates doped with strongly microwave-absorbing materials such as graphite have been used. A 48-position plate made of silicon carbide has been used for parallel synthesis in a microwave unit with success.[13,14] Silicon carbide is an inert, highly microwave-absorbing material and previously has been used as a heating insert for reaction mixtures containing nonabsorbing reagents or solvents. The use of a silicon carbide plate allows for equal heating of the wells, this being confirmed by IR thermal imaging. While it proves useful for rapid library preparation, a drawback of using the plate is that each well has a working volume of only 0.1–0.3 mL. In recent developments, silicon carbide plates capable of holding up to 24 standard glass vials[15] or 20 standard HPLC/GC autosampler vials[16] have become available. Again, uniform heating has been shown by IR thermography. These plates, together with a rotor capable of holding 16 groups of 4 vials, have allowed for efficient and reproducible parallel processing.

4.2.1 Multicomponent Reactions

The use of multicomponent reactions (MCRs) constitutes an attractive synthetic strategy for rapid and efficient library generation because diverse products are formed in a single step.[17] Usually, MCR transformations do not involve the simultaneous reaction of all components. Instead, they are undertaken in a sequence of steps that are determined by the synthetic design. A drawback of many MCR processes is that they can be slow and inefficient, but microwave heating can be used as a tool to overcome these problems, as illustrated here with selected examples.

4.2.1.1 Multicomponent Cyclization/Suzuki Coupling Sequence

A range of 2,6-disubstituted-3-amino-imidazopyridines have been prepared in a one-pot cyclization/Suzuki coupling approach using microwave heating (Scheme 4.1).[18] The motivation behind this comes from the fact that the 5,6-fused heterocyclic core is prevalent in a wide range of synthetic and naturally occurring medicinal compounds. The initial cyclization step involved an Ugi three-component reaction, this being followed by a palladium-catalyzed Suzuki coupling. The Ugi reaction is traditionally performed using a Lewis or Brønsted acid catalyst. However, when screened in this two-step protocol, a number of these acid catalysts had an inhibitory effect on the Suzuki coupling step, a notable exception being $MgCl_2$. As a cyclization partner for the Ugi reaction and as the boron-containing substrate for the Suzuki coupling step, the pinacol ester of 2-aminopyridine-5-boronic acid was used. The boronate was very tolerant to the Lewis acid catalyzed cyclizations, and the Suzuki coupling step proceeded cleanly in the presence of magnesium salts when using $PdCl_2$(dppf) as the catalyst [dppf = 1,1'-bis(diphenylphosphino)ferrocene], potassium carbonate as the base, and water as a cosolvent. With optimized conditions in hand, a library of 2,6-disubstituted-3-amino-imidazopyridines was prepared. Attempts to perform the whole reaction sequence in one step or to carry out the Suzuki coupling followed by Ugi reaction were unsuccessful. The aqueous base required for the Suzuki reaction has the effect of deactivating the Lewis acid catalyst, thereby impeding the Ugi cyclization.

4.2.1.2 Multicomponent Sonogashira Coupling-Cycloaddition Sequence

A Sonogashira coupling/cycloaddition MCR sequence has been reported for the synthesis of isoxazoles (Scheme 4.2).[19] The first step involved reacting acid chlorides

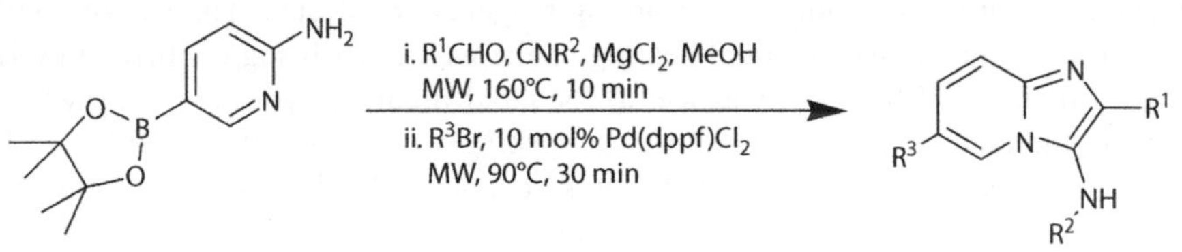

SCHEME 4.1

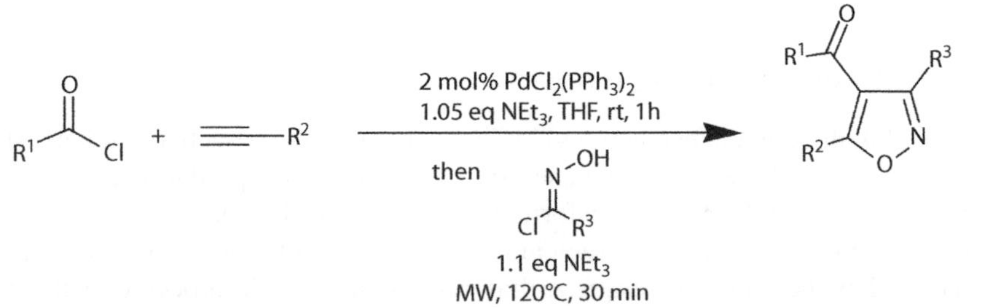

SCHEME 4.2

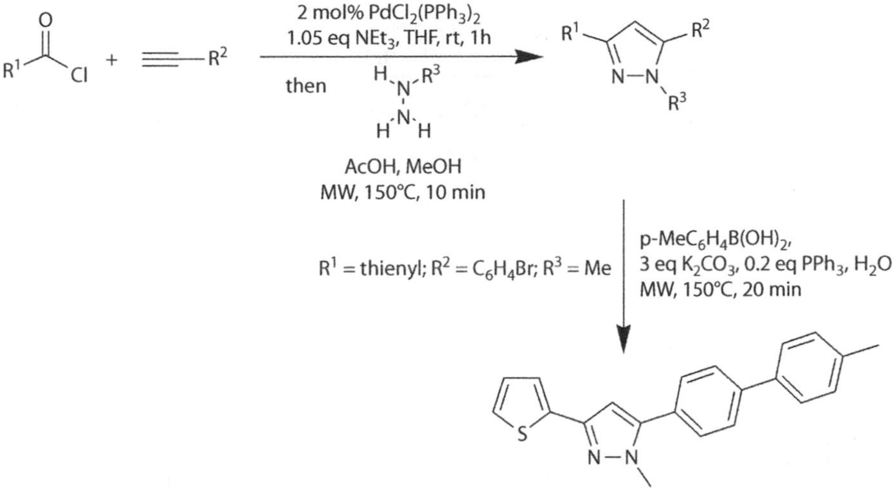

SCHEME 4.3

with terminal alkynes under modified Sonogashira conditions for 1 h at room temperature to furnish the corresponding alkynone intermediates. A hydroximinoyl chloride and triethylamine were added and the reaction mixture heated at 120 °C for 30 min using microwave irradiation. The isoxazoles were obtained in moderate-to-excellent yields. Performing the reaction using conventional heating requires longer reaction times and, as a result, decomposition and side reactions become issues. The cyclization step essentially involves a 1,3-dipolar cycloaddition between the alkyne moiety of the alkynone intermediate and a nitrile oxide, generated in situ from the hydroximinoyl chloride component via a dehydrochlorination reaction with NEt$_3$. As is the case with kinetically controlled 1,3-dipolar cycloadditions, only one of two possible regioisomers of the isoxazole product is observed (Scheme 4.3).

Using a related strategy, a series of pyrazoles have been prepared.[20] The initial Sonogashira step was followed by a Michael addition/cycloaddition with a range of hydrazines. In each case, only one of the two possible regioisomers was formed, the outcome depending on the substituent on the hydrazine component used. When using a bromo-functionalized alkyne as a coupling partner in the first step, the protocol could be further expanded to a four-component one-pot Sonogashira coupling/cyclocondensation/Suzuki coupling process, giving rise to a biphenylyl-substituted pyrazole.

In another variant, 1,5-benzodiazepines have been prepared by reaction of *ortho*-phenylene diamines with the alkynones generated from the Sonogashira step (Scheme 4.4).[21] In this case, microwave heating could be used, but conventional heating at a lower temperature for a significantly longer time (3 d as opposed to 1 h) led to better yields of the products. Interestingly, when using *ortho*-aminothiophenol as a bis-nucleophile, a quinoline product was obtained instead of the expected benzothiazepine. The benzothiazepine initially formed is thought to be unstable and undergoes a sulfur extrusion reaction.

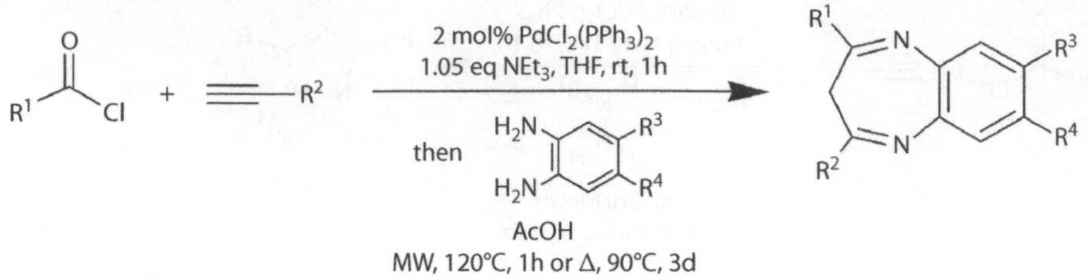

SCHEME 4.4

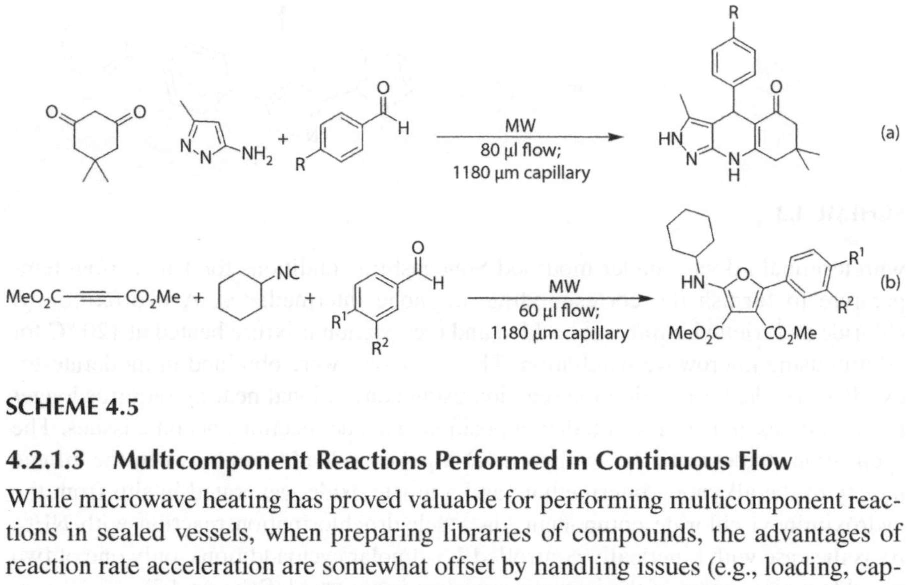

SCHEME 4.5

4.2.1.3 Multicomponent Reactions Performed in Continuous Flow

While microwave heating has proved valuable for performing multicomponent reactions in sealed vessels, when preparing libraries of compounds, the advantages of reaction rate acceleration are somewhat offset by handling issues (e.g., loading, capping, uncapping, and unloading vessels). As an alternative strategy, the use of flow processing has been investigated.[22] By using small capillaries as flow reactors and placing these into the cavity of a microwave unit, it has been possible to prepare a small library of quinolinones from the reaction of a variety of aldehydes with dimedone and 5-amino-3-methyl-1H-pyrazole (Scheme 4.5a). The reagents were passed through the capillary at a rate of 60 µL/min.[23] Using a similar approach, a library of tetrasubstituted furans has been prepared from the reaction of aldehydes with cyclohexyl isocyanide and dimethyl acetylenedicarboxylate (Scheme 4.5b). In both cases, the components were each introduced into the capillary through a separate channel, in equal concentrations and at the same rate. When preparing a library of products, introducing the components separately instead of as a mixture has the potential to expand the combinatorial efficiency of the system.

A flow approach has also been used for the preparation of 1,4-dihydropyridines via a Hantzsch synthesis in a domestic microwave oven.[24] A circular glass tube reactor of 65 mL internal volume was fixed inside the oven along the circumference of the turntable. Two Teflon tubes were attached to the glass reactor, one to each end, and these allowed for material to be passed into and out of the microwave oven via holes cut in the rear. The reaction mixture (125 mL) was flowed from a holding flask through the microwave cavity at a rate of 100 mL/min and back into the original vessel. Four passes were made, and then the product was isolated. Using this approach, it was possible to prepare 50 g of nitrendipine (a calcium channel blocker) as well as other 1,4-dihydropyridines.

4.2.2 DOMINO REACTIONS

Domino reactions are defined as processes in which a consecutive series of transformations take place, each one taking place at functional groups formed in the preceding reaction. They allow chemists to build a large degree of complexity into one transformation. At the same time, they reduce waste and save time since intermediates are not isolated.

4.2.2.1 Copper-Catalyzed Domino Reactions

Fused indoles have been prepared by means of a copper-catalyzed three-component domino coupling-indole formation-N-arylation sequence (Scheme 4.6).[25] The copper salt used salt-catalyzed (a) the Mannich-type coupling of 2-ethynylanilines with formaldehyde and N-substituted *o*-halobenzylamines, (b) indole formation, and (c) arylation of the indole nitrogen. *N*-mesyl-2-ethynylanilines were used as starting materials, the first two steps of the sequence being performed in toluene as a solvent and at 170 °C for 20 min using microwave heating. Upon completion of the indole-forming step, sodium methoxide was added to remove the mesyl-protecting group, and the reaction mixture was heated to 170 °C for a further 20 min to yield the N-arylated product.

A series of 1,2,3-triazoles have been prepared using a Cu-catalyzed Huisgen [2+3] dipolar cycloaddition (click) reaction. The required azide was generated in situ from the corresponding halide and sodium azide (Scheme 4.7).[26] Microwave irradiation dramatically reduced the reaction times from hours to minutes without affecting the 1, 4-regioselectivity of the reaction. Since triazole products often crystallize readily

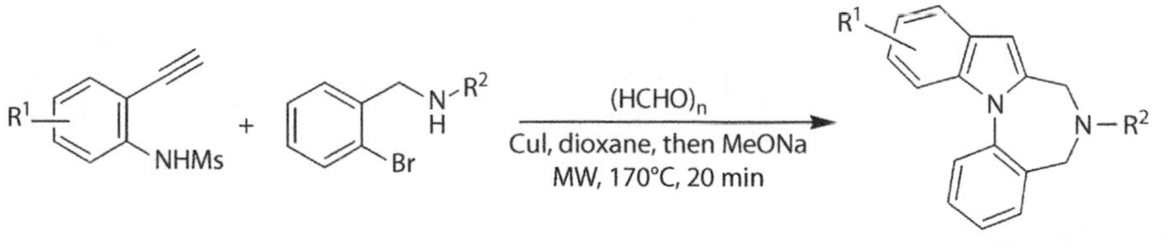

SCHEME 4.6

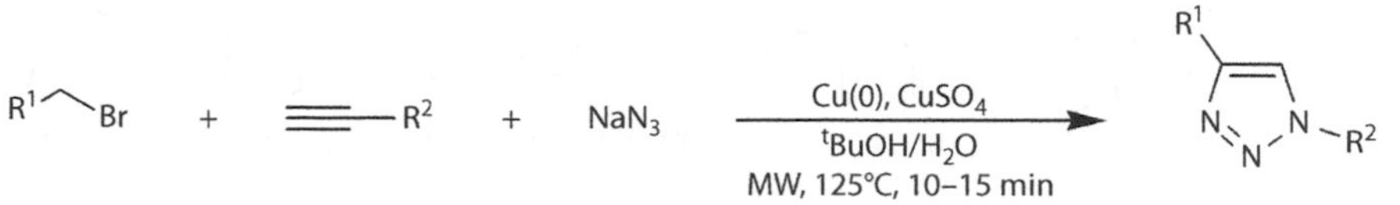

SCHEME 4.7

from solution, purification is greatly facilitated. Enantiomerically pure substituted benzylacetamides have also been prepared using a click protocol in conjunction with microwave heating, 10 min at 125 °C being sufficient to give high yields of the desired products.[27]

4.2.2.2 Organocatalyzed Domino Reactions

An efficient and direct synthesis of polysubstituted indeno[1,2-*b*]quinolines from an aldehyde, 1,3-dione, and enaminone has been performed using *p*-toluenesulfonic acid as an additive, water as the solvent, and microwave heating at 150 °C for 2–7 min (Scheme 4.8).[28] Initial condensation of the aldehyde with a 1,3-indanedione is thought to occur to afford a 2-arylideneindene-1,3-dione intermediate, which undergoes a Michael addition reaction with the enaminone to yield the product.

This procedure has been expanded to the synthesis of 4-azafluorenone (5H-indeno[1,2-*b*]pyridine-5-one) alkaloids, which have significant biological activity (Scheme 4.9).[29] An aldehyde, 1,3-indanedione, aryl ketone, and ammonium acetate are used as reagents. Heating at 120 °C for 6–15 min in a sealed tube and using DMF as solvent proved optimal.

A similar approach has been used for the synthesis of 3-1H-pyrazolo[3,4-*b*]pyridyl-indole[30] and 2-aminochromene[31] libraries, the latter being performed solvent free using a clay as a support and base. In addition, a library of over 100 derivatized substituted 2-pyridones (21 × 5 members) has been prepared by the reaction of 1,3-cyclohexane-diones, dimethylformamide dimethylacetal, and various cyanoacetamides (Scheme 4.10).[32] The outcome of the reaction could be controlled to furnish different products by changing the basicity of the medium to involve either the amide or nitrile functionality of the cyanoacetamide substrate during the 2-pyridone ring-formation step.

A microwave-promoted domino Knoevenagel-hetero Diels–Alder reaction of aldehydes, amides, and dienophiles has been used for the synthesis of a large variety of carbo- and heterocyclic amides (Scheme 4.11).[33] This mechanism involves the initial formation of a 1-(*N*-acylamino)-1,3-butadiene by reaction

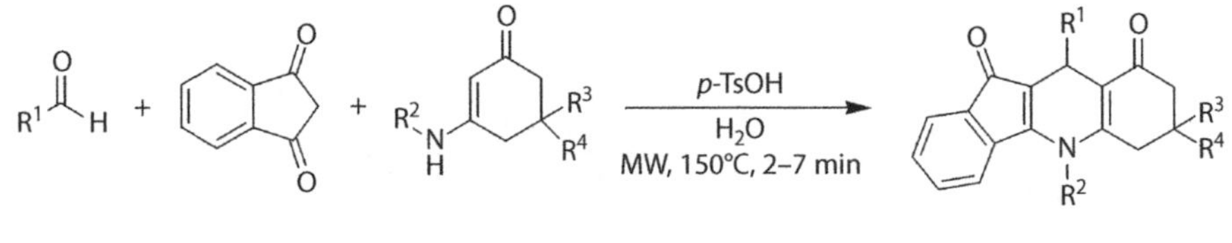

SCHEME 4.8

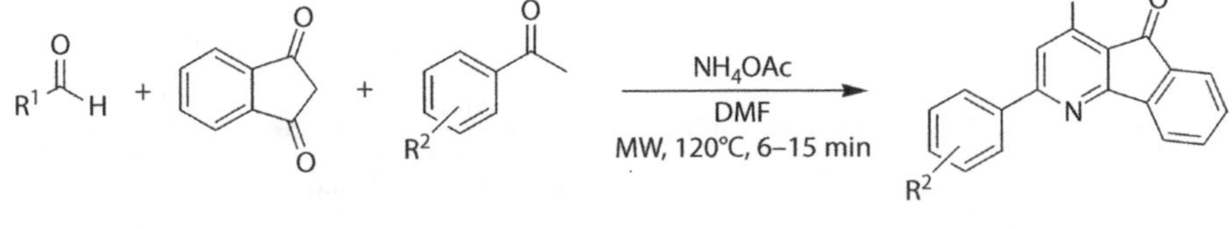

SCHEME 4.9

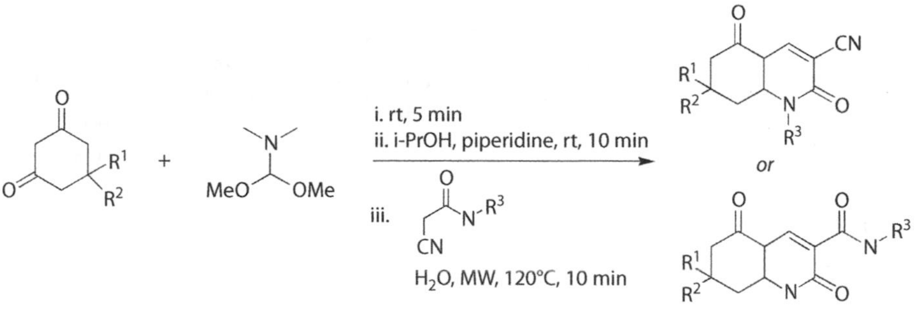

SCHEME 4.10

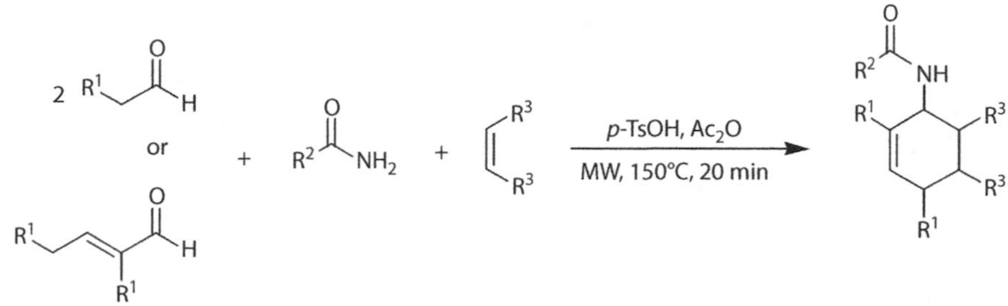

SCHEME 4.11

of the aldehyde and amide substrates, this being followed by a Diels–Alder condensation reaction with an electron-deficient dienophile. Typically, this reaction had been performed at 80–120 °C in dipolar, aprotic solvent such as N-methylpyrrolidone (NMP). However, when trying to obtain a library of products, competitive formation of aldol-type side products limited its scope. Moving to aprotic solvents such as toluene increased the yield of the desired products but required prolonged reaction times (16–120 h). By transitioning to a microwave-promoted solvent-free approach, excellent product yields could be obtained in 20 min at 150 °C, with the addition of acetic anhydride as a water scavenger. Different functionalized amides were reacted with aliphatic as well as α,β-unsaturated aldehydes in the presence of suitable dienophiles, giving a series of 1-acylamino-2-cyclohexene derivatives.

Carbohydrates can be used as biorenewable and versatile starting materials for organic reactions. A number of 1,3-oxazin-2-ones and 1,3-oxazin-2-thiones have been prepared in domino reactions starting from D-glucose and D-xylose (Scheme 4.12).[34] Using montmorillonite K-10 clay as a base catalyst and a support, the reaction of the sugars with semicarbazide initially yielded the corresponding semicarbazone after microwave heating. These rapidly cyclize in the presence of the basic clay to form 1,3-oxazin-2-ones, the whole process taking 2–4 min at 90 °C. The analogous 1,3-oxazin-2-thiones could be prepared using thiosemicarbazide in place of semicarbazide. The 1,3-oxazin-2-ones(thiones) have then been used as scaffolds for

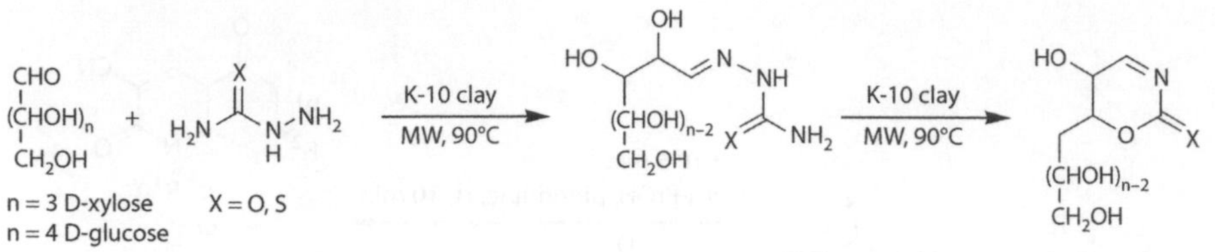

SCHEME 4.12

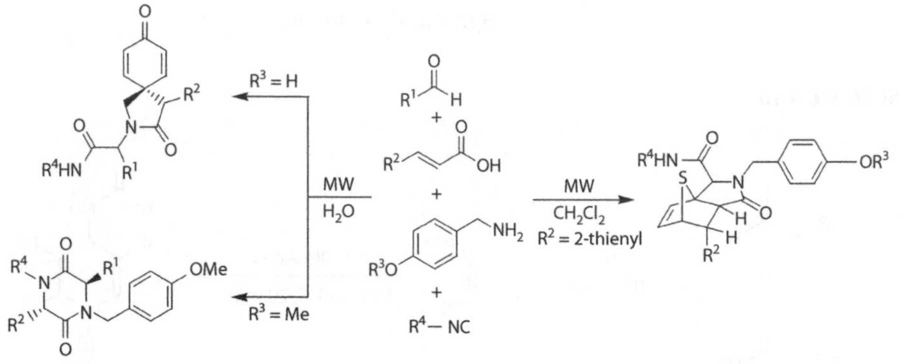

SCHEME 4.13

exploiting chemical diversity and generation of drug-like libraries for screening as potential new leads.

Microwave heating has been used as a tool for preparing a range of heterocycles, all from the same bifunctional class of starting materials. The outcome of the reaction could be controlled by varying the solvent or the functionality of the starting material as shown in Scheme 4.13.[35] In all cases, the first step of the reaction was a 4-component Ugi coupling. Then, when using water as a solvent, a range of 2,5-diketopiperazines were obtained via a subsequent *aza*-Michael reaction or, alternatively, 2-azaspiro[4,5]deca-6,9-diene-3,8-diones were generated through a 5-*exo* Michael cyclization. When dichloromethane was used as a solvent and thiophene-2-carboxaldehyde as a substrate, unique thiophene-derived tricyclic lactams were formed via an intramolecular Diels–Alder pathway.

A library of pyrazoloquinolizinones has been prepared via a multicomponent reaction between 5-aminopyrazoles, cyclic 1,3-diketones, and aromatic aldehydes.[36] Under the strongly basic conditions used, the initial product formed undergoes a subsequent ring-opening/recyclization process to give the pyrazoloquinolizinones (Scheme 4.14).

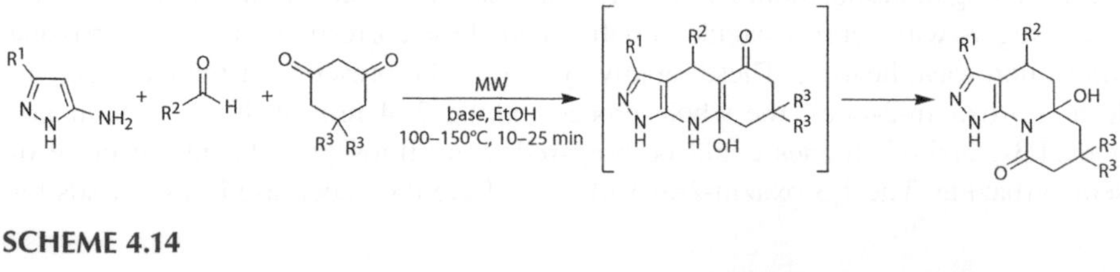

SCHEME 4.14

4.2.3 PARALLEL SYNTHESIS USING SOLID-SUPPORTED STRATEGIES

4.2.3.1 Solid-Supported Synthesis

Solid-supported synthesis offers a number of practical advantages over solution-phase approaches.[37] For example, it is possible to use excess reagent to drive reactions to completion, and then, at the end, it is possible to isolate the supported product easily by filtration. In addition, library synthesis can be facilitated using solid-supported approaches. One disadvantage, however, is that the rate of reactions can be significantly slower when substrates are confined to a solid support. This has been overcome in many cases by using microwave heating.[38]

As an example, in the solid-phase synthesis of di-substituted, tri-substituted, and fully-substituted pyrazolidine-3,5-diones microwave heating proved valuable in a key acylation step.[39] Conditions were first developed for performing the reaction on a solution-phase analog using conventional methods. However, use of controlled microwave heating (90 °C for 1 h) gave superior results. This translated well from solution phase to solid phase. A small library of 25 members was prepared in order to demonstrate the versatility of the chemistry.

In the traceless, solid-phase synthesis of hetero-annulated 1,3-oxazine-6-ones, significant rate enhancement was observed across a range of steps carried out using microwave heating (Scheme 4.15).[40] In a similar traceless, solid-phase approach, a number of 2,4,6-trisubstituted thiazolo[4,5-d]pyridine-5,7-dione derivatives have been prepared using microwave heating as a key step.[41] An amino ester resin was first swollen in DMSO; then an isocyanate was added, and the reaction mixture was heated to 150 °C to yield a thiazolourea, which was converted to the final product in another three steps on the support before a final cleavage protocol (Scheme 4.16).

In an effort to generate methionine aminopeptidase inhibitors, a chromatography-free route for Suzuki coupling reactions has been developed, taking advantage of automation and microwave heating (Scheme 4.17).[42] The approach involves initial immobilization of carboxylic-acid-functionalized heteroaryl halides on a polymer-supported 2-*tert*-Butylimino-2-diethylamino-1,3-dimethylperhydro-1,3,2-

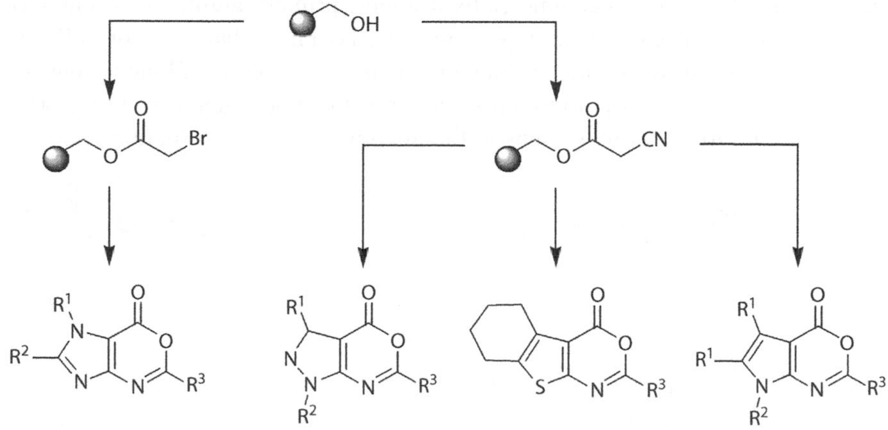

SCHEME 4.15

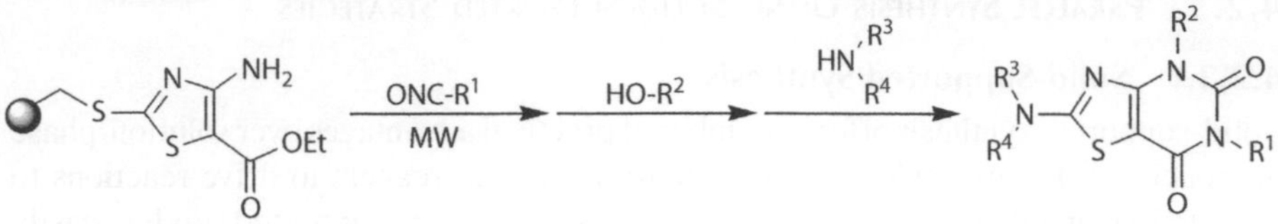

SCHEME 4.16

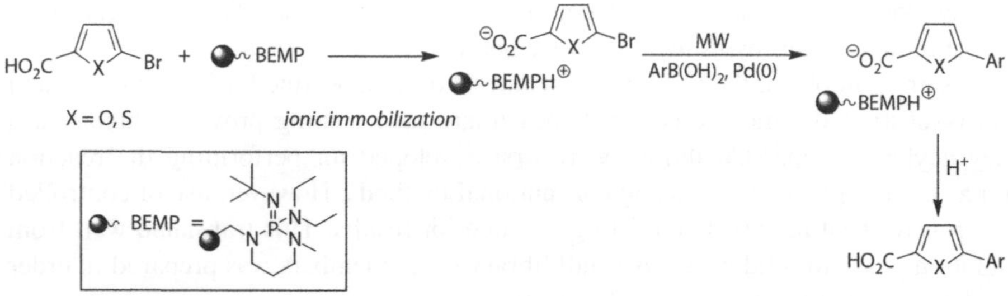

SCHEME 4.17

diazaphosphorine (PS-BEMP). Unlike traditional polymer-supported chemistry, in this case the interaction between the support and the heteroaryl moiety is ionic and not covalent. Reagents were then added to each reaction vessel in order to perform a Suzuki coupling. Loading of the heteroaryl halide onto the resin and dosing of the boronic acid, base, and palladium catalyst were performed in parallel using a Bohdan Miniblock. Each reaction vessel was then sequentially transferred into a microwave unit and heated to 180 °C for 5 min. The supported biaryl products were then filtered and washed in parallel before liberating them from the support in moderate yields but excellent purity.

Microwave heating has been used as a tool for the generation of a library of 625 peptoids (Scheme 4.18).[43] The method, involving the synthesis of N-alkylglycine oligomers, used a set of 10 commercially available primary amines as a source of chemical diversity. The synthetic steps were carried out in tea bags. The overall synthetic strategy involved an initial room temperature deprotection of Fmoc-protected Rink amide resin. This was followed by five repetitive acylation (35 °C for 30 s) and amination (90 °C for 90 s) steps in the microwave unit. Two sublibraries were

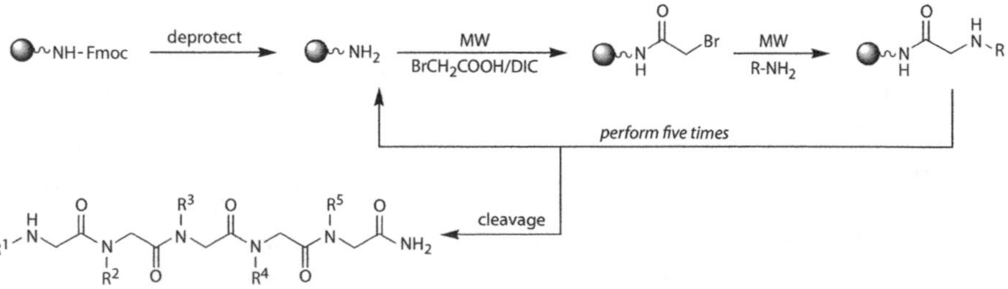

SCHEME 4.18

prepared. In one, a single (different) amine was added in each amination step, giving pentameric products with known composition. In the other, a mixture of all five amines was used in each amination step. This led to a library of randomly substituted pentamers. In this case, the relative reactivity of each of the five amines used was taken into account, and the relative stoichiometry of each varied so as to ensure comparable derivatization in the product mixtures. The overall library of N-acylglycine pentamers was validated by the screening, deconvolution, and identification of active oligomers as trypsin inhibitors.

Other examples of the use of microwave heating in conjunction with solid-supported synthesis strategies include the preparation of a series of 90 N-acylated homoserine lactone-like ligands for testing as modulators of bacteria quorum sensing[44] as well as click chemistry for the preparation of small libraries of galactosyl oligomers from different poly-alkyne DNA-based scaffolds and two galactosyl azide derivatives.[45]

4.2.3.2 Use of Solid-Supported Synthesis Together with Solid-Supported Reagents

Solid-supported reagents have found significant application in organic synthesis.[46] They are easily removed from reaction mixtures by filtration and, as a result, excess reagent can be used to drive reactions to completion without difficulties in purification. They offer ease of handling, this being especially important when using hazardous reagents. Their activity can often be fine-tuned by varying the support used and, in many cases, they react more selectively than their homogeneous counterparts. However, as with performing synthesis on a solid support, a drawback to the use of supported reagents is that reactions are often slow. This is due to issues associated with diffusion of substrates through the polymeric backbone. Microwave heating has been used effectively as a tool to facilitate reactions using supported reagents, dramatic rate enhancements being seen.[47]

Polystyrene-bound carbodiimide together with 1-hydroxybenzotriazole (HOBt) has been used as the final step for the preparation of a library of 3,4-dehydroproline amides (Scheme 4.19).[48] Microwave heating at 100 °C for 10 min was sufficient in most cases, converting the carboxylic acid precursors into the corresponding amides when using a range of amines as substrates. When more volatile amines were used, the reactions were better performed at a lower temperature for a longer time (60 °C for 30 min). Following the reaction, filtration through a silica-bound carbonate solid-phase extraction cartridge to scavenge any unreacted acid and HOBt allowed for easy access to the product. A similar approach has been used for the final step in the preparation of a library of 90 carboxamide-containing oxepines and pyrans.[49]

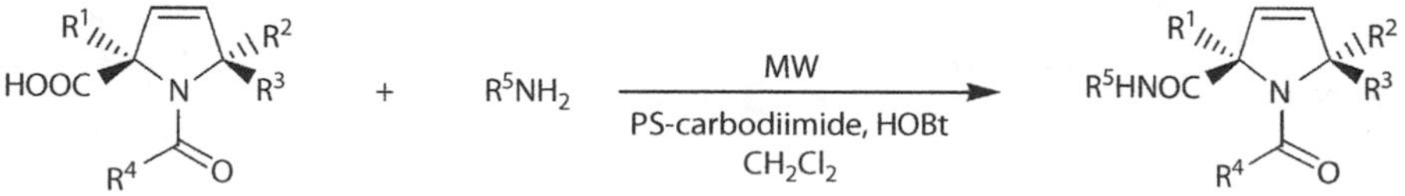

SCHEME 4.19

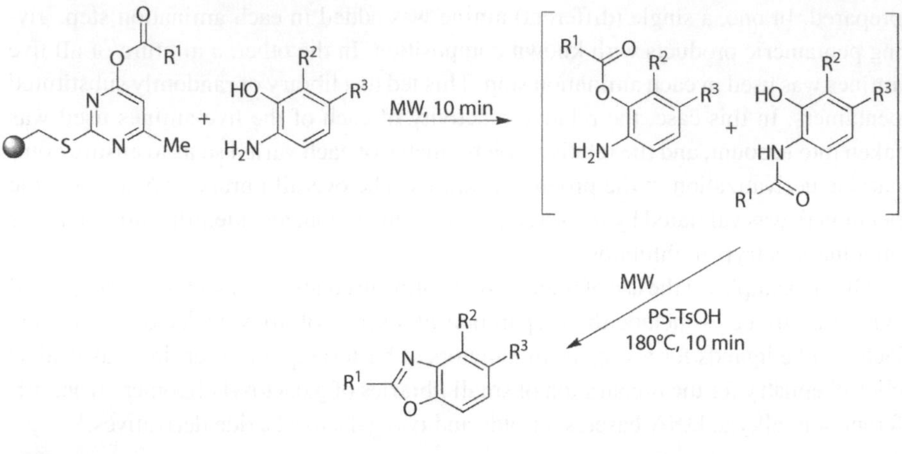

SCHEME 4.20

The combination of solid-supported synthesis and use of a solid-supported reagent has been used successfully in a one-pot two-step approach to the synthesis of substituted benzoxazoles (Scheme 4.20).[50] The protocol employed the combination of a Büchi Syncore parallel synthesizer and a microwave unit. In the first step of the reaction, vessels were each loaded with a polymer-supported activated ester together with a solution of a substituted aminophenol in dichloromethane. The solvent was then removed in parallel using the Syncore. The neat reaction mixtures were then sequentially placed into the microwave unit and heated for 10 min. During the course of the reaction, the uncyclized intermediates were released off the support. Upon cooling, the vessels were opened, polymer-supported p-toluenesulfonic acid (PS-TsOH) and toluene added, and then each reheated for 10 min at 180 °C. Filtration, washing, and evaporation of the solvent were performed in parallel using the Syncore, this allowing for isolation of the benzoxazole products.

A catch-and-release strategy has been used for the synthesis of 2-substituted benzofurans in a two-step approach (Scheme 4.21).[51] Initially, polymer-supported triphosphine (PS-PPh$_2$) was treated with a substituted 2-(bromoalkyl)-phenol to generate the corresponding supported (2-hydroxybenzyl)-triphenylphosphonium bromide. To obtain acceptable conversion, an initial dose of 2-(bromoalkyl)-phenol was added to PS-PPh$_2$ swollen in DMF. This mixture was heated to 85 °C and held for 15 min. Upon cooling, a second dose of the phenol was added and the reaction mixture reheated using the same program. In the second step of the reaction, treatment of the supported (2-hydroxybenzyl)-triphenylphosphonium bromide with an acyl chloride

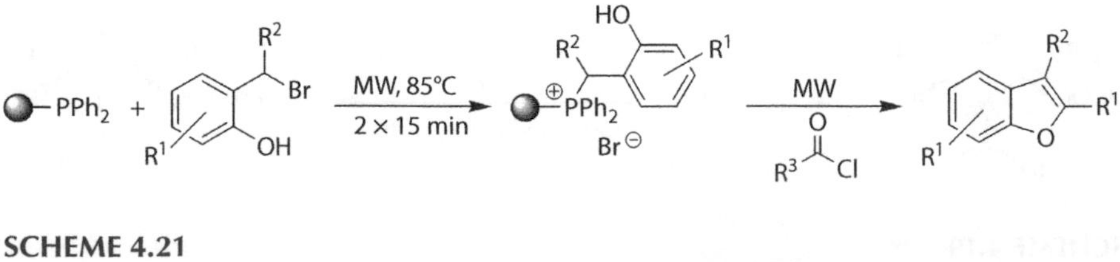

SCHEME 4.21

in the presence of triethylamine generated and liberated the 2-substituted benzo-furan products. The methodology has been further developed for the preparation of chiral α-alkyl-2-benzofuranmethanamides.

Other applications of polymer-supported reagents in conjunction with microwave heating include the preparation of acylhomoserine lactones that selectively inhibit cancer cells using a supported amide coupling reagent[52] and a 66-member C2-aryl pyrrolo[2,1-c][1,4]benzodiazepine library via a Suzuki coupling protocol using a supported palladium catalyst followed by a supported diol for scavenging excess boronic acid.[53]

4.2.3.3 Parallel Synthesis Using Soluble Polymer-Supported Strategies

While insoluble polymer supports have seen great success in small-molecule organic synthesis, the heterogeneous reaction conditions that they require often can compli-cate the direct transfer of traditional solution-phase chemical methodologies to solid-phase synthesis. Other issues include incomplete conversion and the fact that reaction monitoring is very difficult. To make polymer-supported synthesis more "solution-like," there has been increasing interest in the development and use of soluble poly-mers[54] and fluorous chemistry.[55] At its best, soluble polymer-supported synthesis can bring together the advantages of solution-phase and solid-phase chemistry. Using a soluble polyethyleneglycol (PEG) support, a library of benzimidazolyl quinoxalino-nes has been rapidly prepared.[56] Microwave heating proved valuable in three of the five steps to these compounds, starting from PEG-supported ortho-diamino esters (Scheme 4.22). The synthetic approach taken to the products when immobilizing the starting material on a PEG support was much simpler than that required when

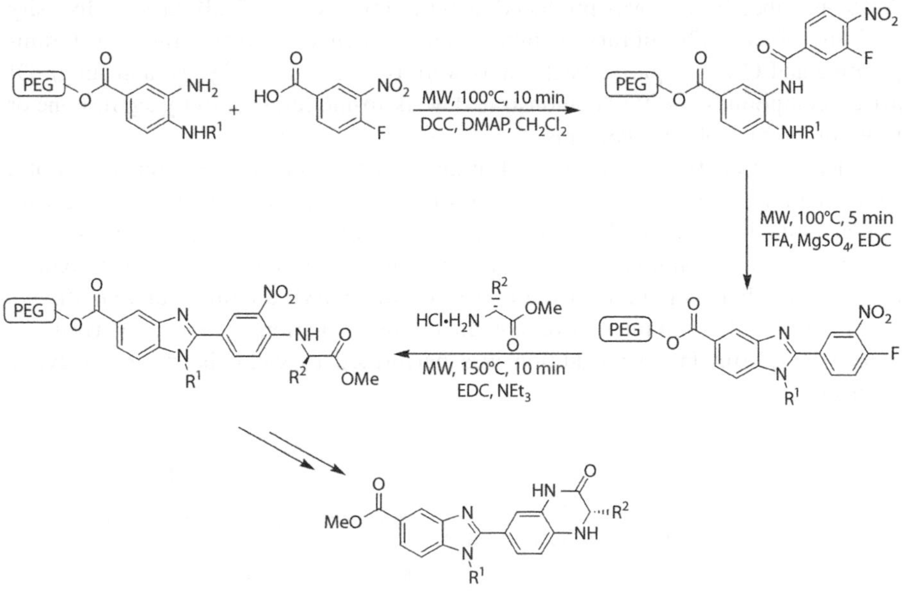

SCHEME 4.22

using traditional solution-phase chemistry. Efficient synthesis of 4-substituted-5-methoxycarbonyl-3,4-dihydropyridone[57] and hydantoin-fused tricyclic tetrahydro-β-carboline[58] libraries on soluble supports has also been reported, microwave heating being central to their success.

4.2.4 SYNTHESIS OF NATURAL PRODUCT-LIKE LIBRARIES

Natural products and their derivatives have historically been invaluable as a source of therapeutic agents.[59] However, in the past decade, research in the pharmaceutical industry directed at natural product chemistry generally saw a decline owing to issues such as the incompatibility of traditional natural product extract libraries with high-throughput screening approaches. However, recent technological advances coupled with unrealized expectations from current lead generation strategies have led to renewed interest in natural products in drug discovery.[60]

Nonactin is a macrotertrolide antibiotic produced by *Streptomyces griseus*. The macrocyclic ring of nonactin is composed of two units of (+)-nonactate and two units of (−)-nonactate. Methanolysis of nonactin followed by resolution provided both enantiomers of methyl nonactate. The methyl nonactate units have been used as complex natural product scaffolds for assembly of combinatorial compound libraries since it is possible to generate a highly diverse library with relatively complex stereochemistry in a relatively simple manner. In one such example, a library of triazoloamides has been generated from an azide-derivatized analog of (−)-nonactatic acid (Scheme 4.23).[61] Initially, a series of amides were formed and then triazoloamides generated using a microwave-promoted, copper-catalyzed 1,3-dipolar cycloaddition of the azido moiety and a series of terminal alkynes. A 161-member library was produced using 5 amines and 32 alkynes as diversity building blocks. The library members were screened against a range of Gram-positive and Gram-negative bacteria as well as a selection of yeast and fungi. Of all the compounds, 14 (9%) demonstrated weak-to-moderate activity against one or more of the microbial organisms.

Microwave heating has been used in many of the steps for the preparation of a small library of aza-analogs of (−)-steganacin, a naturally occurring lactone with potent antileukemic and tubulin polymerization inhibitory activity (Scheme 4.24).[62] To prepare these compounds, a Suzuki coupling was followed by a three-component reaction between an aldehyde moiety on the biaryl, an amine and an alkyne. After functional group interconversion, involving formation of an azide-derivatized analog, a 1,3-dipolar cycloaddition was performed to yield the desired bicyclic products.

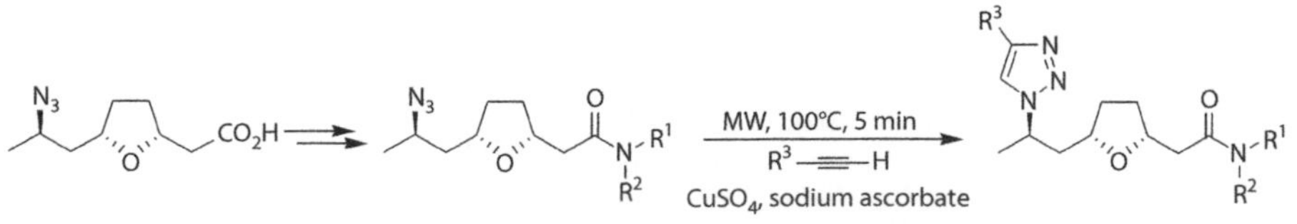

SCHEME 4.23

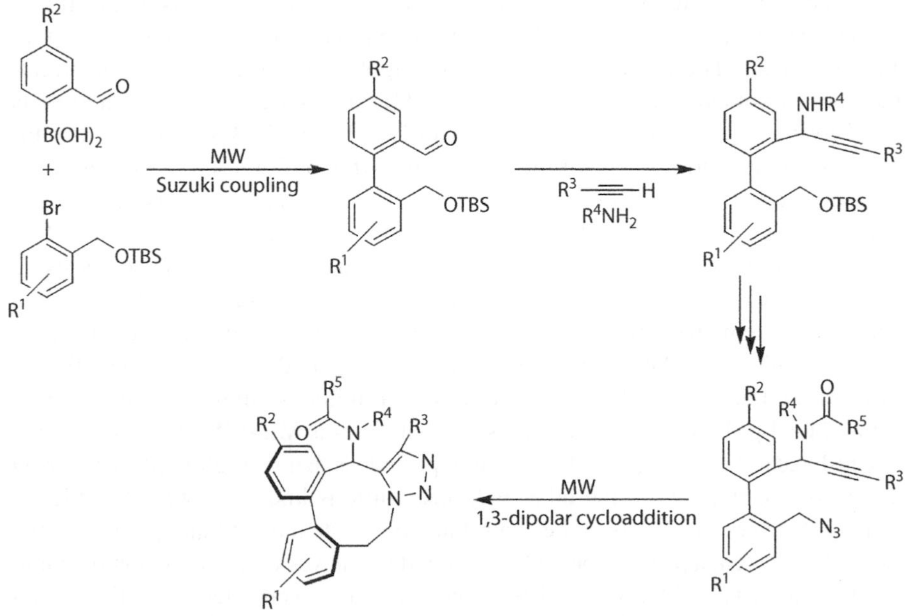

SCHEME 4.24

4.3 MICROWAVE-ASSISTED MEDICINAL CHEMISTRY

The identification of small molecule modulators of protein function and the process of transforming these into useful therapeutic agents are key activities in modern drug discovery. The hit-to-lead phase is usually the follow up of a high-throughput screening (HTS) campaign. The key medicinal chemistry activities at this phase involve the synthesis of analogs to define SARs. The next phase in the drug discovery process is lead optimization to obtain a candidate for preclinical development. An essential component of the lead optimization process is to synthesize new analogs with balanced pharmacological properties such as affinity, safety, pharmacokinetics, ADME (absorption, distribution, metabolism, elimination), and toxicology. In the drug discovery process, both classical iterative organic synthesis and parallel synthesis play significant roles. The proven advantages of microwave heating make it an enabling technology. This section of the chapter will highlight how microwave chemistry has been used as an enabling technology for medicinal chemists as they try to develop drug candidates. Attention will focus on examples from the recent literature and will show the vital role microwave heating plays in the workflow.

4.3.1 HIT-TO-LEAD OPTIMIZATION USING ITERATIVE AND PARALLEL-FOCUSED LIBRARY SYNTHESIS

Muscarinic acetylcholine receptors (mAChRs) belong to the G protein-coupled receptor (GPCR) family. To date, five distinct subtypes of mAChRs (M1–M5) have been cloned and sequenced. They share common orthosteric ligand-binding sites with an extremely high sequence homology across the species. This means that it can be very difficult to identify subtype selective ligands. Research has shown that small molecule M1 antagonists have potential application as new therapeutic treatment for Parkinson's disease and dystonia.[63] Microwave heating has been used as a tool for the rapid generation of a library of [1,2,4]triazolo[4,3-b]pyridazines for screening as potential selective M1 antagonists, this building on the observation that one example had been shown to have weak but selective activity (M1 IC_{50} = 22 µM; M4 IC_{50} > 150 µM).[64] Classical approaches to this class of compound often involve heating at reflux for 16–60 h, and yields are moderate at best. By using microwave heating for both the heterocycle synthesis and subsequent decoration using S_NAr chemistry, the reaction times could be dramatically reduced and product yields enhanced. An iterative library of 60 compounds was generated. Two of these showed an over sixfold increase in M1 inhibitory activity (M1 IC_{50} = 3.99 µM and 6.64 µM) while maintaining selectivity versus M2–M5 (IC_{50} >> 50 µM).

Until recently, no ligands had been reported as being highly M5-preferring or selective. Compared to the other mAChRs, little is known about M5, which is expressed at very low levels in the central nervous system (CNS) and peripheral tissues.[65] In a screen, a p-bromobenzyl-substituted isatin was found to exhibit potential allosteric activity at M1, M3, and M5 receptors but lacked effects at M2 and M4 receptors. In order to explore this in more detail and with the objective of finding an M5-preferring or selective compound, a library was prepared by alkylation reactions of eight commercially available isatins with 9–20 benzyl halides (9–20) using microwave heating (Scheme 4.25). Screening of the compounds led to a hit with an EC_{50} of approximately 1.16 µM at M5 with >30-fold selectivity versus M1 and M3 and with no M2 or M4 activity.[66]

In the literature, there are numerous other reported examples of the use of microwave heating as a tool for iterative parallel library synthesis geared toward HTS campaigns. Pyrazolo[1,5-a]pyrimidine and pyrazolo[3,4-d]pyrimidines have been prepared and screened for cancer therapy.[67] A series of 8-substituted tetrahydrocarbazoles have been prepared in an effort to find potent anti-HPV agents.[68] A microwave-promoted Buchwald–Hartwig amination reaction proved to be an essential transformation for assembling a 2-aminopyrazine library for rapid SAR exploration as Pim Kinase inhibitors.[69] An array of benzothiazole analogs has been prepared and the compounds screened as selective fatty acid amide hydrolase inhibitors

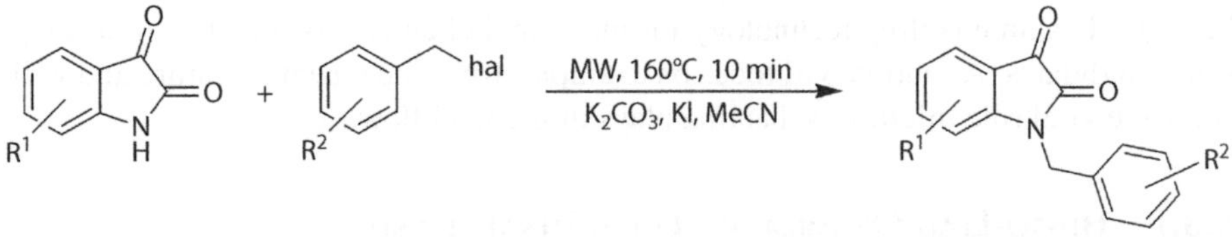

when R^1 = 5-OCF$_3$ and R^2 = 4-OMe: EC_{50} ~ 1.16 µM at M5 with >30-fold selectivity vs M1 and M3 and no M2 or M4 activity

SCHEME 4.25

(FAAH).[70] Pyrido-thieno-pyrimidines have been prepared and screened as potential Cdc7 kinase inhibitors,[71] alkoxythiazoles as isoform-selective RARβ ligands,[72] benzamide tetrahydro-4H-carbazole-4-one as novel inhibitors of Hsp90,[73] and pyrroles and pyrazolones as novel inhibitors of *Mycobacterium tuberculosis*.[74]

4.3.2 STRUCTURE-BASED DESIGN FOR HIT TO LEAD OPTIMIZATION

Structure-based design approaches for analog development have proved very valuable in the drug discovery process. An example is the discovery, SAR, and pharmacokinetics of a novel 3-hydroxyquinolin-2(1H)-one series of potent D-amino acid oxidase (DAAO) inhibitors. D-serine, a coagonist at the glycine site on the *N*-methyl-D-aspartate (NMDA) receptor, is synthesized from L-serine racemates and is metabolized by DAAO.[75] There has been evidence to link low levels of D-serine to schizophrenia. It is possible to ameliorate symptoms by increasing the function of NMDA via blocking of the breakdown of D-serine by DAAO.[76] A number of DAAO inhibitors have been reported, including compounds as simple as benzoic acid as well as a number of heterocycles. Pfizer undertook an HTS of their compound library in search of novel DAAO inhibitors.[77] From this screen, 3-hydroxyquinolin-2(1H)-one was identified as a potent inhibitor of human DAAO in a functional assay. Cocrystallization of 3-hydroxyquinolin-2(1H)-one with the human DAAO enzyme defined the binding site and showed that, in the active site binding pocket surrounding the molecule, the aryl ring is in contact with multiple hydrophobic residues, leaving no room for large multiatom substitution. In light of this, the effect of small functional groups such as F, Cl, methyl, and ethyl around the 5, 6, 7, and 8 positions of the quinolone ring was probed. The preparation of 6-,7-, and 8-substituted analogs consists of an Eistert ring expansion of the appropriate isatin with ethyl diazoacetate to generate an ethyl ester intermediate. This was then saponified and decarboxylated using microwave heating. However, even using extensive microwave heating (150 °C for 2 h), this proved to be challenging, and low (<20%) to moderate (>50%) yields of desired 3-hydroxyquinolin-2(1H)-one analogs were obtained (Scheme 4.26). The 5-substituted analogs did not decarboxylate under these conditions. A series

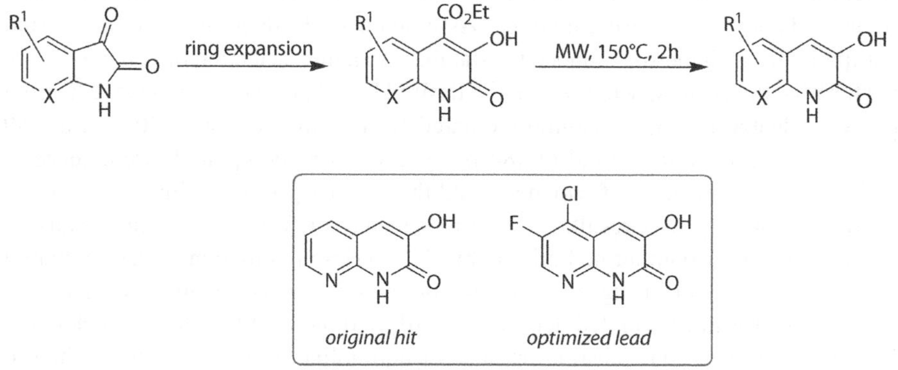

original hit   optimized lead

SCHEME 4.26

of other analogs were also prepared. Overall, placement of small groups around the quinoline ring and/or replacement of ring carbon atoms with nitrogen did not improve the DAAO potency over that of the original lead, 3-hydroxyquinolin-2(1H)-one. However, fluoro substitution at the 6- or 7-position did lead to increased oral bioavailability, chloro substitution at 5-position improved potency, and replacement of the 8-carbon atom with nitrogen improved binding affinity. The combination of these SAR features led to the design of lead molecule 5-chloro-6-fluoro-3-hydroxy-quinolin-2(1H)-one, a 4 nM inhibitor against rat and human DAAO.

4.3.3 KNOWLEDGE-BASED DESIGN APPROACHES TO LEAD OPTIMIZATION

4.3.3.1 Ligand and Pharmacophore-Based Design

The drug discovery process often involves first reviewing genomic and proteomics databases associated with a target together with in-house corporate databases and known ligands for the target. An example of this so-called knowledge-based approach to drug discovery in which microwave heating plays a key role from a synthetic chemistry standpoint is in the generation of a potent and selective series of isoquinoline inhibitors of IκB kinase-β (IKK-β)s. The IκB kinases play key roles as regulators of transcription factors controlling gene expression in innate and adaptive immune responses. Of the four kinases in the family, IKK-β has been the most widely studied, and identification of selective inhibitors has been a particular goal for anti-inflammatory drug discovery. A series of 2-amino-3,5-diarylbenzamides bearing sulfonamide groups have shown inhibitory activity of IKK-β. In order to develop more potent inhibitors, a database was built from over 8000 aromatic halides passing stringent filters for diversification and reagent availability. The motivation for selecting aromatic halides was that they could be readily converted to biaryls containing a sulfonamide moiety in one step using a Suzuki coupling with a sulfonamide-functionalized boronic acid. A 3D pharmacophore was constructed using the conserved interactions between a homology model of IKK-β and docked known inhibitors. Using this pharmacophore as a filter, the potential biaryl products obtained from reaction of the database of aromatic halides with two different sulfonamide-functionalized boronic acids were inputted. Assessment of the docking scores and visual compatibility with the site reduced the database to a final selection of 140 halides. In performing the chemistry to prepare the desired biaryl products, microwave heating proved valuable for the palladium-mediated Suzuki coupling step, 20 min at 150 °C proving successful. A total of 184 of the 280 target compounds were successfully obtained, screening of which showed those bearing isoquinoline groups to be particularly active. As a result, these were selected for further evaluation. Starting from one central isoquinoline building block, a series of compounds were prepared (Scheme 4.27), again using a microwave-promoted Suzuki coupling. In addition, strategic microwave-promoted deprotection allowed the workflow to be accelerated. Products resulting from substitution at the isoquinoline 6- and 7-positions showed substantial enhancement in IKK-β enzyme potency, which translated into significant cellular activity. Interestingly, the same series displayed encouraging selective

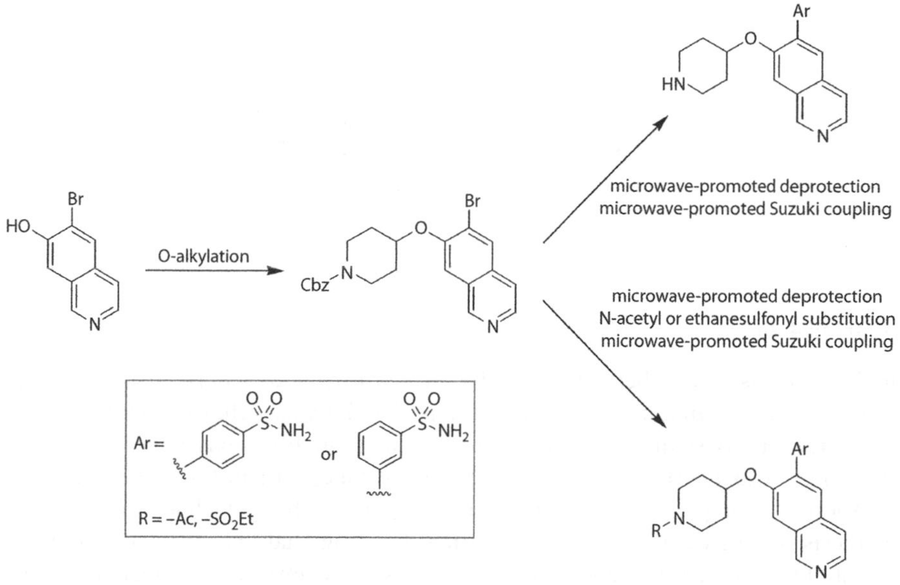

SCHEME 4.27

inhibition within the IκB family against IKK-α and across a wide variety of kinase enzyme and binding assays.

There are numerous other recent reports describing lead generation efforts for various drug targets employing a pharmacophore-based design strategy in combination with microwave-promoted parallel library synthesis as a key synthetic tool.[78] Highlighted examples include the design and synthesis of 9-cyano-1-azapaullone as glycogen synthase kinase-3 (GSK-3) inhibitor,[79] SAR study of 4- and 5-nitroimidazoles as antitubercular agents,[80] the development of substituted pyrazolo[1,5-a]quinazolin-5(4H)-one as potent poly(ADP-ribose)polymerase-1 (PARP-1) inhibitors,[81] use of 1-(piperidin-4-yl)-1H-benzo[d]imidazole-2(3H)-ones as selective phospholipase D (PLD) inhibitors,[82] and in the evaluation of imidazo[1,2-a]quinoxaline as novel inhibitor of human melanoma.[83]

4.3.3.2 Scaffold Hopping

Scaffold hopping is a commonly used strategy by medicinal chemists and can be defined as a process that aims at identifying compounds (chemotypes) with different molecular backbones but the same pharmacological function.[84] Having different chemotypes offer a choice in terms of synthetic accessibility to a target, and having multiple possible lead structures lowers the chance of drug development attrition along the drug discovery pathway. A microwave-promoted Suzuki-coupling reaction has been used to prepare a range of substituted triazolopyridazines in a scaffold hopping approach for exploring their activity as phosphodiesterase-4 (PDE4) inhibitors (Scheme 4.28).[85] While triazolothiadiazines had been shown previously to be active PDE4 inhibitors, triazolopyridazines had not been studied. Not only were selected

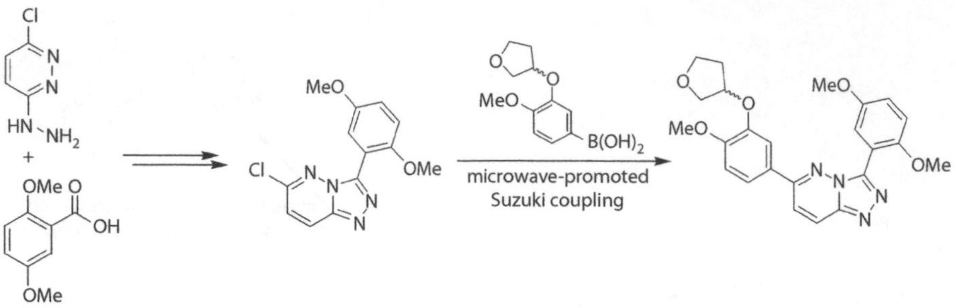

SCHEME 4.28

analogs of this novel chemotype capable of downregulating purified isozymes of PDE4, but they maintained excellent cell-based activity as well. Computational and structure–activity studies also showed that potential modifications at the C3 position of the 1,2,4-triazole ring system could open up avenues for fruitful SAR studies.

Another example of microwave heating as an enabling technology for rapid synthesis of target compounds for a scaffold hopping study is in the generation of potential histamine H4 receptor (H4R) antagonists.[86] The H4R is considered a promising target for the treatment of various chronic inflammatory diseases, including asthma and rheumatoid arthritis.[87] Based on pharmacophore features gleaned from modeling studies using known agonists and antagonists, a six-membered heterocyclic ring was envisioned as a replacement for a chloro-substituted phenyl ring or an indole ring in previously reported scaffolds. However, removal or modification of the *N*-methylpiperidine moiety seen in active compounds was deemed, upon modeling, to have a negative effect. As a result, a small set of compounds with an *N*-methylpiperidine group directly connected to heterocyclic scaffolds such as quinazolines, quinoxalines, quinolines, and isoquinolines were prepared to probe scaffold-hopping possibilities. The compounds were prepared, in one step, from chloro-substituted precursors by reaction with *N*-methylpiperazine using microwave heating (120–140 °C, 5–20 min). Pharmacological evaluation showed the quinoxaline-derived compounds to be the most active and, as a result, further analogs were prepared, using microwave heating in key steps (Scheme 4.29). This led to a series of potent H4R ligands with affinity in the low-nanomolar range. Two of the most potent quinoxaline lead compounds were evaluated in vivo and displayed significant anti-inflammatory activity.

4.3.3.3 Bioisostere Replacement

Another approach to analog development within drug discovery involves exchanging bioisosterically equivalent groups.[88] Bioisosteres can be defined as substituents with similar physical or chemical properties that impart similar biological properties to a molecule. The strategy has been used by medicinal chemists to develop analogs of known commercially successful therapeutics and also as a tool for performing molecular modifications to improve pharmacological activity, gain selectivity for a determined receptor, or reduce adverse effects while keeping target potency. Oxytocin is a peptide hormone that modulates numerous physiological

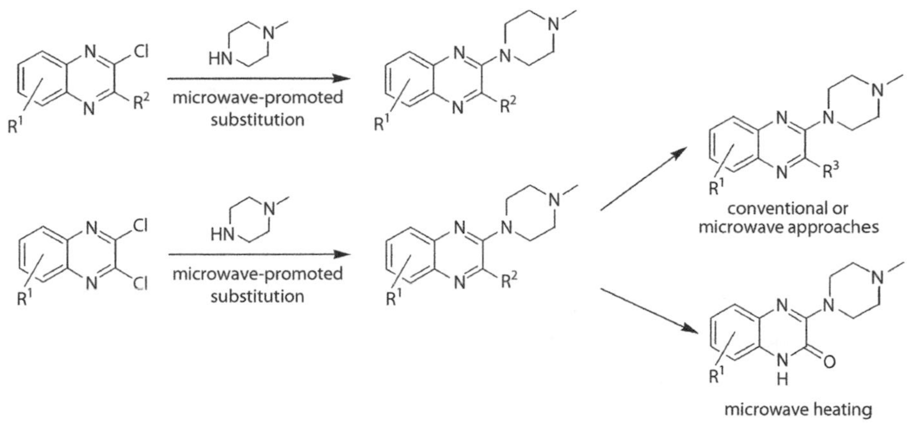

SCHEME 4.29

roles, including the control of uterine contractions. The peptide-based antagonist atosiban offers a potential treatment for threatened preterm birth. However, this is a complex molecule, and development of simpler small molecule drugs would be preferable. A bioisostere replacement approach, together with microwave heating as a tool for synthesis, has been demonstrated by GSK.[89] High-throughput screening of their compound collection identified a tricyclic sulfonamide-containing quinolinone compound as a moderately potent oxytocin receptor antagonist but with the drawbacks of poor solubility and bioavailability. A related, higher-molecular-weight compound with in vivo activity had been reported by Serono Pharmaceutical Research Institute. The team proposed to optimize the potency, solubility, and oral bioavailability of their in-house target without significantly increasing the molecular weight (<500 amu). A number of sulfonamide-containing compounds were prepared with the objective of finding novel isosteres for the tricyclic quinolinone group. Microwave heating was used to facilitate the synthesis of these compounds (Scheme 4.30). Condensation of aldehydes with anilines or an aminopyridine was performed at 150 °C for 10 min, followed by a conventional sodium borohydride reduction. The sulfonamide moiety was attached by reaction with a (hetero)arylsulfonyl chloride. Moderately potent compounds were identified, but mapping them to a pharmacophore based on known active oxytocin receptor antagonists suggested poor alignment of the moiety derived from the aldehyde component in the synthesis. A second range of compounds was prepared using the same synthetic strategy but this time starting from 5-bromo-2-pyridinecarboxaldehyde, upon which a Suzuki coupling could be performed in an additional step (Scheme 4.30). Using microwave heating, a range of biaryls was generated. This resulted in the discovery of two potent compounds, but both compounds had relatively poor pharmacokinetic profiles and solubility. Computational studies suggested that replacement of the aryl group on the sulfonyl moiety could improve the pharmacokinetic profile. This was indeed the case, the preparation of a range of next-generation compounds showing that imidazole sulfonamides not only had improved pharmacokinetic profiles but also good solubility. Overall, by using a combination of high-throughput

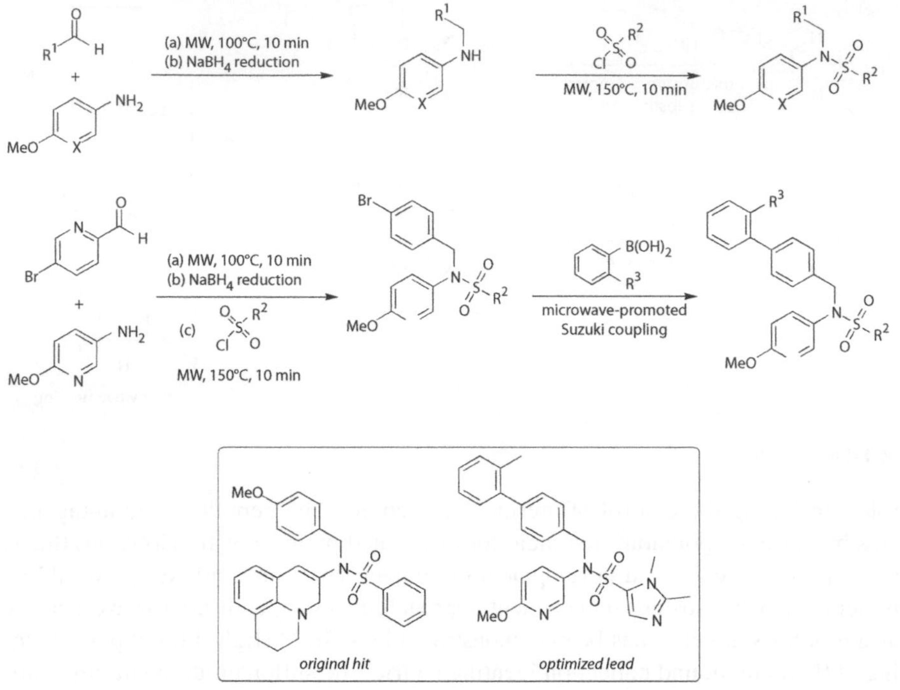

SCHEME 4.30

microwave-promoted chemistry, pharmacophore analysis, and computational pro-filing, bioisosteres of the original moderately potent tricyclic quinolinone were developed and optimized to give a highly potent imidazole-containing sulfonamide antagonist that could proceed from hit to lead development.

A similar medicinal chemistry strategy has been used to design and prepare a series of sulfonamides as potential new HIV-1 entry inhibitors.[90] Known HIV-1 entry inhibitors bearing an α-ketoamide functionality were used as the starting point. Effort was first focused on studying X-ray crystal structures of the α-ketoamides docked in the binding pocket of the viral molecule, looking for notable features. One such feature was that the two carbonyl groups adopt a twisted conformation. With the objective of using a sulfonamide group as an isostere of the α-ketoamide moiety, a series of compounds were prepared using solution-phase parallel library synthesis. Microwave heating was used to facilitate synthetically challenging steps for the preparation of azaindole scaffolds. After pharmacological evaluation on the sulfonamide analogs in a pseudo HIV assay, one compound showed antiviral potency comparable to the original α-ketoamide.

4.3.4 PHENOTYPIC DRUG DISCOVERY

Traditional target-based drug discovery focuses on synthesis and testing of mole-cules designed to interact with a specific target believed to be involved in the disease pathway. While significant R&D funds have been invested in this approach, relatively few new drugs have been approved. Phenotypic drug discovery provides an alternative approach that starts by probing more complex cellular systems instead of specific targets.[91] Using this, the opportunity arises to identify compounds that interact with targets or pathways not previously anticipated. In essence, multiple mechanisms and targets can be screened simultaneously. Also, because data obtained from cellular assays is information rich, the connection of compound action to disease-relevant phenotypes can be established earlier in the drug discovery process. This is increasingly becoming the case with the development of advanced assay technologies and informatics tools. The discovery of carbazoles as antimitotics for cancer treatment and subsequent optimization is a good example of how microwave chemistry can be interfaced into a phenotype-based workflow.[92] Using high-content cellular analysis (HCA) assays, treatment of cells with a carbazole originally developed as part of a heat-shock protein study was seen to elicit a strong apoptotic response like that of small molecule tubulin polymerization-disrupting compounds such as paclixatol and vinblastine. HCA also demonstrated that the resulting cellular phenotype of carbazole-treated cells closely resembled that for vinblastine- and colchicine-treated cells. Further study showed that the carbazole was competitive with colchicine in binding to tubulin but not with labeled vinblastine, this providing the molecular target for this compound. A series of substituted carbazoles were then prepared, the carbazole ring being assembled using a microwave-based modification of the Fischer indole synthesis. The products were tested in a variety of cancer cell lines, and many demonstrated superior activity as compared to the initial hit.

4.3.5 DNA-Based Technology for Drug Discovery

It is fairly well established that a biological HTS campaign examining ~10^6 molecules against a validated drug target could deliver viable lead molecules that may eventually produce one successful drug candidate.[93] If lead discovery efforts rely solely on a blind HTS approach against drug targets, a compound library size well in excess of this would be required. The task of preparing these molecules is daunting, and the infrastructure required to screen them is very costly. An alternative approach, complementary to the conventional screen-based discovery route, is an evolution-based searching method.[94] This has been used with success for the preparation of biopolymers.[95] The key to the methodology lies in the use of nucleic acids as amplifiable blueprints that can be translated into biopolymer sequences. This then allows for iterative functional selections to be carried out with extremely complex molecular libraries. Methods for overcoming the technical challenge of using DNA to direct the chemical assembly of molecules have been developed. Essentially, collections of single-stranded DNA molecules are routed through a split-and-pool combinatorial synthesis. The products are covalent small molecule–DNA hybrids. Microwave heating has been used as a tool for the development of a 100 million-member 8-mer peptoid library, a compound collection two orders of magnitude more complex than that of a large traditional HTS library.[96] Peptoid couplings were performed using either three or six 20 s microwave pulses at 100% power in a domestic oven, the reagents being in a closed system between two syringes. The compounds were subjected to selection for binding to the N-terminal SH3 domain of the proto-oncogene Crk. Over six generations, the molecular population converged to a small number of novel SH3 domain ligands. Of note was that these hits bind with affinities similar to those of natural peptide SH3 ligands.

4.3.6 ENHANCED LEAD OPTIMIZATION

A successful lead compound must fulfill a number of stringent criteria before it can advance along the drug discovery pathway. These include potency (the compound must produce the desirable pharmacological response), oral bioavailability (the compound must be able to pass through multiple of barriers to reach the target), duration (the compound must remain in the system long enough to have the requisite activity), safety (the compound must show selectivity for the targeted response), and pharmaceutical acceptability (the compound must be synthetically accessible, have good chemical stability, show adequate solubility, and possess a reasonable rate of dissolution). Microwave heating has been used as a tool in development processes so that lead compounds pass these strict selection criteria, examples being in the optimization of metabolic stability,[97] pharmacokinetic parameters,[98] and lipophilicity.[99]

4.4 CONCLUSION

This chapter presents an overview of how microwave heating can be integrated efficiently into library synthesis and the drug discovery process as a whole. Alongside other new and innovative technologies, microwave heating has had a real impact on the field. It is often possible to perform synthetic transformations more rapidly, easily, and selectively using microwave heating as opposed to conventional approaches. Also, in some cases, microwave heating opens avenues to new chemistry. It is also impacting other emerging allied areas such as peptide and protein synthesis, proteomics, and extraction of biological material from plants to mention a few. In the future, those working in the drug discovery process will continue to use microwave heating as an enabling technology, and it will see increasing uptake as it becomes more and more the method of choice rather than a last resort.

REFERENCES

1. (a) King, F. D. Ed. *Medicinal Chemistry: Principles and Practice*, 2nd ed.; Royal Society of Chemistry: Cambridge, 2002. (b) Thomas, G. Eds. *Medicinal Chemistry, an Introduction*, 2nd ed.; Wiley-VCH: Weinheim, Germany, 2008.
2. (a) Territt, N. Ed. *Combinatorial Chemistry*, Oxford University Press: Oxford, New York, 1998. (b) MacCoss, M.; Baillie, T. A. *Science*, 2004, *303*, 1801–1803. (c) Colombo, M.; Peretto, I. *Drug Discov. Today*, 2008, *13*, 677–684.
3. (a) Lidström, P.; Tierney, J.; Wathey, B.; Westman, J. *Tetrahedron*, 2001, *57*, 9225–9283. (b) Loupy, A., Ed. *Microwaves in Organic Synthesis*; Wiley-VCH: Weinheim, Germany, 2002. (c) Hayes, B. L., Ed. *Microwave Synthesis: Chemistry at the Speed of Light*; CEM Publishing: Matthews, NC, 2002. (d) Kappe, C. O. *Angew. Chem., Int. Ed.* 2004, *43*, 6250–6284. (e) Tierney, J. P.; Lidstrom, P., Eds. *Microwave Assisted*

Organic Synthesis; Blackwell Scientific: Boca Raton, FL, 2005. (f) Van der Eycken, E.; Kappe, C. O. Eds. *Microwave-Assisted Synthesis of Heterocycles,* Springer-Verlag: Berlin, 2006. (g) Kappe, C. O. *Chem. Soc. Rev.*, 2008, *37*, 1127–1139. (h) Kappe, C. O.; Dallinger, D.; Murphree, S. Eds. *Practical Microwave Synthesis for Organic Chemist,* Wiley-VCH: Weinheim, Germany, 2009. (i) Caddick, S.; Fitzmaurice, R. *Tetrahedron,* 2009, *65*, 3325–3355. (j) Kappe, C. O.; Dallinger, D. *Mol. Divers.* 2009, *13*, 71–193.

4. Koppitz, M.; Eis, K. *DDT*, 2006, *11*, 561–568.
5. (a) Kolb, H. C.; Finn, M. G.; Sharpless, K. B. *Angew. Chem. Int. Ed.* 2001, *40*, 2004–2021. (b) Kolb, H. C.; Sharpless, B. *DDT*, 2003, *8*, 1128–1137. (c) Olesen, P. H.; Sorensen, A. R.; Ursö, B.; Kurtzhals, P.; Bowler, A. N.; Ehrbar, U.; Hansen, B. F. *J. Med. Chem.* 2003, *46*, 3333–3341. (d) Lebsack, A. D.; Gunzner, J.; Wang, B.W.; Pracitto, R.; Schaffhauser, H.; Santini, A.; Aiyar, J.; Bezverkov, R.; Munoz, B.; Liu, W. S.; Venkatraman, S. *Bioorg. Med. Chem.* 2004, *14*, 2463–2467. (e) Zhu, X. M.; Schmidt, R. R. *J. Org. chem.* 2004, *69*, 1081–1085.
6. (a) Larhed, M.; Moberg, C.; Hallberg, A. *Acc. Chem. Res.*, 2002, *35*, 717–727. (b) Nilsson, P.; Olofsson, K.; Larhed, M. *Topics in Current Chemistry: Microwave-Assisted and Metal-Catalyzed Coupling Reactions,* Springer: Berlin, Heidelberg, 2006.
7. Rozenman, M.; McNaughton, B. R.; Liu, D. *Curr. Opin. Chem. Bio.* 2007, *11*, 259–268.
8. Trnka, T. M.; Grubbs, R. H. *Acc. Chem. Res.* 2001, *34*, 18–29.
9. (a) Larhed, M.; Hallberg, A. *DTT*, 2001, *6*, 406–416. (b) Ersmark, K.; Larhed, M.; Wannberg, J. *Curr. Opin. Drug Discov. Dev.* 2004, *7*, 417–427. (c) Kappe, C. O.; Alexande, S. Eds. *Microwaves in Organic and Medicinal Chemistry*; Wiley-VCH: Weinheim, Germany, 2005. (d) Kappe, C. O.; Dallinger, D. *Nat. Rev. Drug Discov.* 2006, *5*, 51–63. (e) Pilotti, A.; Mavandadi, F. *DDT*, 2006, *11*, 165–174. (f) Larhed, M.; Olofsson, K. Eds, *Topics in Current Chemistry: Microwave Methods in Organic Synthesis*; Springer: Berlin, New York, 2006. (g) Alcázar, J.; Diels, G.; Schoentjes, B. *Mini-Reviews in Med. Chem.* 2007, *7*, 345–369. (h) Alcazar, J.; Diels, G.; Schoentjes, B. *Comb. Chem. High Throughput Screening,* 2007, *10*, 918–932. (i) Polshettiwar, V.; Varma, R.S. *Chem. Soc. Rev.* 2008, *37*, 1546–1557.
10. Fergus, S.; Bender, A.; Spring, D. R. *Curr. Opin. Chem. Biol.* 2005, *9*, 304–309.
11. Stadler, A.; Kappe, C. O. *J. Comb. Chem.* 2001, *3*, 624–630.
12. (a) Kappe, C. O.; Matloobi, M. *Comb. Chem. High Throughput Screening,* 2007, *10*, 735–750. (b) Dai, W. M.; Shi, J. Y. *Comb. Chem. High Throughput Screening,* 2007, *10*, 837–856. (c) Nüchter, M.; Ondruschka, B. *Mol. Divers.*, 2003, *7*, 253–264.
13. (a) Kremsner, J. M.; Stadler, A.; Kappe, C. O. *J. Comb. Chem.* 2007, *9*, 285–291. (b) Pisani, L.; Prokopcová, H.; Kremsner, J. M.; Kappe, C. O. *J. Comb. Chem.* 2007, *9*, 415–421.
14. Stencel, L. M.; Kormos, C. M.; Avery, K. B.; Leadbeater, N. E. *Org. Biomol. Chem.* 2009, *7*, 2452–2457.
15. Treu, M.; Karner, T.; Kousek, R.; Berger, H.; Mayer, M.; McConnell, D. B.; Stadler, A. *J. Comb. Chem.* 2008, *10*, 863–868.
16. (a) Damm, M.; Kappe, C. O. *J. Comb. Chem.*, 2009, *11*, 460–468. (b) Baghbanzadeh, M.; Molnar, M.; Damm, M.; Reidlinger, C.; Dabiri, M.; Kappe, C. O. *J. Comb. Chem.* 2009, *11*, 676–684.
17. (a) Dömling, A. *Chem. Rev.* 2006, *106*, 17–89. (b) Tempest, P. A. *Curr. Opin. Drug Discov. Devel.* 2005, *8*, 776–788. (c) Hulme, C.; Gore, V. *Curr. Med. Chem.* 2003, *10*, 51–80. (d) Weber, L. *Curr. Med. Chem.* 2002, *9*, 2085–2093. (e) Dömling, A.; Ugi, I. *Angew. Chem. Int. Ed.* 2000, *39*, 3168–3210.
18. DiMauro, E. F.; Kennedy, J. M. *J. Org. Chem.* 2007, *72*, 1013–1016.
19. Willy, B.; Rominger, F.; Müller, T. J. J. *Synthesis*, 2008, 293–303.
20. Willy, B.; Müller, T.J. J. *Eur. J. Org. Chem.* 2008, 4157–4168.

21. Willy, B.; Dallos, T.; Rominger, F.; Schönhaber, J.; Müller, T. J. J. *Eur. J. Org. Chem.* 2008, 4796–4805.
22. Baxendale, I. R.; Hayward, J. J.; Ley, S. V. *Comb. Chem. High Throughput Screening*, 2007, *10*, 802–836.
23. Bremner, W. S.; Organ, M. G. *J. Comb. Chem.* 2007, *9*, 14–16.
24. Khadilkar, B. M.; Madyar, V. R. *Org. Process Res. Dev.* 2001, *5*, 452–455.
25. Ohta, Y.; Chiba, H.; Oishi, S.; Fujii, N.; Ohno, H. *Org. Lett.* 2008, *10*, 3535–3538.
26. Appukkuttan, P.; Dehaen, W.; Fokin, V.V.; Van der Eycken, E. *Org. Lett.* 2004, *6*, 4223–4225.
27. Castagnolo, D.; Dessi, F.; Radi, M.; Botta, M. *Tetrahedron Asymm.* 2007, *18*, 1345–1350.
28. Tu, S. J.; Jiang, B.; Zhang, J. Y.; Jia, R. H.; Zhang, Y.; Yao, C.-S. *Org. Biomol. Chem.* 2006, *4*, 3980–3985.
29. Tu, S. J.; Jiang, B.; Jia, R. H.; Zhang, J. Y.; Zhang, Y. *Tetrahedron Lett.* 2007, *48*, 1369–1374.
30. Zhu, S. L.; Ji, S. J.; Zhao, K.; Liu, Y. *Tetrahedron Lett.* 2008, *49*, 2578–2582.
31. Surpur, M. P.; Kshirsagar, S.; Samant, S. D. *Tetrahedron Lett.* 2009, *50*, 719–722.
32. Yermolayev, S. A.; Gorobets, N. Y.; Desenko, S. M. *J. Comb. Chem.* 2009, *11*, 44–46.
33. Strübing, D.; Neumann, H.; von Wangelin, A. J.; Klaus, S.; Hübner S.; Beller, M. *Tetrahedron*, 2006, *62*, 10962–10967.
34. Yadav, L.D.S.; Srivastava, V.P.; Rai, V.K.; Patel, R. *Tetrahedron*, 2008, *64*, 4246–4253.
35. Santra, S.; Andreana, P. R. *Org. Lett.* 2007, *9*, 5035–5038.
36. Chebanov, V.A.; Saraev, V. E.; Desenko, S. M.; Chernenko, V. N.; Shishkina, S. V.; Shishkin, O.V.; Kobzar, K. M.; Kappe, C. O. *Org. Lett.* 2007, *9*, 1691–1694.
37. Dörwald, F. Z. *Organic Synthesis on Solid Phase: Supports, Linkers, Reactions,* 2nd ed.; Wiley-VCH: Weinheim, 2002.
38. (a) Moos, W. H.; Hurt, C. R.; Morales, G. A. *Mol. Divers.*, 2009, *13*, 241–245. (b) Dai, W. M.; Shi, J. Y. *Comb. Chem. High Throughput Screening*, 2007, *10*, 837–856.
39. He, R. J.; Lam, Y. L. *Org. Biomol. Chem.* 2008, *6*, 2182–2186.
40. Che, J.; Raghavendra, M. S.; Lam, Y. L. *J. Comb. Chem.* 2009, *11*, 378–384.
41. Lee, T.; Park, J. H.; Lee, D. H.; Gong, Y. D. *J. Comb. Chem.* 2009, *11*, 495–499.
42. Vedantham, P.; Guerra, J.M.; Schoenen, F.; Huang, M.; Georg, G. I.; Lushington, G. H.; Mitscher, L.A.; Ye, Q. Z.; Hanson, P. R. *J. Comb. Chem.* 2008, *10*, 185–194.
43. Messeguer, J.; Cortés, N.; García-Sanz, N.; Navarro-Vendrell, G.; Ferrer-Montiel, A.; Messeguer, A. *J. Comb. Chem.* 2008, *10*, 974–980.
44. Geske, G. D.; O'Neil, J. C.; Miller, D. M.; Mattmann, M. E.; Blackwell, H. E. *J. Am. Chem. Soc.* 2007, *129*, 13613–13625.
45. Pourccau, G.; Meyer, A.; Vasseur, J.J.; Morvan, F. *J. Org. Chem.* 2008, *73*, 6014–6017.
46. (a) McNamara, C. A.; Dixon, M. J.; Bradley, M. *Chem. Rev.* 2002, *102*, 3275–3300. (b) Clapham, B.; Reger, T. S.; Janda, K. D. *Tetrahedron* 2001, 57, 4637–4662. (c) Ley, S. V.; Baxendale, I. R.; Bream, R. N.; Jackson, P. S.; Leach, A. G.; Longbottom, D. A.; Nesi, M.; Scott, J. S.; Storer, R. I.; Taylor, S. J. *J. Chem. Soc., Perkin Trans. 1* 2000, 3815–4195.
47. Solinas, A.; Taddei, M. *Synthesis* 2007, 2409–2453.
48. Werner, S.; Kasi, D.; Brummond, K. *J. Comb. Chem.* 2007, *9*, 677–683.
49. Mao, S.L.; Probst, D.; Werner, S.; Chen, J. Z.; Xie, X, G.; Brummond, K. *J. Comb. Chem.* 2008, *10*, 235–246.
50. Radi, M.; Saletti, S.; Botta, M. *Tetrahedron Lett.* 2008, *49*, 4464–4466.
51. De Luca, L.; Giacomelli, G.; Nieddu, G. *J. Comb. Chem.* 2008, *10*, 517–520.
52. Oliver, C. M.; Schaefer, A. L.; Greenberg, E. P.; Sufrin, J. R. *J. Med. Chem.* 2009, *52*, 1569–1575.

53. Antonow, D.; Cooper, N.; Howard, P. W.; Thurston D. E. *J. Comb. Chem.* 2007, *9*, 437–445.
54. (a) Lu, J. N.; Toy, P. H. *Chem Rev.* 2009, *109*, 815–838. (b) Dickerson T. J.; Reed, N. N.; Janda, K. D. *Chem. Rev.* 2002, *102*, 3325–3344. (c) Toy, P. H.; Janda, K. D. *Acc. Chem. Res.* 2000, *33*, 546–554.
55. (a) Zhang, W. *Chem. Rev.* 2009, *109*, 749–795. (b) Zhang, W. *Tetrahedron* 2003, *59*, 4475–4489.
56. Chanda, K.; Kuo, J.; Chen, C. H.; Sun, C. M. *J. Comb. Chem.* 2009, *11*, 252–260.
57. Fu, G.Y.; Zhang, X.L.; Sheng, S.R.; Wei, M. H.; Liu, X. L. *Synthetic Comm.* 2008, *38*, 1249–1258.
58. Yeh, W. P.; Chang, W. J.; Sun, M. L.; Sun, C. M. *Tetrahedron,* 2007, *63*, 11809–11816.
59. (a) Koehn, F. E.; Carter, G. T. *Nat. Rev. Drug Discovery* 2005, *4*, 206–220. (b) Newman, D. J.; Cragg, G. M. *J. Nat. Prod.* 2007, *70*, 461–477.
60. (a) Kanafani, Z. A.; Corey, G. R. *Expert. Rev. Anti-Infect. Ther.* 2007, *5*, 177–184. (b) Harvey, A. L. *DDT,* 2008, *13*, 894–901. (c) Challis, G. L. *J. Med. Chem.* 2008, *51*, 2618–2628.
61. Luesse, S. B.; Wells, G.; Nayek, A.; Smith, A. E.; Kusche, B. R.; Bergmeier, S. C.; McMills, M. C.; Priestley, N. D.; Wright, D. L. *Bioorg. Med. Chem. Lett.* 2008, 18, 3946–3949
62. Mont, N.; Mehta, V. P.; Appukkuttan, P.; Beryozkina, T.; Toppet, S.; Van Hecke, K.; Van Meervelt, L.; Voet, A.; DeMaeyer, M.; Van der Eycken, E. *J. Org. Chem.* 2008, *73*, 7509–7516.
63. (a) Hulme, E. C.; Birdsall, N. J. M.; Buckley, N. J. *Annu. Rev. Pharmacol. Toxcol.* 1990, *30*, 633–673. (b) Eglen, R. M.: *Progress in Medicinal Chemistry*, King, F. D.; Lawton, G. Eds.; Elsevier: San Diego, 2005, *43*, 105–106.
64. Aldrich, L. N.; Lebois, E.P.; Lewis, M.; Nalywajko, N. T.; Niswender, C. M.; Weaver, C. D.; Conn, P. J.; Lindsley, C. W. *Tetrahedron Lett.* 2009, *50*, 212–215.
65. Langmead, C. J.; Watson, J.; Reavill, C. *Pharmacol. Ther.* 2008, *117*, 232–243.
66. Bridges, T. M.; Mario, J. E.; Niswender, C. M.; Jones, C. K.; Jadhav, S. B.; Gentry, P. R.; Plumley, H. C.; Weaver, D. C.; Conn, P. J.; Lindsley, C. W. *J. Med. Chem.* 2009, *52*, 3445–3448.
67. (a) Daniel, R. N.; Kim, K.; Lebois, E.; Muchalski, H.; Hughes, M.; Lindsley, C. W. *Tetrahedron Lett.,* 2008, *49*, 305–310. (b) Rodriguez, A. L.; Williams, R.; Zhou, Y.; Lindsley, S. R.; Le, U.; Grier, M. D.; Weaver, C. D.; Conn, P. J.; Lindsley, C. W. *Bioorg. Med. Chem. Lett.* 2009, *19*, 3209–3213.
68. Gudmundsson, K. S.; Sebahar, P. R.; Richardson, L. D.; Catalano, J. G.; Boggs, S. D.; Spaltenstein, A.; Sethna, P. B.; Brown, K. W.; Harvey, R.; Romines, K. R. *Bioorg. Med. Chem.* 2009, *19*, 3489–3492.
69. Qian, K.; Wang, L.; Cywin, C.; Farmer B. T.; Hickey, E.; Homon, C.; Jakes, S.; Kashem, M. A.; Lee, G.; Leonard, S.; Li, J.; Magboo, R.; Mao, W.; Pack, E.; Peng, C.; Prokopowicz, A.; Welzel, M.; Wolak, J.; Morwick, T. *J. Med. Chem.* 2009, 52, 1814–1827.
70. Wang, X. Q.; Sarri, K.; Kage, K.; Zhang, D.; Brown, S. P.; Kolasa, T.; Surowy, C.; El Kouhen, O. F.; Muchmore, S. W.; Brioni, J. D.; Stewart, A. O. *J. Med. Chem.* 2009, *52*, 170–180.
71. Zhao, C. L.; Tovar, C.; Yin, X. F.; Xu, Q.; Todorov, I. T.; Vassilev, L. T.; Chen, L. *Bioorg. Med. Chem. Lett.* 2009, *19*, 319–323.
72. Lund, B. W.; Knapp, A. E.; Piu, F.; Gauthier, N. K.; Begtrup, M.; Hacksell, U.; Olsson, R. *J. Med. Chem.* 2009, *52*, 1540–1545.
73. Barta, T. E.; Veal, J. M.; Rice, J. W.; Partridge, J. M.; Fadden, R.P.; Ma, W.; Jenks, M.; Geng, L. F.; Hanson, G. J.; Huang, K. H.; Barabasz, A. F.; Foley, B. E.; Otto, J.; Hall, S. E. *Bioorg. Med. Chem. Lett.* 2008, *18*, 3517–3521.

74. Manetti, F.; Magnani, M.; Castagnolo, D.; Passalacqua, L.; Botta, M.; Corelli, F.; Saddi, M.; Deidda, D.; De Logu, O. *ChemMedChem* 2006, *1*, 973–989.
75. (a) Javitt, D. C.; Zukin, S. R. *Am. J. Psychiatry* 1991, *148*, 1301–1308. (b) Madeira, C.; Freitas, M. E.; Vargas-Lopes, C.; Wolosker, H.; Panizsutti, R. *Schizophrenia Res.* 2008, *101*, 76–83.
76. Millan, M. J. *Psychopharmacology*, 2005, *179*, 30–53.
77. Duplantier, A. J.; Becker, S. L.; Bohanon, M. J.; Borzilleri, K. A.; Chrunyk, B. A.; Downs, J. T.; Hu, L. Y.; El-Kattan, A.; James, L. C.; Liu, S. P.; Lu, J. M.; Maklad, N.; Mansour, M. N.; Mente, S.; Piotrowski, M. A.; Sakya, S. M.; Sheenhan, S.; Steyn, S. J.; Strick, C. A.; Williams, V. A.; Zhang, L. *J. Med. Chem.* 2009, *52*, 3576–3585.
78. (a) Piscitelli, F.; Coluccia, A.; Brancale, A.; La Regina, G.; Sanson, A.; Giordano, C.; Balzarini, J.; Maga, G.; Zanoli, S.; Samuele, A.; Cirilli, R.; La Torre, F.; Lavecchia, A.; Novellino, E.; Silverstri, R. *J. Med. Chem.* 2009, *52*, 1922–1934. (b) Niculescu-Duvaz, D.; Gaulon, C.; Dijkstra, H. P. Niculescu-Duvaz, I.; Zambon, A.; Ménard, D.; Suijkerbuijk, B. M. J. M.; Nourry, A.; Davies, L.; Manne, H.; Friedlos, F.; Ogilvie, L.; Hadley, D.; Whittaker, S.; Kirk, R.; Gill, A.; Taylor, R. D.; Raynaud, F. I.; Moreno-Farre, J.; Marais, R.; Springer, C. J. *J. Med. Chem.* 2009, *52*, 2255–2264. (c) Kattar, S. D.; Surdi, L. M.; Zabierek, A.; Methot, J. L.; Middleton, R. E.; Hughes, B.; Szewczak, A. A.; Dahlberg, W. K.; Kral, A. M.; Ozerova, N.; Fleming, J. C.; Wang, H. M.; Secrist, P.; Harsch, A.; Hamill, J. E.; Cruz, J. C.; Kenific, C. M.; Chenard, M.; Miller, T. A.; Berk, S.; Tempest, P. *Bioorg. Med. Chem. Lett.* 2009, 19, 1168–1172. (d) Nadarraj, V.; Selvi, S. T.; Mohan, S. *Eur. J. Med. Chem.* 2009, *44*, 976–980. (e) Kennedy, J. P.; Conn, P. J.; Lindsley, C. W. *Bioorg. Med. Chem. Lett.* 2009, *19*, 3204–3208.
79. Stukenbrock, H.; Mussmann, R.; Geese, M.; Ferandin, Y.; Lozach, O.; Lemeke, T.; Kegel, S.; Lomow, A.; Burk, U.; Dohrmann, C.; Meijer, L.; Austen, M.; Kunick, C. *J. Med. Chem.* 2008, *51*, 2196–2207.
80. (a) Kim, P.; Zhang, L.; Manjunatha, U. H.; Singh, R.; Patel, S.; Jiricek, J.; Keller, T. H.; Boshoff, H. I.; Barry III., C. E.; Dowd, C. S. *J. Med. Chem.* 2009, *52*, 1317–1328. (b) Kim, P.; Kang, S.; Boshoff, H.; Jiricek, J.; Collins, M.; Sighn, R.; Manjunatha, U. H.; Niyomrattanakit, P.; Zhang, L.; Goodwin, M.; Dick, T.; Keller, T. H.; Dowd, C. S.; Barry C. E. III.; *J. Med. Chem.* 2009, *52*, 1329–1344.
81. Orvieto, F.; Branca, D.; Giomini, C.; Jones, P.; Koch, U.; Ontoria, J. M.; Palumbi, M. C.; Rowley, M.; Toniatti, C.; Muraglia, E. *Bioorg. Med. Chem. Lett.* 2009, *15*, 4196–4200.
82. Scott, S.; Selvy, P.; Buck, J. R.; Cho, H. P.; Criswell, T. L.; Thomas, A. L.; Armstrong, M. D.; Arteaga, C. L.; Lindsley, C. W.; Brown, H. A. *Nat. Chem. Biology*, 2009, *5*, 108–117.
83. Deleuze-Masquefa, C.; Moarbess, G.; Khier, S.; David, N.; Gayraud-Paniagua, S.; Bressolle, F.; Pinguet, F.; Bonnet, P. A. *Eur. J. Med. Chem.* 2009, *44*, 3406–3411.
84. (a) Schneider, G.; Schneider, P.; Renner, S. *QSAR Comb. Sci.* 2006, *25*, 1162–1171. (b) Oyarzabal, J.; Howe, T.; Alcazar, J.; Andrés, J. I.; Alveraz, R. M.; Dautzenberg, F.; Iturrino, L.; Martinez, S.; Van der Linden, I. *J. Med. Chem.* 2009, *52*, 2076–2089.
85. Skoumbourdis, A.P.; LeClair, C.A.; Stefan, E.; Turjanski, A. G.; Maguire, W.; Titus, S. A.; Huang, R. L.; Auld, D. S.; Inglese, J.; Austin, C. P.; Michnick, S. W.; Xia, M. H.; Thomas, C. J. *Bioorg. Med. Chem. Lett.* 2009, *19*, 3686–3692.
86. Smits, R. A.; Lim, H. D.; Hanzer, A.; Zuiderveld, O. P.; Guaita, E.; Adams, M.; Coruzzi, G.; Leurs, R.; de Esch, I. J. *J. Med. Chem.* 2008, *51*, 2457–2467.
87. Hill, S.J. *Pharmacol. Rev.* 1990, *42*, 45–83. (b) Nguyen, T.; Shapiro, D.A.; George, S.R.; Setola, V.; Lee, D. K.; Cheng, R.; Rauser, L.; Lee, S. P.; Lynch, K. R.; Roth, B. L.; O'Dowd, B. F. *Mol. Pharmacol.* 2001, *59*, 427–433.
88. (a) Cheeseright, T. *Innovations in Pharmaceutical Technology*, 2009, *28*, 22–26 (b) Moreira, L.; Barreiro, E. J. *Curr. Med. Chem.* 2005, *12*, 23–49.

89. Barton, N. P.; Bellenie, B. R.; Doran, A. T.; Emmons, A. J.; Heer, J. P; Salvagno, C. M. *Bioorg. Med. Chem. Lett.* 2009, *19*, 528–532.
90. Lu, R. J.; Tucker, J. A.; Zinevitch, T.; Kirichenko, O.; Konoplev, V.; Kuznetsova, S.; Sviridov, S.; Pickens, J.; Tandel, S.; Brahmachary, E.; Yang, Y.; Wang, J.; Freel, S.; Fisher, S.; Sullivan, A.; Zhou, J. Y.; Stanfield-Oklaey, S.; Greenberg, M.; Bolognesi, D.; Bray, B.; Koszalka, B.; Jeffs, P.; Khasanov, A.; Ma, Y.-A.; Jeffries, C.; Liu, C. H.; Proskurina, T.; Zhu, T.; Chucholowski, A.; Li, R. S.; Sexton, C. *J. Med. Chem.* 2007, *50*, 6535–6544.
91. Rydzewski, R. M. *Real World Drug Discovery: A Chemist's Guide to Biotech and Pharmaceutical Research*; Elsevier: Oxford, 2008.
92. Barta, T. E.; Barabasz, A. F.; Foley, B. E.; Geng, L. F.; Hall, S. E.; Hanson, G. J.; Jenks, M.; Ma, W.; Rice, J. W.; Veal, J. *Bioorg. Med. Chem. Lett.* 2009, *19*, 3078–3080.
93. Oprea, T. I. *J. Comput. Aided Mol. Des.* 2002, *16*, 325–334.
94. (a) Halpin, D. R.; Harbury, P. B. *PLoS Biol.* 2004, *2*, 1015–1021. (b) Halpin, D. R.; Harbury, P. B. *PLoS Biol.* 2004, *2*, 1022–1030. (c) Halpin, D. R.; Lee, J. A.; Wrenn, S. J.; Harbury, P. B. *PLoS Biol.* 2004, *2*, 1031.
95. (a) Joyce, G. F. *Annu. ReV. Biochem.* 2004, *73*, 791–836. (b) Roberts, R. W.; Ja, W. W. *Curr. Opin. Struct. Biol.* 1999, *9*(4), 521–529.
96. Wrenns, S. J.; Weisinger, R. M.; Halpin, D. R. *J. Am. Chem. Soc.* 2007, *129*, 13137–13143.
97. Kinzel, O.; Llauger-Bufi, L. Pescatore, G.; Rowley, M.; Schultz-Fademrecht, C.; Monteagudo, E.; Fonsi, M.; Gonzalez, P. O.; Fiore, F.; Steinkühler, C.; Jones, P. J. *Med. Chem.* 2009, *52*, 3453–3456.
98. (a) Wu, W. L.; Burnett, D.; Domalski, M.; Greenlee, W. J.; Li, C.; Bertorelli, R.; Fredduzzi, S.; Lozza, G.; Veltri, A.; Reggiani, A. *J. Med. Chem.* 2007, *50*, 5550–5553. (b) Kempson, J.; Spergel, S. H.; Guo, J. Q.; Quesnelle, C.; Gill, P.; Belanger, D.; Dyckman, A. J.; Li, T. L.; Watterson, S. H.; Langevine, C. M.; Das, J.; Moquin, R. V.; Furch, J. A.; Marinier, A.; Dodier, M.; Martel, A.; Nirschl, D.; Van Kirk, K.; Burke, J. R.; Pattoli, M. A.; Gillooli, K.; McIntyre, K. W.; Chen, L. S.; Yang, Z.; Marathe, P. H.; Wang-Iverson, D.; Dodd, J. H.; Mckinnon, M.; Barrish, J. C.; Pitts, W. J. *J. Med. Chem.* 2009, *52*, 1994–2005. (c) Saulnier, M. G.; Frennesson, D. B.; Wittman, M. D.; Zimmermann, K.; Velaparthi, U.; Langley, D. R.; Struzynski, C.; Sang, X. P.; Carboni, J.; Li, A. X.; Greer, A.; Yang, Z.; Balimane, P.; Gottardis, M.; Attar, R.; Vyas, D. *Bioorg. Med. Chem. Lett.* 2008, *18*, 1702–1707. (d) Pettus, L. H.; Xu, S. M.; Cao, G. Q.; Chakrabarti, P. P.; Rzasa, R. M.; Sham, K.; Wurz, R. P.; Zhang, D. W.; Middleton, S.; Henkle, B.; Plant, M. H.; Saris, C. J. M.; Sherman, L.; Wong, L. M.; Powers, D. A.; Tudor, Y.; Yu, V.; Lee, M. R.; Syed, R.; Hsieh, F.; Tasker, A. S. *J. Med. Chem.* 2008, *51*, 6280–6292. (e) Herberich, B.; Cao, G. Q.; Chakrabarti, P. P.; Falsey, J. R.; Pettus, L.; Rzasa, R. M.; Reed, A. B.; Reichelt, A.; Sham, K.; Thaman, M.; Wurz, R. P.; Xu, S. M.; Zhang, D. W.; Hsieh, F.; Lee, M. R.; Syed, R.; Li, V.; Grosfeld, D.; Plant, M. H.; Henkle, B.; Sherman, L.; Middleton, S.; Wong, L. M.; Tasker, A. S. *J. Med. Chem.* 2008, *51*, 6271–6279. (f) Bailey, N.; Bamford, M.J.; Brissy, D.; Brookfield, J.; Demont, E.; Elliott, R.; Garton, N.; Farre-Gutierrez, I.; Hayhow, T.; Hutley, G.; Naylor, A.; Panchal, T. A.; Seow, H-X ; Spalding, D.; Takle, A. K. *Bioorg. Med. Chem. Lett.* 2009, *19*, 3602–3606.
99. Gentile, G.; Di Fabio, R.; Pavone, F.; Sabbatini, F. M.; St-Denis, Y.; Zampori, M. G.; Vitulli, G.; Worby, A. *Bioorg. Med. Chem. Lett.* 2007, *17*, 5218–5221.

5 Microwave Heating as a Tool for Process Chemistry

Jonathan D. Moseley

CONTENTS

5.1   INTRODUCTION AND OVERVIEW

The focus of process chemistry is toward pilot- and full-scale production. As a result, this chapter will address the topic of scale-up of microwave-promoted chemistry. Microwave heating is used on very large scales in industries such as drying, polymer and rubber preparation, and food processing,[1] as well as on more moderate scales for niche applications in analytical, environmental, and biomedical chemistry. However, the examples presented here will be drawn mainly from pharmaceutical process chemistry[2] derived from traditional organic synthesis and particularly the chemistry of drug discovery, since this is from where process chemists get their target molecules and often their first synthetic route. Microwave heating is relatively expensive compared to other types of conventional, conductive heating methods. Bogdal has listed some of the attractive features of microwave heating and concluded that to be commercially competitive, it must be applied to high-value products.[3] This fits well with the manufacture of high-value, low-tonnage pharmaceuticals rather than low-value, high-tonnage agrochemicals and commodity chemicals. Other high-value products are polymers and peptides (effectively biopolymers), which also benefit from microwave heating and may be scaled up in a limited sense.

After a brief historical overview, the various approaches to the scale-up of microwave chemistry will be discussed. This will be followed by a section detailing the current commercially available microwave reactors for scale-up. Practical applications and concerns will be addressed together with a discussion of reaction classes that work particularly well in microwave reactors. Chemistry that has been performed in the different available reactors will be presented in order to exemplify particular techniques and practices, but the discussion will not attempt to be exhaustive since several other books and reviews have already attempted this.[4–7] Attention here will be drawn to special features and uses. Further applications to demonstrate how microwave heating can be used as an investigative tool at the small scale for process chemists and others will follow. In accordance with the main theme of sustainability of this book, the current hot topic of the energy efficiency of microwave heating will also be covered in this chapter. Recent developments in the area of pilot-scale microwave chemistry will be noted before the final conclusions.

There is much chemistry that has and continues to be conducted using microwave heating under solvent-free or dry media solid-supported conditions,[8] often in domestic ovens, the use of which is strongly discouraged by this and other authors.[9] Although reactions carried out under these conditions claim to be "green" by avoiding solvent in the reaction step, pre- and postsolvent treatment is often considerable. Such techniques may in fact be less "green" overall than conventional chemistry. This chapter will focus on conventional "wet" organic chemistry conducted in standard organic solvents as used by most academic, medicinal, and process chemists.

5.2   SCALE-UP

5.2.1   Historical Perspective

Microwave heating was first used for organic synthesis in 1986, but it was not until the mid-1990s that the issue of scale-up was first investigated. Strauss pioneered the technology, building a prototype microwave batch reactor (MBR)[10] and a continuous-flow microwave reactor (CMR)[11] for large-scale synthesis, testing 15 and 26 reactions in each one, respectively.[12] Since this period, a significant number of other prototypical microwave scale-up reactors have been reported, both batch and continuous in operation, usually from chemical engineering groups more so than organic chemistry ones. Examples are discussed in other reviews of the field[13,14] but not here, since they are not available for general use.

From around 2000, small-scale commercial scientific microwave units became available, and their use was quickly taken up by medicinal chemists keen to capitalize on short reaction times to generate their desired compounds more rapidly. Integrated robotic handling in several instruments also aided this uptake. It is now estimated that nearly all new potential pharmaceuticals are first synthesized in a microwave reactor. Once medicinal chemists began to incorporate microwave chemistry steps routinely into their synthetic routes, the possibility to scale up this chemistry began to have an obviously visible benefit. As a result, the scientific microwave instrument manufacturers began to apply themselves to the problem of microwave scale-up.

5.2.2 THE NEED FOR SCALE-UP

Scaling up chemistry has been a topic of keen interest at conferences on the topic of microwave chemistry over the last 5–6 years[15-17] but, despite considerable investigation of potential scale-up reactors, it has not progressed as well as other areas such as peptide synthesis using microwave heating. The question arises as to why is this is the case. It could be that there is simply no need; or that there is a need, but it is difficult to achieve; or that irrespective of need, it cannot be achieved (due to physical limitations, for example). To gain a deeper understanding, it is worth reviewing the advantages of microwave heating for organic synthesis on the smaller scale, which are typically given as follows:

- Faster reactions, due to the higher reaction temperatures achievable.
- Increased yield; this is probably because it is easier to reach the end point of a reaction that is heated to high temperatures for a short period, compared to waiting for an uncertain end point at more modest temperatures over a prolonged period.
- Reduced impurities, which are usually attributed to reduced "wall effects";[18] fewer impurities also mean more product and, hence, higher yields.
- Enlarged reaction space through the use of superheated solvents in sealed-vessel systems (i.e., autoclave-type conditions), this may allow for new chemistries to be accessed, such as reactions in near-critical water.[19]
- Linear reproducibility on scale-up, potentially avoiding traditional scale-up problems.
- Possible energy savings.

All of these features are of as much interest to process chemists as to medicinal chemists. Faster reactions mean that more product can be processed more quickly, which

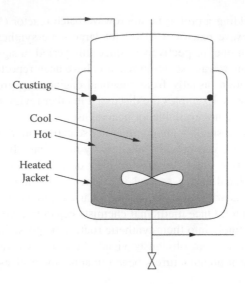

Crusting

Cool

Hot

Heated
Jacket

FIGURE 5.1   Heating effects in a conventionally heated large-scale batch reactor.

may obviate some of the inherent scale-up issues with microwave chemistry. Increased yield on scale-up is usually more a matter of efficient isolation than conversion (irrespective of heating method), but reduced impurities can be very beneficial, especially when chromatography is generally avoided and/or impossible. Conventional plant-scale batch reactors heat conductively through their walls, often resulting in temperature gradients even if well stirred,[18] which may result in charring and decomposition (Figure 5.1). These effects are more marked at higher temperatures, where longer reaction times may be needed and most molecules are simply spectators in the reaction during this period. The cool walls during microwave heating avoid these problems on scale-up, which not only improves the chemistry (yield, selectivity, etc.), but also the sustainability, since isolation and purification often have the greatest impact on the economic and environmental factors of pharmaceutical processes.

A further advantage of microwave heating in process chemistry is the potentially enlarged reaction space that may be accessed due to the ability to heat readily to high temperatures. Most conventional plant is heated by superheated steam, usually through some other heat transfer fluid circulating through the walls of the reactor vessel. However, ~150 °C is typically the highest accessible temperature for much pharmaceutical plant equipment. Chemistry that requires higher temperatures may need specialized plant that is not available or it may have to be contracted out to a specialist, resulting in added time and cost to the project. Alternatively, the chemistry can be performed at a lower temperature for a longer period, but problems with "overcooking" may result.

An often-quoted advantage of small-scale microwave chemistry has been linear scale-up from small tubes to larger tubes, cylindrical vessels, or other large reactors.[16] This has been demonstrated numerous times to emphasize the ease and applicability of microwave scale-up.[20–24] However, a key milestone in pharmaceutical development requires ~0.5–5 kg of the desired compound for initial clinical trials.

Many potential new pharmaceuticals fail in clinical trials after this point, but some process development effort will already have been invested in the project for future scale-up. If the (microwave-heated) medicinal chemistry method can be scaled to supply these quantities, without any additional development, both time and resource would be saved on the project. This would take the project to a critical decision point more quickly, and save considerable resources.

5.2.3 THE PROBLEMS WITH SCALE-UP

The scale-up of microwave chemistry clearly has several benefits to offer over conventional heating. These benefits largely overlap those at small scale, but also go beyond them, particularly regarding sustainability issues. However, there are some problems that make scale-up of microwave chemistry difficult to achieve.

The key limiting factor is the penetration depth of microwave irradiation, which is only a few centimeters in most solvents at 2.45 GHz.[25] An issue therefore arises in getting sufficient microwave power into the reaction mixture to achieve the desired heating effect. The core of a large reactor vessel will not receive any microwave radiation as it will all have been absorbed by the outer layers. As a result, the center is effectively conductively or convectively heated, and the potential benefits of microwave heating will be lost. Penetration depth does, however, vary with frequency. Only a limited number of Industrial, Scientific, and Medical (ISM) frequencies are allowed[25] so as not to interfere with military and civil aviation frequencies and telecommunications.[26] Alternative frequencies are used for other large-scale applications and thus may provide an alternative solution to the scale-up of microwave chemistry.[1]

A potential solution to the issue of limited penetration depth could be to add a microwave reactor externally to a large batch reactor and cycle the reaction mixture through a loop continuously.[15] However, this is effectively using the conventional reactor as a reservoir while processing small quantities of reaction mixture in the microwave field and, overall, it offers no advantage. Another solution is to either process smaller volumes in batch mode or else use a continuous-flow microwave reactor. These are the options that most investigators of microwave scale-up have used.

All the major scientific microwave manufacturers have designed microwave reactors capable of providing scale-up from 10 g to possibly low-kilogram-scale quantities, usually without the need to change the chemistry, thus confirming a key advantage of microwave chemistry on scale-up. These reactors have markedly different operating principles, and the scale-up achievable in each varies. The need to avoid competitors' patents and to have a competitive advantage may also contribute to the diversity of equipment for scale-up.[27] It also indicates that no consensus has been reached as to best practice.[17] The approaches taken, shown in Figure 5.2, can be loosely categorized as: scale-up in size, scale-up in parallel, scale-up in sequence, and scale-up by continuous flow. Hybrid approaches have also been devised.

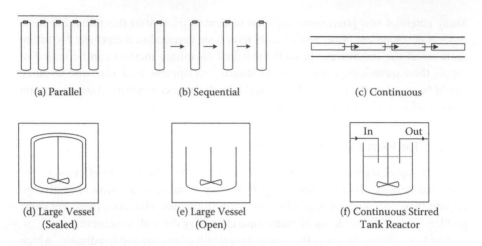

(a) Parallel	(b) Sequential	(c) Continuous

(d) Large Vessel (Sealed)	(e) Large Vessel (Open)	(f) Continuous Stirred Tank Reactor

FIGURE 5.2   Strategies for microwave scale-up.

5.3   COMMERCIALLY AVAILABLE MICROWAVE APPARATUS

Commercially available microwave reactors will be used to exemplify the different approaches to scale-up. Alongside this, an assessment of the relative advantages and disadvantages of each approach (and where relevant, each reactor) will be made, this also being the subject of previous reports.[28,29] Key technical data on the equipment to be discussed are given in Table 5.1. More comprehensive technical specifications have been given previously[13,30] or are available from the manufacturers.[27] Scale-up in parallel is discussed initially because it is the first logical extension of small-scale microwave chemistry.

5.3.1   SCALE-UP IN MULTIPLE SEALED VESSELS (PARALLEL)

Small-scale chemistry is often performed in microwave reactors such as the Biotage Initiator and the CEM Discover, both shown in Chapter 1. These units do have a limited scale-up option using sealed glass tubes of 20–50 mL working volume, but an obvious, more substantial scale-up approach is to process multiple larger reaction tubes in parallel (cf. Figure 5.2a). This is exemplified by the Anton Paar Synthos 3000 (Figure 5.3), the CEM MARS (Figure 5.4), and the Milestone MicroSYNTH (Figure 5.5).[27] These reactors typically have 6–20 tubular vessels of between 45 and 300 mL individual volume, giving up to 1 L of useful reaction volume. The narrow tubular reaction vessels avoid the potential limitation of penetration depth by ensuring the whole vessel can be irradiated. However, charging and emptying multiple vessels can be tedious without automation, especially if solids are present. Solutions are easier to process but, even so, manual handling of many reaction tubes requires time, and the maximum volume is only ~1 L, which, at typical process concentrations of around 10% w/v, equates to 100 g of product per run. Such reactors can often reproduce small-scale microwave reactions without requiring additional development. They are used successfully by several small and medium-sized companies which offer outsourced medicinal chemistry programs to larger organizations, particularly

TABLE 5.1

Operating Parameters of Commercial-Scale-Up Microwave Reactors

Type	Microwave Reactor	Power (W)	Vessel Size (mL)	Max Fill^a (mL)	Max Temp (°C)	Max Pressure (bar)	Solid Handling^b
Parallel	Synthos (HF100)	1400	16 × 100	1000	240	40	(Yes)
	Synthos (XQ80)	1400	8 × 80	400	300	80	(Yes)
	MARS	1600	14 × 75	700	200	40	(Yes)
	MicroSYNTH Q20	1000	20 × 45	600	250	80	(Yes)
	MicroSYNTH SK-10	1000	10 × 100	700	250	80	(Yes)
Single, sealed	Advancer	1200	350	250	250	20	Yes
	BatchSYNTH^c	1000	1000^d	700^d	230^d	8^d	(Yes)
	UltraCLAVE^e	1000	3500	2000^f	300	200	(Yes)
Single, open	MARS	1600	5000	3000	Solvent bp	1	Yes
	MicroSYNTH	1000	4000	2500	Solvent bp	1	Yes
Sequential	Voyager	300	80	n × 50	250	20	No
	Kilobatch	1200	350	4 × 250^g	250	20	Yes
Continuous	FlowSYNTH	1000	200	Unlimited	200	30	No

a   Usable reactor volume per cycle.
b   (Yes) = solids handled under most conditions dependent on density, loading, and agitation.
c   Based on MicroSYNTHPlus.
d   Smaller single vessels possible with different operating parameters.
e   Multiple parallel vessel configurations also possible.
f   In single vessel use.
g   n × 250 if homogeneous reaction mixture.

FIGURE 5.3 The Anton Paar Synthos 3000. (Reproduced with permission from Anton Paar.)

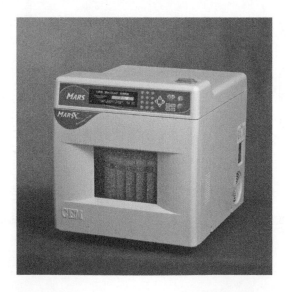

FIGURE 5.4 The CEM MARS in sealed-vessel mode. (Reproduced with permission from CEM Corp.)

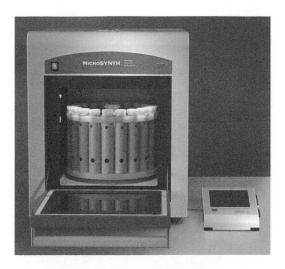

FIGURE 5.5 The Milestone Micro SYNTH in sealed vessel mode. (Reproduced with permission of Milestone s.r.l.)

to scale up initial syntheses of advanced key intermediates on the 100–300 g scale, or to make multiple analogs on 5–10 g scale in a single run.[31]

5.3.2 SCALE-UP IN A SINGLE LARGE SEALED VESSEL

When considering batch scale-up, another approach is to use a single, larger, reaction vessel (Figure 5.2d). This avoids the problem of tedious charging and discharging of multiple reaction vessels. However, penetration depth and/or power density could begin to become issues as the vessel size increases. Even so, the limiting factor if the vessel is to be run in sealed-vessel mode is more likely to be safety concerns related to the vapor pressure generated by a superheated solvent. While moderately high pressures (~20 bar) are readily contained in small scale microwave reactors, using larger vessel sizes requires more safety features and greater engineering expertise in the design. This results in more complex reactors that are less easy to use and more expensive. The Milestone BatchSYNTH, which is based on the MicroSYNTHPlus, can accommodate a 1 L vessel, which achieves a compromise between single-vessel operation and pressure containment for limited scale-up. The Milestone UltraCLAVE (Figure 5.6) is a much larger instrument with currently the largest reaction vessel for this type of microwave reactor (3.5 L with 2 L useful reaction volume). This can handle substantial pressures, but drawbacks are greater cost and an instrument that does not fit in a chemist's conventional fume cupboard.

A further problem arises with large single-vessel microwave reactors. To contain the pressures likely to be generated, strong materials must be used for the reaction vessels, but they must also be microwave transparent. This rules out metals on this scale and leaves thick-walled (quartz) glass, ceramics, and polytetrafluoroethene (PTFE, better known by the DuPont brand name Teflon), or combinations thereof.

FIGURE 5.6   The Milestone UltraCLAVE. (Reproduced with permission from Milestone s.r.l.)

However, these are all thermal insulators. Furthermore, as the scale increases, the surface area-to-volume ratio drops, so while heat (microwave energy) can readily be put directly into the reaction sample, it is not so easily removed. Active cooling of such reaction vessels is also not easy. Generally, compressed or fan air is used, but air has a low specific heat capacity. Liquids can be used, but they must be microwave transparent (or drained from the microwave cavity between batches). In either case, cooling is further compromised by the necessarily thick-walled vessels such that, for larger microwave reactors, the cooling time is often notably longer than the combined heat-up and reaction periods. This increases the cycle time per batch, and reduces the overall output, lessening the benefit of rapid microwave heating.

The Biotage Advancer (Figure 5.7) attempts to bridge these requirements by using a 350 mL sealed-vessel. The slim PTFE vessel design ensures that penetration depth is not an issue. Flash evaporation using the mechanical pressure of superheated solvent in the vessel can discharge the reaction mixture while it is still hot into a collection vessel, thus allowing the cooling to occur offline from the reactor. However, like the UltraCLAVE, it is also large and expensive.

5.3.3   SCALE-UP IN A SINGLE LARGE OPEN VESSEL

If superheated solvents are not required, or if the process can be performed successfully below the solvent boiling point, then large reactions can be run in open-vessel mode (Figure 5.2e). This is possible in both the CEM MARS (Figure 5.8) and the

FIGURE 5.7   The Biotage Advancer. (Reproduced with permission from Biotage.)

Milestone MicroSYNTH (Figure 5.9). The advantage of this configuration is that standard laboratory glassware can be used, even up to the 5 L scale in the CEM MARS. This is much more attractive to the average organic chemist since it looks and handles like a normal laboratory setup. Furthermore, using a thin-walled glass vessel allows for a faster cooling time, although of course the surface area-to-volume ratio will still be reduced. However, penetration depth or power density may be an issue, and of course the possible advantages of accessing elevated temperatures and pressures cannot be realized. However, considerable scale-up can be, and has been, achieved in such reactors.

5.3.4   Scale-Up in Sequence

A way to retain the advantages of sealed-vessel operation while not incurring the performance penalties of larger-vessel operation is to process many small batches in sequence (Figure 5.2b). However, to be viable for production of sizable quantities of material, the operation must be possible in rapid succession. The CEM Voyager can process small batches of up to 50 mL in an 80 mL glass vessel (Figure 5.10). Automation is achieved by pumping reaction solutions in and out of the reaction vessel through a number of PTFE lines. In this respect, the stop-flow Voyager is a hybrid between sequential batch and true continuous-flow operation. The vessel can be sealed between operations so that elevated temperatures and pressures can be reached. Penetration depth is not an issue, although power density can sometimes

FIGURE 5.8 The CEM MARS in open-vessel mode. (Reproduced with permission from CEM Corp.)

FIGURE 5.9 The Milestone MicroSYNTH in open-vessel mode. (Reproduced with permission from Milestone s.r.l.)

FIGURE 5.10 The CEM Voyager SF. (Reproduced with permission from CEM Corp.)

be a limitation when trying to use low-absorbing solvents at higher temperatures (>200 °C) since the instrument has only a 300 W output available. The cycle time is fast because the small vessel heats and cools very quickly. This allows many small batches to be processed in sequence so that a moderate scale-up can be obtained (typically a few hundred grams per day). Alternatively, multiple microwave reactors could be used to achieve a larger scale-up (the scale-out or numbering-up approach).[32] A further advantage is that once a process is established, it needs very little operator input, unlike the much more user-intensive parallel processing approach. However, a significant limitation of the stop-flow apparatus is that mixtures must generally be homogeneous, otherwise materials cannot be pumped into or out of the reaction vessel. Unfortunately, most pharmaceutical processes involve the use of solids at some point in their operation.

The recently introduced Biotage Advancer Kilobatch (Figure 5.11), a development of the Advancer, is perhaps the only true example of a microwave reactor that processes small batches in sequence while retaining solid charging. It incorporates automated loading of solvents and slurries and automated discharging of the product mixture. A limited cycle of four batches can be run before reagent restocking is required, giving a degree of semiautomated operation. The Kilobatch retains the larger vessel size (350 mL) and rapid cooling of its predecessor so that larger volumes can be processed quickly, but it also retains the size and cost.

5.3.4 SCALE-UP USING CONTINUOUS FLOW

There is considerable interest in continuous-flow chemistry at the present time, for rapid analog synthesis in medicinal chemistry in addition to scale-up.[33] A true continuous-flow microwave reactor is available in the Milestone FlowSYNTH

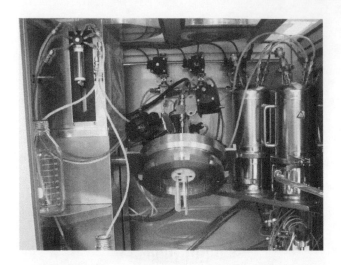

FIGURE 5.11 The Biotage Advancer Kilobatch, showing liquid- and solid-handling functionality together with product collection receptacles. (Reproduced with permission from Biotage.)

(Figure 5.12). This effectively achieves scale-up by heating the reaction mixture as it passes through the reactor column, which is 200 mL in volume. Throughput is dependent simply on flow rate, which is itself dependent on the residence time; the more energy that can be put into the reaction mixture, the shorter the residence time and the faster the flow rate. Such reactors can produce kilograms per day at typical process concentrations. The cooling time is also short since only a small portion of reaction mixture needs to be cooled at any time. A limitation is that both the reaction mixture going into and the product mixture coming out of the reactor must be in essence homogeneous for efficient operation. While the FlowSYNTH (similar to the Voyager), can in principle handle slurries, this is often not possible. This is especially problematic when using solvents above their boiling points because the back pressure regulator on the unit effectively acts as a filter for solids (even fine particulates) and quickly blocks them.

A number of prototype and bespoke continuous-flow microwave reactors of varying scales have also been fabricated and chemistry reported in them.[34] However, because these are specialized instruments that are not readily available, they will not be discussed further here.

Given that commercially available conventionally heated continuous-flow reactors are also available,[35] the question arises as to the advantages that microwave heating could offers in the case of flow chemistry. Conventional flow reactors are especially good at removing heat from exothermic reactions, but can also provide heat to endothermic ones. However, very high temperatures are often less accessible, resulting in lengthened reaction times and reduced throughputs. Microwave heating may facilitate access to the higher-temperature window, thus making some continuous-flow processes viable that otherwise would not be so.

An alternative continuous-flow approach has been reported, using a CEM MARS unit operating in open-vessel mode to mimic a continuous stirred tank reactor

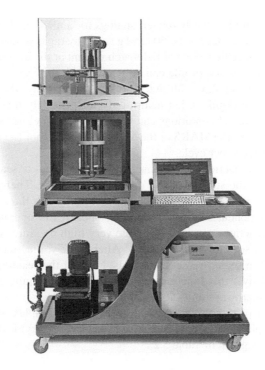

FIGURE 5.12   The Milestone FlowSYNTH. (Reproduced with permission from Milestone s.r.l.)

(CSTR; Figure 5.2f). As long as the residence time for diffusion of molecules from the input to the exit is long enough to achieve an acceptable conversion of reactants to products, a standard batch reactor can be turned into a CSTR. As an example, a 4 L flask in a MARS has been used for the preparation of biodiesel via a transesterification reaction with a flow rate of 7.2 L/min (33 s residence time).[36]

5.4   PRACTICAL CONSIDERATIONS

5.4.1   General Points

The first point that needs to be addressed when scaling up a reaction is the quantity of material that is required. For many medicinal and academic chemists, the use of a 20 or 35 mL reaction vessel as opposed to a 2–5 mL tube will constitute adequate scale-up. If more product is required, the same procedure is run five or ten times, preferably by automation, and the individual batches combined for workup. This may provide typically 5–25 g of product. No special expertise or redevelopment of the chemistry is required, and the user can begin immediately.

Working on larger scales in sealed-vessel mode will probably require the use of a multimode instrument such as the MARS, the MicroSYNTH, or the Synthos 3000, which can provide typically 50–250 g quantities per cycle (about 1 h) in sealed-vessel operation at typical concentrations. These are also ideal for preparing in a single cycle

5–10 g each of multiple, closely related analogs for a medicinal chemistry program. The Advancer can also access the 50–250 g range by rapid sequential operation, and larger scales can be achieved in the Kilobatch variant or the Milestone UltraCLAVE. If a homogeneous reaction profile can be guaranteed, the Voyager can deliver these types of quantities per day with automation. Alternatively, the FlowSYNTH can provide much more (kg/day), but again, only for generally homogeneous reaction mixtures. If open-vessel conditions can be used, then an easier and larger scale-up can be achieved using the MARS or the MicroSYNTH (from 1–3 L), giving typically 100–300 g of material per cycle.

Many classes of reaction lend themselves to either open- or sealed-vessel processing, depending on the substrates and operator preference. There are some obvious exceptions. If extrusion of a small molecule is required in the reaction, giving rise to a volatile component such as a gas, then open-vessel operation is likely to be essential. Such reactions can often generate very high pressures or else be inhibited when operated in sealed-vessel mode.[29,37] Alternatively, reactions using gases have to be performed in sealed-vessel conditions. Most of the reactors discussed can be pressurized to deliver this functionality, although some require slight modification. In essence, the microwave units act as autoclaves.

An indication of typical operational cycle times, complexity of loading, batch numbers, and typical throughputs achievable per day for a typical 10% w/v pharmaceutical process is given in Table 5.2.

5.4.2 APPLICABILITY TO SYNTHETIC TRANSFORMATIONS

Microwave heating is convenient in practice for most small-scale reactions, and hence its wide uptake in small-scale organic synthesis. In principle, any reaction requiring heating can be performed in a microwave reactor, but not all such reactions are suitable. As noted, microwave reactors are especially good at heating reactants and containing pressure, but rather poor at cooling. Furthermore, many reactions require only moderate heating, or heating for short periods. As a result, the benefit of using microwave heating in these cases is limited, although certainly convenient on the small scale.

A general analysis of the reactions used in the preparation of drug candidate molecules across three large pharmaceutical companies was conducted recently and provides a good starting point for discussion of the viability of microwave heating as a tool for scale-up for different chemistries.[38] Some very general observations are summarized in Table 5.3. Classes of reactions that may be expected to gain from microwave heating are alkylations/acylations (all types, C/N/O/S), additions, condensations, S_NAr reactions, and heterocycle formation. All these tend to require elevated temperatures and lengthy reaction times conventionally. Less major reaction classes that can also be identified are high-temperature rearrangements such as the Claisen rearrangement, cycloadditions such as the Diels–Alder reaction, and Friedel–Crafts-type acylations. Other reactions requiring autoclave-like conditions or gas pressure can be very conveniently and safely scaled up in microwave reactors. As an example, hydrogenations, treated separately from reduction or deprotection steps, make up a large group and can be performed successfully in microwave reactors.[39–41]

TABLE 5.2
Reactor Productivity for a Typical 10% w/v Pharmaceutical Process

Type	Microwave Reactor	Approximate Cycle Time	Typical Throughput (g)	Manual Handling	Achievable Number of Batches Per Day[a]	Typical Daily Throughput (g)
Parallel	Synthos (HF100)	45 min	100	High	2 (4)[b]	250
	Synthos (XQ80)	45 min	40	High	2 (4)[b]	100
	MARS	1 h	70	High	2 (4)[b]	200
	MicroSYNTH[c]	1 h	70	High	2 (4)[b]	200
Single, sealed	Advancer	30 min	25	Low	8–16[d]	300
	BatchSYNTH	1 h	70	Low	4–6	500
	UltraCLAVE	1.5 h	200	Low	4	800
Single, open	MARS	1 h	300	Medium	4–6	1200
	MicroSYNTH	1 h	250	Medium	4–6	1000
Sequential	Voyager	15 min	n × 5	Automated	30 (90)[e]	200 (600)[e]
	Kilobatch	20 min	4 × 25	Automated	32 (90)[e]	800 (2400)[e]
Continuous	FlowSYNTH	2 L/h[f]	200 g/h[f]	Automated	Continuous	1600 (4800)[e,f]

a　Assuming typical loading, unloading, and cleaning times.
b　Assumes use of two rotors.
c　For several vessel configurations.
d　For heterogeneous reactions; for homogeneous, see Kilobatch below.
e　With 24 h continuous operation.
f　Data from Reference 28.

TABLE 5.3
General Summary of Reaction Classes Suitable for Microwave Scale-Up

	Beneficial/Suitable	No Benefit/Unsuitable
Major reaction classes	Additions condensations	Amide bond formation
	Alkylations/acylations	Deprotections (excluding
	Heterocycle formation	hydrogenations)
	Hydrogenations	Functional group additions
	S_NAr reactions	Functional group
		interconversions
		Protection reactions
Minor reaction classes	Cycloadditions	Grignard reactions
	Friedel–Crafts reactions	Low-temperature organometallic
	Metal-catalyzed reactions	reactions (e.g., lithiation)
	(e.g., Heck and Suzuki	Oxidations
	couplings)	Reductions (metal hydrides,
	[Peptide synthesis][a]	excluding hydrogenations)
	[Polymer synthesis][b]	
	Thermal rearrangements	
Other reaction parameters	Autoclave/pressure reactions	
	Reactions with gases	
	Reactions with solid-support	
	reagents	
	Reactions with water as	
	solvent	
	Where thermodynamic product	
	required	

[a] See Chapter 9.
[b] See Chapter 3.

Since metal complexes interact well with microwave energy, metal-catalyzed reactions (e.g., Heck and Suzuki couplings) are also good candidates for scale-up.[42] However, even conventionally, scale-up of such reactions is still relatively rare in process chemistry despite huge academic interest. Polymer chemistry also performs well under microwave heating and can be scaled up. Amide bond formations for peptides can also be conveniently and efficiently prepared by microwave heating but, because a few grams of material classes as scale-up in these cases, discussion is not so relevant here where attention is on production of larger quantities of material. Direct conversion of esters to amides with excess volatile amines under forcing conditions is a useful application of microwave heating, and it should be noted that many heterocycle formations are effectively amide-bond-forming reactions followed by condensations/dehydrations.

Reactions that are unsuited to microwave heating are transformations that are usually performed at moderate or low temperatures, oxidations, and reductions, as well as functional group additions and interconversions. However, it is freely acknowledged that examples of all these types of reactions have been reported using microwave heating.

5.4.3 Nonlinear Scale-Up Parameters

A widely made claim of microwave chemistry is predictable, linear scale-up, and a number of studies have supported this assertion for both monomode and multimode microwave reactors.[20–24,28,29,43,44] Generally, this claim is valid, and certainly when scaling up from a small (2–5 mL) to a medium (100 mL, Synthos 3000) to a large sealed tube (350 mL, Advancer). Scaling from sealed tubes into open round-bottomed flasks in multimode reactors is also generally reliable.[28,29] This assumes that identical conditions such as temperature and reaction time can be obtained in all cases, so that predictable scale-up can be achieved. However, some factors do not scale linearly. As already noted, microwave power density drops with scale and, if this becomes a limiting factor, it will be revealed in reduced heat-up rates or temperatures achievable. Pressure should remain constant if the same relative fill is used in each vessel, so it is easily controlled (or calculated). While most aspects of microwave heating are indeed predictable on scale-up, it is not immune to some conventional scale-up effects. The application of common sense and forethought is still required. Two factors that certainly do not remain constant on scale-up of microwave-promoted chemistry and that are not so easily controlled are agitation and cooling.

Agitation is achieved in small-scale microwave reactors by magnetic stirring, but this is often ineffective on larger scales, and process chemists generally favor mechanical stirrers, which can put much more mechanical energy into a reaction mixture.[2b] This is especially important for reactions using solids, such as the especially dense cesium carbonate, or those that produce heavy precipitates. If agitation fails or is ineffective on any scale, hot spots or temperature gradients may be generated in parts of the reaction mixture as several reports have shown.[45–47] Magnetic stirrer bars are particularly unsuited for stirring solids on >1 L scale, and in heavily loaded tubular cylindrical vessels used in the parallel scale-up approach. Ineffective agitation may have implications for safety, but is more likely to result in incomplete reaction. Note too that the combination of particle size and agitation can also affect reaction success.

Cooling is also not directly scalable in microwave heating due to the drop in surface-area-to-volume ratio. Although a large sample can be heated as quickly as a small one in a microwave reactor due to the volumetric heating (assuming the power density is sufficient for the task), the heat cannot be removed as quickly from the larger sample. The issue of rapid, efficient cooling is not well treated at present, with the exception of continuous-flow systems and the Biotage Advancer with its rapid removal of product mixture from the vessel to allow cooling off-line.

5.4.4 Safety

All current scientific microwave units are robust instruments designed with many safety features and are capable of containing vessel ruptures (explosions). Even if the instrument is inoperative as the result of a vessel failure, the operator should be protected from harm. This being said, attention to safety is still essential. Basic safety checks such as ensuring that reaction vessels are in good condition and that all other aspects of instrumentation are in good working order is just the beginning.

It is important that microwave vessels be clean, especially those made of glass, since residues on the surface may absorb microwaves strongly and lead to hot spots that can rupture vessels that are under pressure. This is particularly a problem when using heterogeneous metal catalysts, which can melt vessels if grains of strongly microwave-absorbing catalyst are left above the solvent line. This can also happen with strongly microwave-absorbing, viscous liquids. For such reactions, it is important to load vessels carefully and ensure that all residues are washed off the walls. However, such reactions are relatively rare. More common is deposition of metal films from dissolved metal catalysts and reagents onto the walls of the reaction vessel during the run. This may present a problem for recurring use and, of course, care should be taken in cleaning the vessels to avoid scratching the surface or otherwise damaging the walls.

Accurate temperature measurement should be made, as in any chemical reaction, but microwave heating does present different problems. The reactors described here each have their own in-built solutions to providing temperature measurement. One further problem can arise with biphasic reaction. Depending on which phase the temperature measurement is taken from, results can be very different since the phases may heat at very different rates. This problem has not limited the use of biphasic conditions such as in phase-transfer catalysis,[48] but does require some forethought.

Elevated pressure in sealed systems poses a significant hazard. The pressure of super-heating a solvent can be determined beforehand by a simple calculation. Generally, the more microwave-absorbing solvents are also those with the higher boiling points; lower-boiling solvents tend to be more difficult to heat to higher temperatures but, not surprisingly, also tend to generate more pressure if they are. If gases or other volatile components are produced stoichiometrically during the course of a reaction, this will add to the internal pressure. Of more concern are solvents that may decompose to gaseous components under strong heating (e.g., DMF, formic acid), as these may lead to rapid and often uncontrollable pressure rises.

As noted already, the degree of agitation may vary from small to larger scales. When working on large scales, significant superheating of a solvent in open-vessel use should be avoided by the use of a mechanical stirrer.[45] A magnetic stir bar placed into the vessel is much less efficient and may be compromised if solids are present, leading to an eventual and sudden boiling of the solvent. In sealed-vessel use, failure in stirring is unlikely to cause a safety issue, but the reaction may fail as a consequence. One additional precaution related to magnetic stirrer bars which may apply at intermediate scales is that they should not be one quarter of the wavelength of the radiation in length (i.e., ~3 cm), otherwise they can function as an antenna and may lead to destruction of the stirring bar and even rupture of the vessel.[49]

A further safety warning is necessary when running open-vessel reactions under reflux. The solvent being refluxed can in some cases function as an antenna for microwave radiation and so conduct microwaves out of the cavity into the external environment. This is likely to depend on the solvent and other parameters. When running microwave reactors in open-vessel reflux mode, a microwave leakage test should be performed to ensure that no microwaves are escaping from the cavity

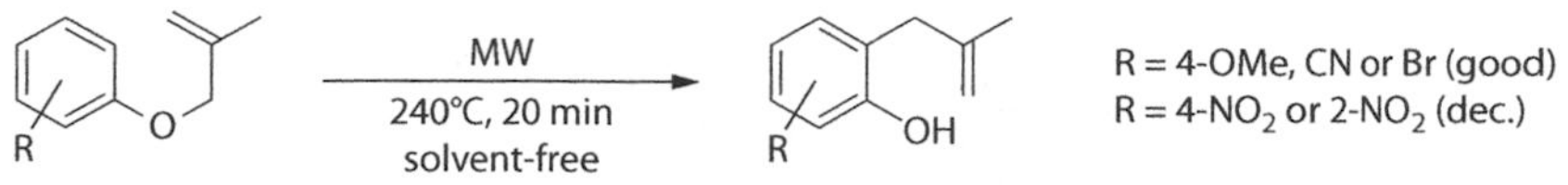

R = 4-OMe, CN or Br (good)
R = 4-NO$_2$ or 2-NO$_2$ (dec.)

SCHEME 5.1

before continuing. Commercially available leakage detectors as used for checking domestic microwave ovens are adequate for this purpose.

When performing a reaction on a small scale using a monomode microwave unit, a temptation is to heat the reaction mixture using the full 300–400 W available. However, this can sometimes lead to uncontrollable temperature and pressure rises and, as a result, vessel failure. When scaling up reactions using larger multimode microwave units, the ability to heat the reaction mixture as intensively is much reduced since the power density is significantly lower.

Microwave reactions are invariably "all-in" processes and are often endothermic, so runaway reactions are rare. However, exothermic decompositions can occur in some cases. The Claisen rearrangement of some nitro-functionalized arylvinyl ethers offers a case in point (Scheme 5.1). Upon heating, they decomposed exothermically despite the fact that related analogs were all well behaved.[50] Fortunately, the chemistry was first screened on a small scale. Although microwave chemistry can potentially be scaled directly in a linear manner, a gradual approach is prudent, a test reaction at intermediate scales being a wise precaution. The small investment in time is worth the effort to discover occasionally unsuitable chemistry. This is particularly important when performing chemistry using concentrated or solvent-free protocols.

5.5 EXAMPLES OF SCALE-UP

The examples that have been selected are intended to illustrate applications using the different approaches to microwave scale-up presented above and are not intended to be comprehensive. Examples have been taken from the more recent literature since these are consistent with the current state of equipment. Earlier examples have been covered elsewhere.[4–7] The focus will be on scale-up, but alternative applications will also be highlighted.

Two recent studies have reported a range of reactions performed in all the commercial reactors discussed above. The first of these deals with the high-temperature Newman–Kwart rearrangement (NKR) (Scheme 5.2) performed in all the microwave units, thus giving a direct comparison between them.[28] The second covers a suite of reactions performed using different microwave units, several of which will be discussed further below.[29] In both cases, scale-up to 1 mol or beyond was accomplished with commercially available equipment. Earlier, partial comparisons between several scale-up microwave reactors had also been reported.[20–22,43,44]

5.5.1 Parallel Scale-Up (MARS, MicroSYNTH, Synthos 3000)

The Anton Paar Synthos 3000 is one of a number of parallel sealed-vessel multimode microwave reactors designed for scale-up. The first results using this unit

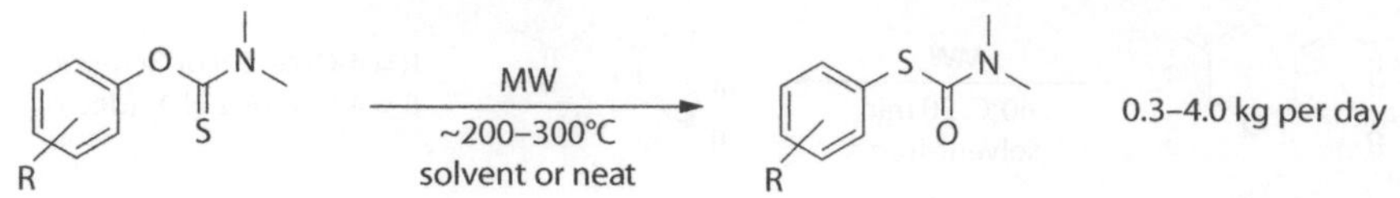

0.3–4.0 kg per day

SCHEME 5.2

were in 2003 for a range of reactions, including the Biginelli dihydropyrimidinone synthesis and a solid-phase procedure, together with Kindler, Heck, Negishi, and Diels–Alder reactions.[22] The focus of this study was more on proving comparable results from small-scale monomode reactors to multimode parallel batch reactors rather than pushing the limits of scalability, since not all 8 or 16 tubes were used every time. Even so, the Biginelli reaction was scaled to 0.64 mol using just eight tubes (Scheme 5.3). This raises an important point. If the quantity of material available for reactions or required at the end does not necessitate loading all the vessels in this type of microwave unit, a smaller scale-up can still be achieved using fewer tubes, as long as they are symmetrically placed in the rotor to keep it evenly distributed. A dummy solvent load can also be used for the same purposes. If working at less than full load, it is usually more efficient to use fewer tubes at full volume than more tubes partially filled since this reduces the number of weighing, charging, discharging, and cleaning operations required.

The Synthos 3000 has also been used in a comparative study with two other microwave units (and conventional heating) for three S_NAr reactions under varying reaction conditions.[44] Eight tubes at 50 mL fill were used for substitution of a pyrimidine core by heating at 120 °C for 10 min (Scheme 5.4). A yield of 76% was obtained, but with only 90% purity, although this was better than the conventionally heated equivalent (66% yield of 80% purity), or an autoclave (74% yield of 85% purity). This was also somewhat lower than the 95–96% obtained with smaller-scale microwave heating, the differences being attributed to the difference in heating profile between the different scales.

Another early example of parallel scale-up in sealed vessels was reported using the Milestone MicroSYNTH unit, attention being focused on an alkylation reaction that was first optimized on a small scale (Scheme 5.5).[21] The objective of the study was to probe variations in outcome across multiple reaction vessels, especially since the temperature is accurately monitored in only one tube, the others being measured by an external IR sensor (which is typical of all these instruments). The reproducibility was found to be good both across the vessels and also from small to larger scale. Indeed, several analogs were scaled up in a single run in a good demonstration of parallel scale-up in support of a medicinal chemistry program.

Several of the parallel-type Milestone microwave reactor systems have been used for scale-up of reactions, including the pharmaceutically valuable Biginelli and Hantzsch multi-component reactions.[51] The work showed reproducibility throughout a set of small reaction vessels run in parallel, and also the applicability of one set of conditions across a range of substrates.

Working with multiple parallel vessels can be advantageous in some circumstances. For example, a protocol for the solvent-free *aza*-Michael reaction requires

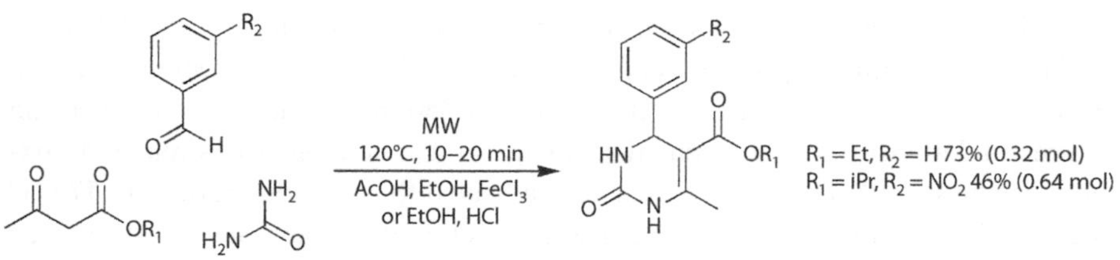

SCHEME 5.3

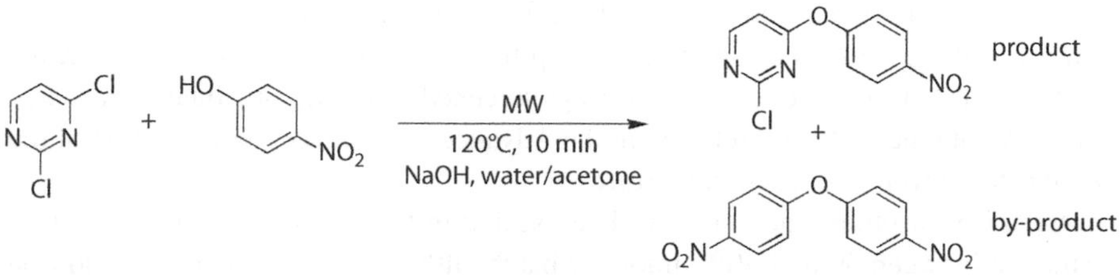

R₁ = Et, R₂ = H 73% (0.32 mol)
R₁ = iPr, R₂ = NO₂ 46% (0.64 mol)

SCHEME 5.4

99–100% conversion
81–99% yield

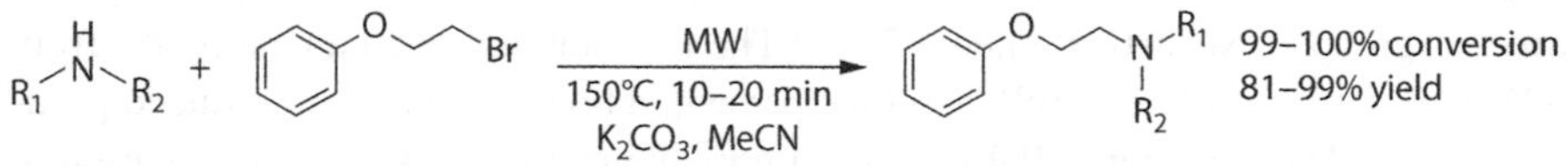

SCHEME 5.5

heating to 200 °C for 20 min (Scheme 5.6).[52] The reaction is performed neat using methyl acrylate as a substrate, generating considerable pressure at the elevated temperatures reached (~14 bar). As a scale-up strategy, a MicroSYNTH with a rotor comprising 20 quartz reaction vessels was used as opposed to performing the reaction in one larger vessel.[29] IR temperature monitoring in each of the 45 mL quartz tubes was possible, with a control vessel monitored by a fiber-optic probe inside a thermowell that was directly placed into one of the reaction vessels. Scale-up to 0.4 L (2.8 mol) was possible, a product yield of 78% being obtained for the combined reaction mixture with little variation between the vessels. Use of a parallel small-vessel system was felt to be better in this case for containing the high pressure generated, while charging a single stock solution to each vessel was not too onerous.

5.5.2 SEALED-VESSEL SCALE-UP (ADVANCER, BATCHSYNTH, ULTRACLAVE)

Scale-up in a single sealed vessel represents perhaps the least utilized scale-up option, being limited to a few reports in the Advancer and the BatchSYNTH, and even less using the UltraCLAVE. This may be because both the Advancer and the UltraCLAVE are large and expensive instruments, and require more expertise in their use, and also because the cycle time tends to be long for the scale achievable.

A Heck reaction on 0.1 mol scale (100 mL) has been performed in the Advancer at 170 °C for 15 min to give an isolated yield of 93% of 4′-methoxycinnamic acid (Scheme 5.7).[29] The vessel used in this unit is of 350 mL capacity, with a working volume of 250 mL. It has a mechanical stirrer so that heterogeneous reaction mixtures can be stirred. The 1200 W magnetron heated the reaction mixture to 170 °C in 5 min. The adiabatic flash cooling also means that the cycle time is fast, so even though the scale is not large, 0.25 molar batches could be run in rapid succession, perhaps 2–3 per hour.

In a more recent report, the same Heck coupling has been studied in more detail.[53] The chemistry is performed on a small scale in sealed glass tubes using very low catalyst loadings. However, upon scaling up the chemistry using the Advancer, the reaction failed at the lower catalyst loadings screened. This was attributed to adsorption of Pd into the PTFE vessel used in the Advancer and, hence, sequestration of the metal catalyst from the reaction mixture.

A number of other reactions have been scaled in the Advancer, most being performed at between 50 and 400 mmol per batch, although the *aza*-Michael addition discussed above could be performed on the 1.5 mol scale. One particular example of pharmaceutical interest was the Grandberg synthesis of 2-methyltryptamine.[53] A conventional approach to this reaction had already been scaled up to 20 kg by a process group at Novartis (Scheme 5.8).[54] The reaction exhibited an exotherm from ~75 °C if not carefully controlled, which presented both a hazard and reduced product quality. The very controllable heating possible using the Advancer was used to moderate the initial exothermic event before continuing to heat to 150 °C using a programmed heating ramp. Results comparable to those from the Novartis group were achieved on 0.2 mol scale.

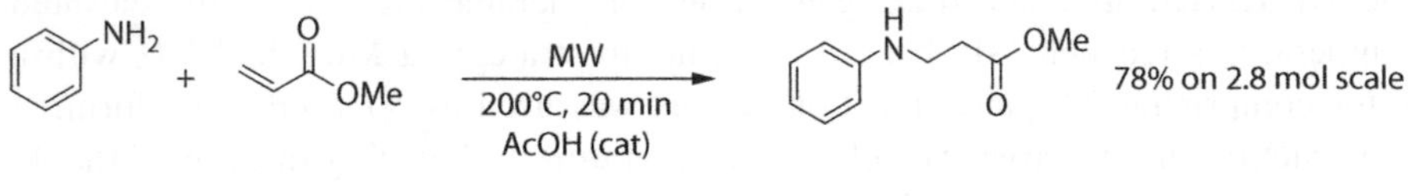

SCHEME 5.6

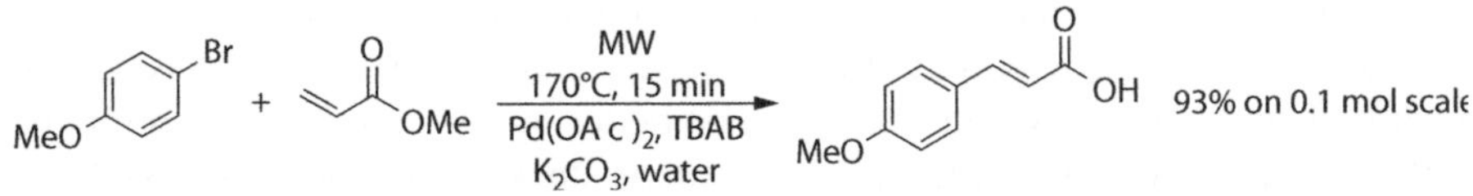

SCHEME 5.7

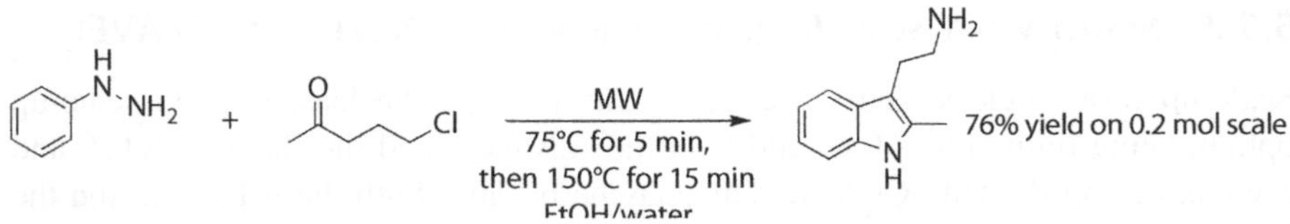

SCHEME 5.8

The Milestone UltraCLAVE is the microwave unit with the largest single vessel (3.5 L total, ~2 L working volume) for operation under elevated pressures. It is essentially a microwave digestion unit and can accommodate a range of vessel sizes from many tubes up to a single PTFE pot. Typical uses are analytical, environmental, food technology, and medical applications. The NKR has been scaled up using this unit.[28] It has also been used for the palladium-catalyzed ethoxycarbonylation of iodobenzene on the 1 mol scale under a pressure of carbon monoxide.[55] The reactions required heating to 125 °C for 30 min, a slight modification from previous smaller-scale protocols, yielding 120 g (80%) of ethyl benzoate product. The unit allowed for 27 bar of CO (and 23 bar of N_2, 50 bar total) to be safely loaded. However, the cooling time at the end of the reaction was lengthy (45 min) due to the size of the vessel and the thickness of the walls.

Building on this chemistry, the UltraCLAVE has been used in parallel sealed-vessel mode for alkoxycarbonylation reactions.[55] Six 150 mL vessels were charged at 50 mmol each and heated at 125 °C for 30 min to give 91–99% conversion for six different aryl iodide substrates run simultaneously (Scheme 5.9). Although each vessel is open to the pressurized CO/N_2 atmosphere inside the 3.5 L reactor, they are effectively operating as individual pressurized vessels in parallel. One advantage over the other pressurized parallel reactors was evident in this case: the temperature probe measures the ethanol bath in which the individual reaction vessels sit. This is more reliable than measuring just one control vessel and assuming that all other vessels absorb microwaves in a similar fashion.

One moderately large-scale microwave reactor is potentially available for hydrogenations using hydrogen gas. A Milestone ETHOS 1600 has been modified to take up to 30 bar H_2 on up to 200 mL working volume.[39] Several hydrogenation reactions have been conducted using the apparatus, including aromatic and azide reductions, an amine debenzylation, and reduction of the trisubstituted double bond in strychnine. All reductions used Pd/C as the catalyst and were reportedly faster than conventionally heated analogs.

Vessels in a Synthos 3000 have been prepressurized with ethene in order to perform a Diels–Alder reaction with a range of 2(1H)-pyrazinones and prepare bis-lactam products in a single run on modest multiple gram scale.[56] A similar approach has been used for Pd-catalyzed alkoxy- and hydroxycarbonylations of substituted aryl iodides using gaseous carbon monoxide.[57] In principle, all these reactions can be scaled to useful medicinal chemistry quantities without needing to change the chemistry, either by filling the vessels further, increasing the gas pressure, or using a solid/liquid equivalent source of the desired gas.

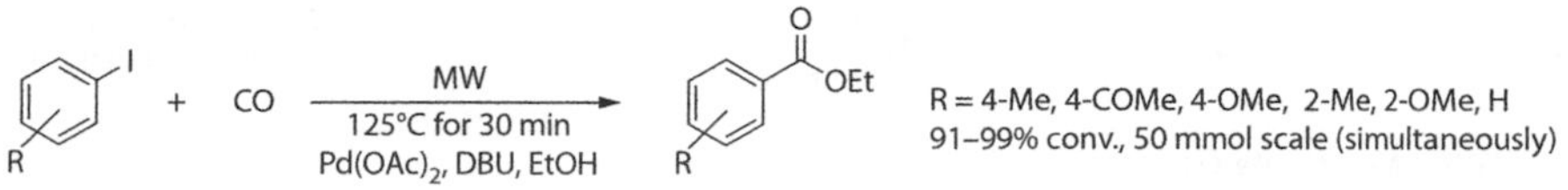

R = 4-Me, 4-COMe, 4-OMe, 2-Me, 2-OMe, H
91–99% conv., 50 mmol scale (simultaneously)

SCHEME 5.9

5.5.3 OPEN-VESSEL SCALE-UP (MARS, MicroSYNTH)

A classic pharmaceutical reaction involves dialkylation of a primary amide with an alkyl dihalide to form simple nitrogen heterocycles (Scheme 5.10). In an early example of the scale that can be achieved in open-vessel mode, *N*-heterocyclizations were performed on scales up to 1 mol using a MARS microwave unit, with several azacycloalkanes and isoindolines being prepared.[58] The reactions were conducted in water as solvent with K_2CO_3 as base and gave high yields and a clean product profile, which further aided the already simple workup procedure. The reactions were scaled from 1 mmol in sealed tubes directly to the 1 mol level in a 5 L open flask. Although the small-scale conditions (120 °C for 20 min) could not be reproduced in open-vessel mode, it was possible to achieve temperatures in the range of 100–110 °C, which proved adequate over a 20 min period to drive the reaction to completion. Running the reaction in open-vessel mode also facilitated removal of the byproducts (HBr, HCl) from the reaction mixture. Overall cycle time per 1 mol batch could be estimated to be about 1 h, allowing 10 min to heat the reaction mixture to reflux, 20 min at reflux, and approximately 30 min for cool down. This was much superior to the 5 h required to perform the reaction conventionally.

A Suzuki coupling reaction has been scaled to the 1 mol level for several substrates in a MARS unit, which was preceded by an optimization study (Scheme 5.11).[59] The reaction time had to be increased from 5 min at 150 °C in a sealed tube on a small scale to 20 min at solvent reflux (86 °C) on the larger scales. In addition, the Pd loading had to be increased to 1–5 ppm from ppb levels, but this was hardly inconvenient. Curiously, the conventionally heated reaction required much longer reaction times and, even so, gave poor conversion at these very low Pd loadings, maybe due to agglomeration of Pd nanoparticles.

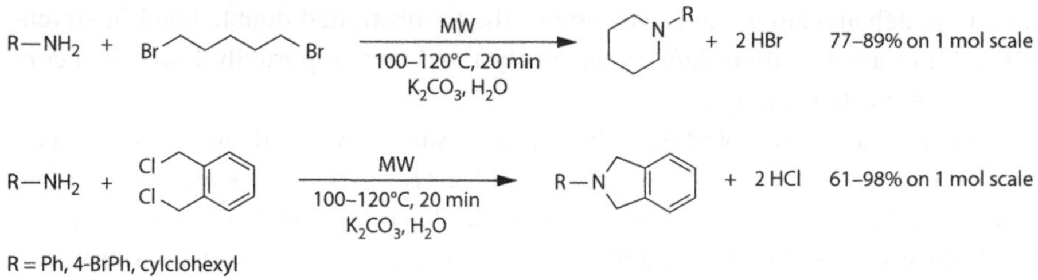

$$R-NH_2 + Br \sim\sim\sim Br \xrightarrow[\substack{100-120°C, 20 min \\ K_2CO_3, H_2O}]{MW} \quad N-R \quad + \; 2\,HBr \quad 77\text{–}89\% \text{ on 1 mol scale}$$

$$R-NH_2 + \substack{Cl \\ Cl} \xrightarrow[\substack{100-120°C, 20 min \\ K_2CO_3, H_2O}]{MW} \quad R-N \quad + \; 2\,HCl \quad 61\text{–}98\% \text{ on 1 mol scale}$$

R = Ph, 4-BrPh, cylclohexyl

SCHEME 5.10

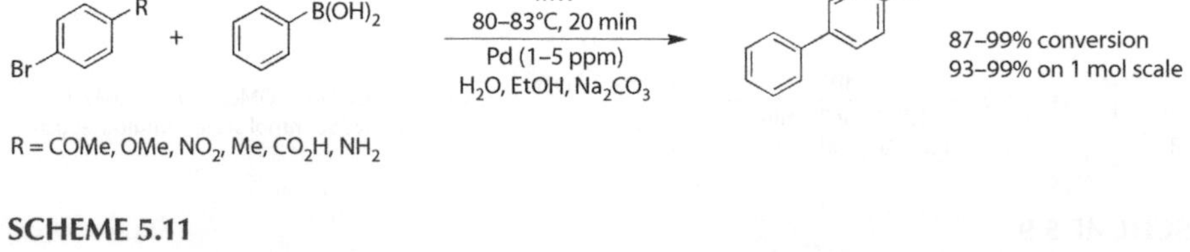

$$\substack{Br} \quad R + \quad B(OH)_2 \xrightarrow[\substack{80-83°C, 20 min \\ Pd (1-5 ppm) \\ H_2O, EtOH, Na_2CO_3}]{MW} \quad R \quad \substack{87\text{–}99\% \text{ conversion} \\ 93\text{–}99\% \text{ on 1 mol scale}}$$

R = COMe, OMe, NO_2, Me, CO_2H, NH_2

SCHEME 5.11

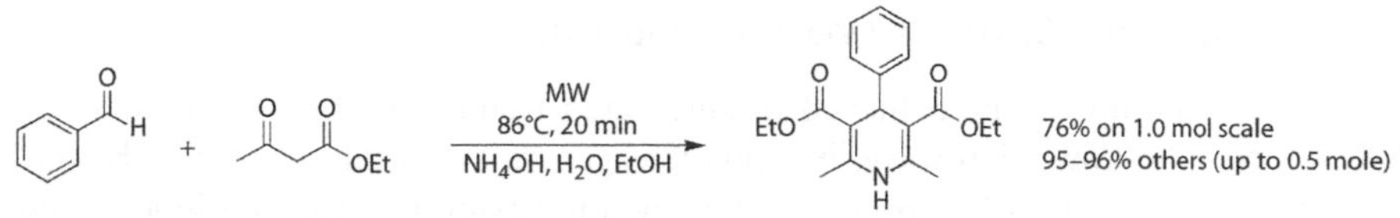

SCHEME 5.12

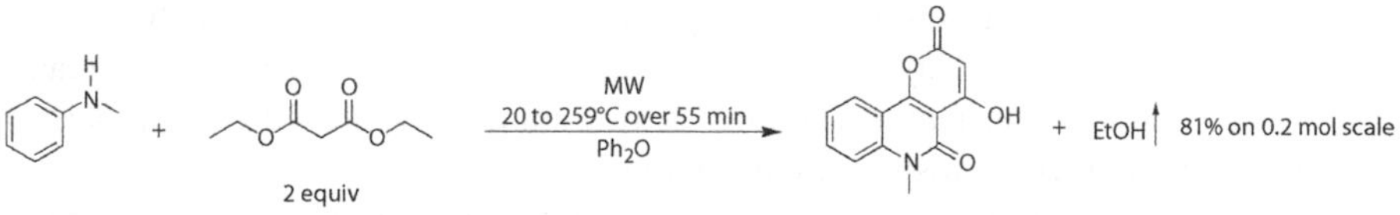

SCHEME 5.13

A solvent-free approach has been used to scale up a Heck reaction to the 1 mol level.[60] Again, this study was preceded by optimization and investigation of substituent effects. However, on 0.5–1.0 molar scale under solvent-free conditions, the reaction was markedly exothermic, self-heating to ~150 °C from the 100 °C set point. It is important to stress the importance of taking every precaution when performing reaction under solvent-free conditions on larger scales. In this case, the reaction mixture was heated slowly, and an over-sized flask was used.

The Hantzsch 1,4-dihydropyridine synthesis was also scaled successfully to the 0.5–1 mol level in the MARS unit, requiring heating up to 1 L of reaction mixture (Scheme 5.12).[29] One batch provided 250 g of dihydropyridine product in less than 1 h.

Open-vessel conditions prove advantageous for driving equilibrium reactions to completion when the product or by-product is a more volatile component than the starting materials. A Milestone MicroSYNTH has been used to perform a fractional distillation in the synthesis of a tricyclic heterocycle.[37] N-Methylaniline was reacted with two moles of diethylmalonate in a cyclocondensation reaction with elimination of four moles of ethanol to give the product pyranoquinoline on 0.2 molar scale (Scheme 5.13). Stepwise heating and fractional distillation were required because too fast a distillation resulted in codistillation of the aniline and/or malonate starting materials with ethanol. On this scale, the distillation took only 82 min as compared to 3–5 h conventionally, and reaction progress could be monitored by still-head temperature or volume of ethanol collected. Performing the reaction in a sealed-vessel microwave resulted in <5% product formation.

An open-vessel approach using a Dean–Stark trap placed outside the microwave cavity has been used to facilitate esterifications of acetic acid and acid-catalyzed dehydrations of octanol on the 2–6 mol scale.[29] Open-vessel microwave heating is ideal for high-temperature distillations compared to conventional approaches due to the rapid noncontact heating possible, making the procedure easy and safe.

5.5.4 SEQUENTIAL SCALE-UP (VOYAGER, KILOBATCH)

A limited scale-up can be achieved in automated fashion by the use of the CEM Voyager SF stop-flow microwave.[27c] A previous generation of the unit used a Kevlar-lined PTFE coil of 10 mL capacity as the reaction vessel in a true continuous-flow manner (Voyager CF).[23] This was used to perform a [3+2] cycloaddition reaction promoted by a suspended clay to form a triazole,[61] but further Voyager CF examples have not been forthcoming, and this variant is no longer commercially available. Using the Voyager SF, a number of groups have reported scale-ups in the region of typically 50–500 g per day; depending on concentration, the process was made easier by the automation features. Larger scales could be achieved by numbering up,[32] although in practice this has not been done.

The Voyager SF offers a useful scale-up capability for a small investment in cost, space, and training. It is mainly suitable for processing homogeneous reaction mixtures and reaching scales more suitable for medicinal chemistry scale-up (~50–500 g). However, a range of reactions has been conducted, yielding between 200 g and 1400 g per 24 h period.[62] These included the Heck coupling, two high-temperature rearrangement reactions (Claisen and NKR) performed under high-concentration conditions, a simple heterocycle formation, and a hydrolysis and an alkylation. The latter pair were combined into a two-step sequence (Scheme 5.14) which required use of all three input lines on the microwave unit. This reaction was run at a concentration that would be considered dilute by process chemistry standards (30 L/kg), but would still have yielded nearly 200 g of product over a 24 h period.

Sometimes, it is necessary to modify conditions from those used on a small scale in order to be able to process reaction mixtures using the Voyager SF. A case in point is the scale-up of Suzuki and Heck reactions using the unit. On small scales, these reactions were performed using water as the solvent, tetrabutylammonium bromide as a phase-transfer agent, and either sodium or potassium carbonate as base. When moving to the Voyager SF, in order to make the reaction mixture homogeneous, either an ethanol–water (Suzuki) or DMF–water (Heck) mix was used as the solvent.[63] The organic substrates, mineral base, and palladium catalyst dissolved in this. Both reactions worked well under these conditions, one further modification in the case of the Heck protocol being to extend the reaction time slightly. However, another problem was encountered at the end of the reaction. The products were crystalline and blocked the exit lines even at 90 °C. The problem was solved by first venting excess pressure at 110 °C, and then pumping a small charge of either ethyl acetate (Suzuki) or DMF (Heck) into the reaction vessel to help solubilize the products before discharging them. Ten cycles of each reaction were then run.

Two successful examples of processing heterogeneous reaction mixtures using the Voyager SF have been reported. In one, a fine slurry of micronized zinc cyanide was

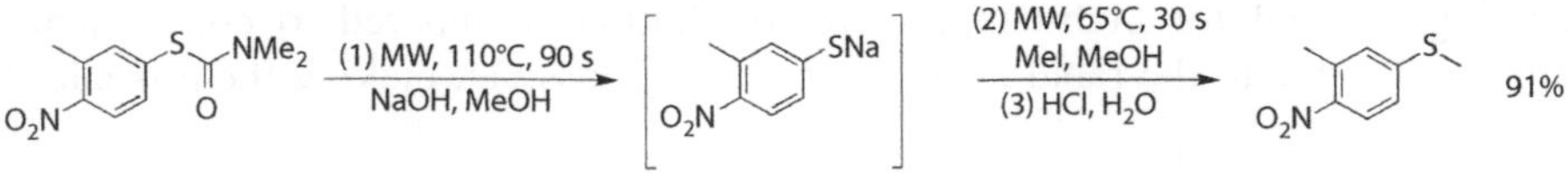

SCHEME 5.14

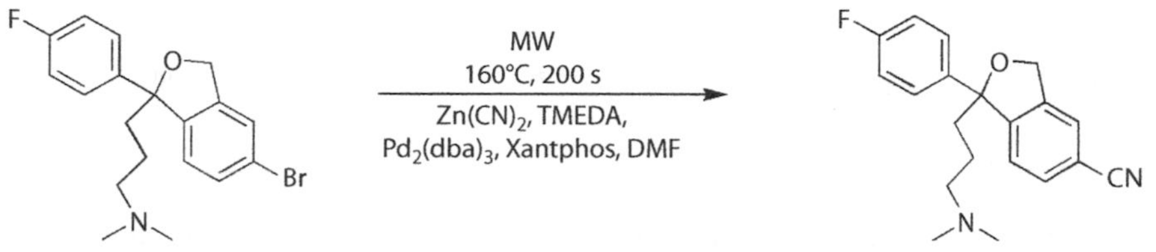

SCHEME 5.15

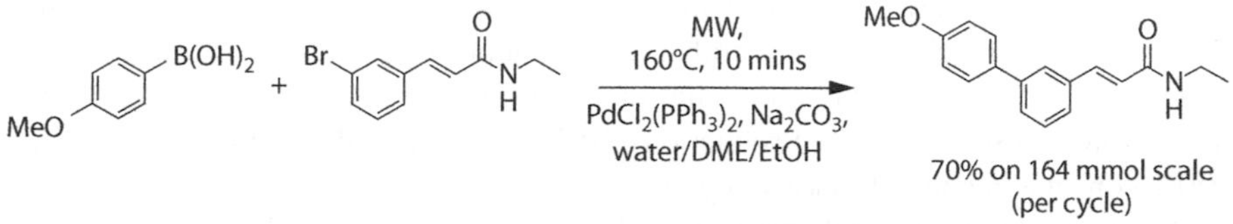

70% on 164 mmol scale
(per cycle)

SCHEME 5.16

used as the cyanide source for the final Pd-mediated cyanation step in the synthesis of escitalopram (Scheme 5.15).[64] Eleven cycles at 50 mL capacity with a 10 min cycle time gave 150 g of product in under 2 h with excellent purity and greatly improved work-up over the existing process used. In the other report, three Buchwald–Hartwig aminations were performed using NaO^{t}Bu as the base suspended in trifluoromethylbenzene as solvent.[43] If run repeatedly, this would have yielded ~260 g (1.35 mol) in a 24 h period.

No formal reports of chemistry in the Advancer Kilobatch have been made, although the manufacturer has shown the application of the unit in a series of trials performed in-house.[65] A Suzuki coupling was performed (Scheme 5.16), and the results from each of the first four batches (the first cycle) were analyzed together with the combined batches from four sequential cycles run by automation (i.e., 16 batches). Excellent reproducibility was seen between batches and across the cycles. Each cycle represented 160 mmol of starting materials. Since many palladium-catalyzed coupling reactions require the use of an inert atmosphere, as a demonstration all the solid charging vessels were maintained under nitrogen prior to and during addition. A simple Diels–Alder reaction between maleic anhydride and 1,3-cyclohexadiene was also performed in 56 batches (160 °C for 10 min), both reagents being added as solutions in DMF, thus making this effectively a stop-flow reaction. The total processed volume was 13.5 L, the whole procedure taking 14.5 h and yielding 1.24 kg of product (~7 mol, 99% yield).

5.5.5 Continuous Scale-Up (FlowSYNTH)

The only commercially available continuous-flow microwave available is the Milestone FlowSYNTH, based on their earlier ETHOS system.[27d,66] The functionality of the FlowSYNTH has been studied using the NKR as a test reaction.[67] In addition, a number of reactions that were initially performed in batch mode have been translated for use in the FlowSYNTH.[29] The simple homogeneous esterification of acetic acid and butanol was repeated, but at 150 °C and with an adjusted

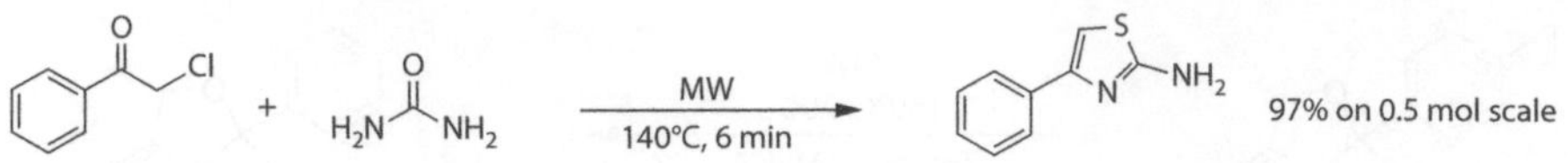

SCHEME 5.17

stoichiometry to drive the equilibrium forward since water could not be removed during the course of the reaction. Even so, only a 1 min residence time was required, which allowed a 200 mL/min flow rate to be attained with 78% conversion to the butyl ester. A production rate of ~1 mol/min was achieved over 22 min, consuming 4.5 L of reagent stock solution. Another homogeneous reaction studied was the synthesis of an aminothiazole in ethanolic solution with a 6 min residence (34 mL/min) time at 140 °C (Scheme 5.17). Production of 0.5 mol was possible in just over 1 h with 100% conversion and 97% isolated yield.

Attempts to prepare a 1,4-dihydropyridine heterocycle using a Hantzsch reaction (Scheme 5.12) were only partially successful because the product began to crystallize in the exit line on cooling, resulting in eventual blockage. A Suzuki reaction was also performed under continuous-flow conditions, but modified to ensure a completely homogeneous reaction mixture. Modeling from small-scale studies indicated that 5 min at 140 °C (40 mL/min flow rate) would give complete conversion with only 500 ppb Pd catalyst, and 0.5 mol was processed. However, only a 63% conversion was achieved. This is a common problem in converting small-scale batch chemistry to flow mode and, inevitably, some minor adjustments must be made before starting the main run.[62] This is probably because small-scale microwave units slightly overheat the contents compared to larger-scale (batch or flow) microwave systems. However, small-scale sealed-tube reactions do provide good starting points for developing flow approaches and can be used as such.

While in principle continuous-flow reactors can process many types of reactions, in practice they cannot always manage even the small number of homogeneous reactions available. Furthermore, some additional optimization is usually required, with alterations leading to other problems. Adjusting the reaction conditions loses one of the benefits of microwave heating, namely, linear scale-up. Reoptimizing the process is usually simple, but does take some effort and time and is likely to require other problems to be solved along the way. For early stage scale-up in a pharmaceutical context, this may be counterproductive, especially if an adequate (microwave) batch process can be scaled sufficiently.

5.6 OTHER APPLICATIONS IN PROCESS CHEMISTRY

5.6.1 REACTION OPTIMIZATION

When attempting to optimize reaction conditions, it is often prudent to increase the temperature in steps while keeping the time short. This allows for the optimum reaction time and temperature to be rapidly identified more efficiently than when fixing the temperature and increasing the time. In addition, it is important to compare incomplete reactions, since there is little useful differentiation between reactions at or near 100% conversion. In a medicinal chemistry context, these principles have been clearly demonstrated in the optimization of the alkylation reaction shown in Scheme 5.5 using the parallel vessel capability of a Milestone MicroSYNTH.[21]

A Design of Experiments (DoE) approach can be used for rapid optimization of reaction conditions. It is used extensively by process chemists to optimize large-scale reactions, but can just as easily be applied to small-scale (medicinal chemistry) examples. In either case, microwave heating is an excellent technology to pair it with, as has been demonstrated in the optimization of Ugi three component[68] and Biginelli reactions.[69] The statistical power of DoE combined with the automation and speed of microwave synthesis followed up by rapid HPLC analysis enabled the optimum conditions to be identified quickly. In the case of the Biginelli reaction, the optimum catalyst–solvent and temperature–time–catalyst concentration combinations had been identified after three iterations of optimization, this taking only 29 programmed microwave reactions.

The Milestone MultiSYNTH, which can operate in both monomode and multimode function, has been used to optimize reaction conditions rapidly and then screen a range of substrates.[20] Three reactions have been studied using this approach; a Suzuki coupling, aza-Michael reaction, and Williamson etherification. When the microwave absorptivity of a reaction mixture is dictated by the solvent, there is little effect on the heating characteristics of a reaction mixture when the substrate is varied in a screening run. However, this is not the case when reaction mixtures do not couple well with the microwave energy. In this case, the characteristics of the substrate can significantly affect the outcome of the reaction in a screen. Careful consideration needs to be taken when deciding into which vessel the fiber-optic probe should be placed and thus is used as the control.

Another particularly thorough example of reaction optimization relates to the transfer hydrogenation/debenzylation conditions for a model substrate, as shown in Scheme 5.18.[40] Five parameters were studied: temperature, solvent, reaction time, catalyst loading, and equivalents of hydrogen. Each of these was independently varied two or three times, with reactions being performed on the 45 mg/300 μL scale. The optimized conditions were then applied to the deprotection of 14 differentially substituted O-benzyl aromatic ethers to test the generality of the method. For compatible functionalities, conversions and yields of 90–100% were typically obtained.

5.6.2 Reaction Understanding

When probing the kinetics of a reaction, multiple data points (50–100) are required for an accurate comparative Arrhenius plot to be obtained.[70] An advantage of modern small-scale automated microwave apparatus is that it can be used for this very

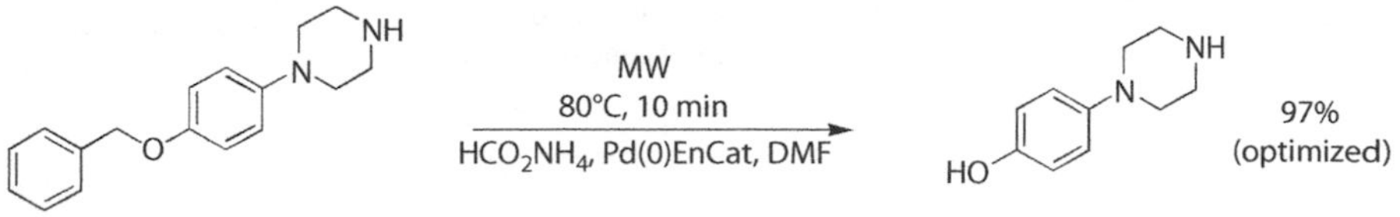

MW
80°C, 10 min
HCO$_2$NH$_4$, Pd(0)EnCat, DMF

97%
(optimized)

SCHEME 5.18

function, producing data points very quickly on small samples and in a programmed fashion. A thorough reinvestigation of the kinetics of the NKR has been undertaken using microwave heating to facilitate the acquisition of data.[71] The work confirmed that the reaction is first order with known substituent and solvent effects. The Arrhenius parameters and activation energies were also determined. No specific microwave effects were found, this being a prerequisite for this type of study if microwave heating is to substitute for conventional heating in kinetic studies.

A kinetic study has been undertaken with the objective of uncovering possible nonthermal microwave effects in the reductive elimination from a Pt(IV) complex using a low-polarity solvent.[72] The highly polar transition state through which the reaction is known to proceed should interact strongly with microwaves if there were any nonthermal effects in operation, thus changing the kinetics of the reaction. Determining the reaction rate constants under conventional and microwave heating in several solvents at several concentrations and microwave powers revealed no such phenomena. Lack of agitation markedly affected the comparison between microwave and conventional heating, further underlining the importance of this parameter, whether for kinetic studies or scale-up of reactions.

A problem associated with using microwave heating is the general inability to sample the reaction while it is in progress. One technique that can be used in real time, however, is Raman spectroscopy.[73,74] As well as being used for monitoring reaction progress, Raman spectroscopy in conjunction with microwave heating can be used to obtain kinetic data, an example being the study of the formation of 3-acetylcoumarin from salicylaldehyde and ethyl acetoacetate (Scheme 5.19).[75] From this data, the reaction order and relative reaction order of the various components could be determined, and the Arrhenius parameters and the activation energy for the reaction obtained. A deeper mechanistic understanding of the reaction was also gleaned.

In situ Raman monitoring has also been used as a tool to probe for microwave effects on a microscopic level.[76] The results show no evidence for localized superheating, an often-cited specific microwave effect. While the microwave energy may interact with the polar molecules more so than with nonpolar ones, the conversion of electromagnetic energy into kinetic energy is slower than conversion of kinetic energy into thermal energy. As a result, more polar molecules are not at a temperature greater than that of the bulk.

The value to the process chemist (and others) in each of these examples is the advantage rapid microwave heating brings to understanding a reaction in a short time while employing small quantities of material. The use of automated small-scale

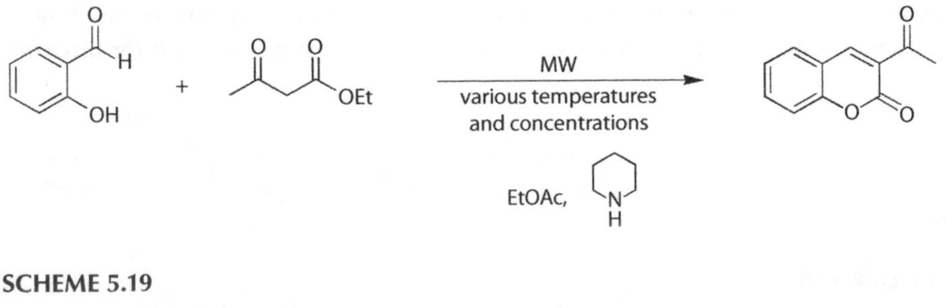

SCHEME 5.19

microwave reactors greatly facilitates this type of study and suggests they also have a valuable role to play in the process chemistry laboratory.

5.7 ENERGY EFFICIENCY OF MICROWAVE HEATING

When considering large-scale processes involving microwave heating, energy efficiency becomes an important topic for consideration. Kitchen microwave ovens can cook, or more usually reheat, much more quickly than conventional stoves and ovens, with concomitant energy savings. According to one source, conventional cooking times may be reduced by 60–75% with only a quarter of the power required.[77] Many proponents of using microwave heating as a tool for synthetic chemistry have glibly assumed by analogy that such energy savings will transfer to the laboratory and onto the pilot plant. Clearly, heating a single sealed tube in a microwave reactor is more energy efficient than heating the same-sized reaction mass in an oil bath containing 1–2 L of heating fluid constantly radiating heat to the environment, but this is not a fair comparison. Energy savings on this scale, while beneficial, are unlikely to be significant or economically worthwhile.

Many other industries do use microwave heating on a large scale for heating, so the cost-effective application of microwave heating is of proven commercial value.[1] The concern, however, lies first in the conversion of electrical energy to microwave energy, and then in the efficient uptake of microwave energy to provide heating. For microwave systems operating at 2.45 GHz, efficiencies of 50–72% have been suggested, assuming all the microwave energy can usefully heat the reaction mass.[3,78] The conclusions drawn are that such inefficiencies can be tolerated only for high-value products (such as pharmaceuticals), or where special processing is achieved that would be difficult or impossible to achieve by alternative methods (some of the drying methods used in the food industry, for example). As a point of comparison, the efficiency of a steam-heated pilot or production plant has been quoted as 90% if heated by electricity and 65% if by fossil fuels;[79] either figure is better than microwave heating. Of interest is that efficiencies for microwave heating can be higher at lower frequencies (80–87% at 915 MHz), another reason for investigating their use, perhaps.[3]

Surprisingly, and despite these figures, very few studies have been conducted to investigate the overall heating efficiency of microwave heating as applied to organic synthesis. The first investigation was in 2005, probing three reactions performed using three different reaction methods (microwave heating, oil bath heating, and use of supercritical CO_2).[80] The conclusion was that there was a "significant gain in energy efficiency" using microwave heating. However, this was a very limited study, and the data were based on small-scale experiments. A more recent report, focusing on small- to moderate-scale chemistry, draws the conclusion that microwave heating is not very energy efficient,[81] being worse at small scale because not all the microwave energy generated can be absorbed by the sample, and better at larger scale in multimode instruments, but still with energy efficiencies of only ~30% overall (although the microwave used in this case may give particularly low figures[82]). In another study, focusing on four reactions, the conclusions reached were that small-scale reactions are more energy efficient, especially when the reaction time is short and elevated temperatures can be accessed; but for larger-scale reactions conducted over the same or longer time periods, microwave heating became progressively less efficient.[83]

More recently, a thorough comparison of four reactions conducted in the commercial microwave reactors described above and a conventionally heated jacketed reactor run on the same scale has been made.[82] At the scale of 1–3 L, it was found that microwave reactors varied from comparable to much more energy efficient, depending on their type and method of operation. Throughputs were also generally higher in the microwave reactors, particularly so for the continuous-flow reactors, where energy efficiencies were also greater. However, as with the other studies,[80,81] energy efficiencies fell with longer reaction times, lower temperatures, and low-absorbing solvents.

A direct comparison with pilot plant scales is ongoing at the time of writing, but a point raised by a number of authors is that true pilot plant or production comparison must probably be made on a case-by-case basis.[81–83] One such example quoted in the literature is for the synthesis of the cosmetic Laurydone (presumably a high-value product), which used a batch-loop reactor where an energy savings of 40% had been claimed.[15] However, the original comparison has now been identified as flawed, and the reactor setup dismantled.[31]

In the biofuels arena, the energy efficiencies of microwave and conventional heating have been compared in the preparation of biodiesel.[36] To be viable, biodiesel production must be conducted on a very large scale and also must be energy efficient over the entire process. Calculations indicate that while microwave heating in batch mode was essentially equivalent in energy consumption to using conventionally heated reactors, the process had the potential to be much more efficient when run in continuous-flow microwave reactors.

Pharmaceutical production facilities generally have easy access to cheaply generated steam heating and generous plant capacity. Given the unavoidable inefficiency in converting electrical into microwave energy, successful applications can be expected only for high-value products in combination with some additional benefit of microwave heating or in very large-scale applications where marginal savings become significant. A specific pharmaceutical reaction may benefit in these circumstances, but it seems likely that the energy debate will remain unresolved until a genuine side-by-side comparison at pilot plant scale is reported.

Aside from direct energy usage, it is worth remembering that other benefits of microwave heating on the large scale may also contribute to a more sustainable future. Small improvements in product quality through quicker heating, or lower catalyst loadings, for example, may contribute toward an easier and less solvent intensive purification. Processing times, solvent and effluent volumes, and remediation requirements may all be reduced, which is likely to have a much greater impact on the overall energy balance of the process than on direct energy usage. Microwave heating can be expected to benefit some processes in unexpected ways. It is worth restating that microwave heating is used on very large scales for economically viable processes involving low-value products,[1,3] so it should be expected that it will also be viable for pharmaceuticals and other high-value chemicals.

5.8   RECENT DEVELOPMENTS IN SCALE-UP

Within the context of pharmaceutical organic chemistry, many observers have concluded that continuous flow is the best way to scale up microwave chemistry. The general advantages and disadvantages of this technology have been presented above, including a discussion of whether microwave heating benefits a continuous-flow device. Details of several reactors have been presented: the true continuous-flow FlowSYNTH, the stop-flow Voyager SF, and a CSTR adaptation in the CEM MARS.

The main advantage of continuous flow is high throughput; the main disadvantage is inability to handle slurries in most cases. Recently, another microwave flow reactor has become available from pharmaceutical supplier Cambrex,[84] working in collaboration with microwave experts C-Tech Innovation.[85] This is a bespoke microwave reactor (GENESIS) that is potentially available for purchase and certainly for contract manufacture. It has already been used successfully for in-house applications in Cambrex projects, and some generic chemistries have been reported.[86] The reactor consists of a thick-walled quartz glass tube vertically mounted in the microwave wave guide serviced by feed and receiver vessels (Figures 5.13 and 5.14). Temperatures of 200 °C and pressures of 20 bar can be attained with flow rates up to 200 mL/min. The reactor can process homogeneous and light heterogeneous reaction mixtures, even up to 10% w/v in practice, and sufficient to cope with heterogeneous catalysts (e.g., Pd/C) and fines, which often occur in many processes. Computer control, dual feed, and gas inertion are additional features. The reactor can operate for many hours and is limited only by the size of the service vessels.

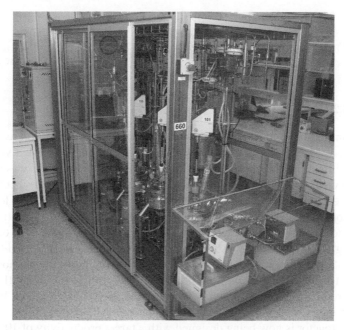

FIGURE 5.13   Cambrex GENESIS continuous-flow microwave reactor. (Reproduced with permission from Cambrex. Copyright 2009 Cambrex. All rights reserved.)

FIGURE 5.14 Receiver vessels for Cambrex GENESIS continuous-flow microwave reactor. (Reproduced with permission from Cambrex. Copyright 2009 Cambrex. All rights reserved.)

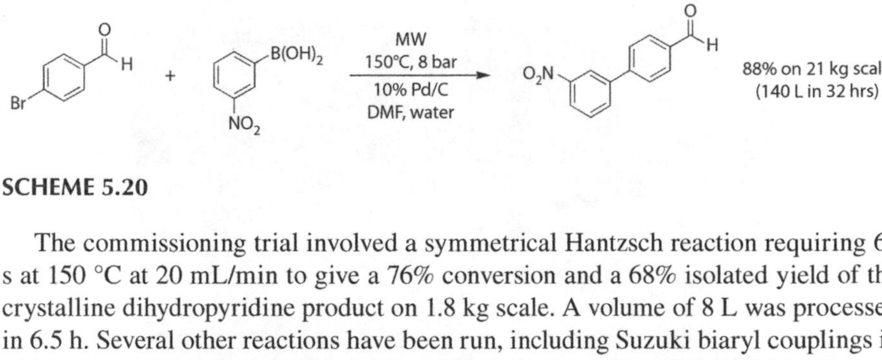

SCHEME 5.20

The commissioning trial involved a symmetrical Hantzsch reaction requiring 60 s at 150 °C at 20 mL/min to give a 76% conversion and a 68% isolated yield of the crystalline dihydropyridine product on 1.8 kg scale. A volume of 8 L was processed in 6.5 h. Several other reactions have been run, including Suzuki biaryl couplings in which the reaction time could be reduced from 2 min to 15 s, and the Pd loadings to 5 ppm in some cases. The largest-scale preparation so far was a 32 h continuous run for the model Suzuki reaction (Scheme 5.20) at 150 °C with a flow rate of 60–70 mL/ min, giving a total processed volume of 140 L. Heterogeneous 10% Pd/C catalyst was used (~2 mol%), requiring 460 g slurried in DMF on this scale, but no problems with blockage or sedimentation were observed. Overall, this provided 21 kg of the biphenyl product in 88% yield and >99% purity. This would equate to 100 kg per week or >5 metric tons per year if run continuously.

A larger reactor is now being designed with a target productivity of 100–200 kg/ day and that can handle heterogeneous reactions with up to 5–10 wt% solids. This will greatly widen the range of chemistries that can be attempted. The whole reactor

FIGURE 5.15 (a) CAD rendering of the prototype AccelBeam Synthesis reactor; (b) CAD rendering of the AccelBeam Synthesis unit that will soon be commercially available. (Reproduced with permission from AccelBeam Synthesis.)

will be mounted on a skid that can be wheeled between appropriately sized plant vessels to serve as feed and workup vessels.

AccelBeam Synthesis, a U.S. company working in conjunction with Ukrainian microwave engineers, has recently developed a batch unit that allows for reactions to be performed on scales from 2 to 12 L.[88] Engineering renderings of the prototype unit, together with the unit to be commercially available soon, are shown in Figure 5.15. The reactor employs three 2.45 GHz magnetrons with an accessible power of 2.5 kW each for a total maximum allowed output of 7.5 kW. There are three interchangeable reaction vessels: 5, 9, and 13 L glass vessels with working volumes of 2–4, 4–8, and 7–12 L, respectively. The desired reaction vessel, equipped with a PTFE cover, is placed in a mechanically sealed stainless steel reaction chamber capable of operating at pressures up to ~24 bar. To run a reaction, the chamber is pre-pressurized to 17–21 bar. Reaction parameters such as time, temperature, pressure, stirring level, and magnetron power are monitored and collected using software. Upon completion of a reaction, the contents can be ejected through a water-cooled counter-flow heat exchanger. Alternatively, the contents can be ejected directly into a receiving chamber at ambient pressure, this affording rapid cooling with concomitant evaporation of solvent.

A number of solvents have been heated in the unit, the results showing that penetration depth was not an issue when heating larger volumes of solvents. A range of reactions has been scaled up using the unit, including homogeneous condensations, heterogeneous reactions, and phosphine-free palladium-catalyzed transformations carried out in water. Two examples are shown in Scheme 5.21.

At the time of this writing, French microwave and radio frequency experts Sairem has two large-scale microwave reactors, LABOTRON synthesis and LABOTRON extraction, in advanced development.[89] One is a batch reactor (volume from 1.5 to 17 L), and the other is a continuous-flow reactor capable of flow rates of 10 mL up to 1 L per minute, which can also process solids up to 10 wt% (Figure 5.16). Both reactors have significant cooling functionality built in and are designed to use the same 6 kW microwave generator at 2450 MHz, key features of which are the U-shaped

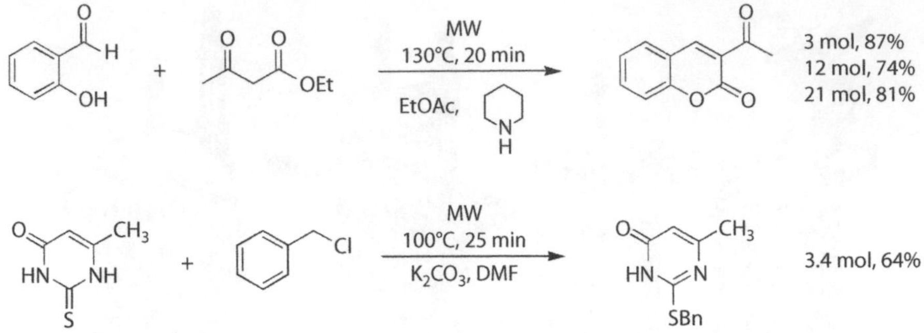

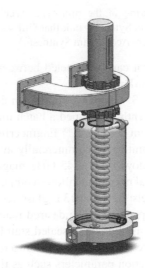

SCHEME 5.21

FIGURE 5.16 Schematic of Sairem 2450 MHz microwave continuous-flow reactor. (Reproduced with permission from Sairem.)

wave guide and metallic vessels (which will help with pressure containment but also faster thermal transfer). Patents are pending on both. Also in development is a truly pilot-plant-capable batch reactor (up to 100 L) using a 30 kW generator at 915 MHz (Figure 5.17).

Scale-up of microwave chemistry has resisted resolution for some time. The continuous-flow strategy seems to be prevailing at the larger scale, which is to be expected. This has required synthetic and pharmaceutical chemists to understand and use another technology with which they were unfamiliar, although of course, continuous-flow technology is widely used in the chemical industry. In combination with the relative novelty of microwave chemistry, this means putting two new technologies together, which is more than twice the challenge.

Microwave heating does have advantages to offer at the large scale compared to conventional heating. However, heating volatile, often-flammable, organic solvents in combination with suspended solids to make delicate organic molecules under

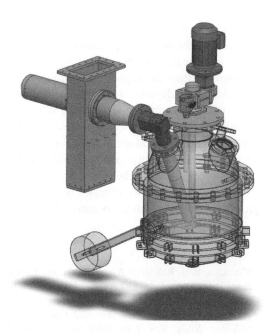

FIGURE 5.17 Schematic of Sairem 915 MHz microwave 100 L batch reactor. (Reproduced with permission from Sairem.)

well-controlled conditions is not trivial on the large scale, whatever the heating methodology. Previous efforts to scale up microwave synthesis have been driven by chemists, engineers, or microwave manufacturers, but not generally in combination. It is notable that the companies discussed above which now have pilot-plant-capable microwave technology for synthesis have engaged a threefold combination of chemistry, engineering, and microwave expertise.

Lastly, the Sairem 915 MHz batch reactor is also a good illustration of a more fundamental change in strategy to microwave scale-up through the use of a different wavelength, since penetration depths, dielectric constants, and loss factors vary with wavelength as well as with solvent and temperature.[90,91] Developments at other companies may also be forthcoming.

5.9 CONCLUSIONS

The central tenet of this chapter has been the scalability of microwave heating. Academics and medicinal chemists can number up on small-scale instruments to a few grams. If more material is required, any one of several parallel or open-vessel microwave reactors can be purchased at modest expense and a few hundred grams prepared relatively easily in a day or two. Larger quantities up to 1 kg are likely to require the larger, more costly reactors and an acceptance of higher batch numbers (which will be tedious to load in the parallel reactors). For multiple kilograms, existing microwave reactors are not really equal to the task, except for very concentrated reactions. However, in the less common cases where continuous-flow chemistry can be used, kilogram quantities can be accessed relatively easily. Pilot-scale continuous and batch reactors are available for use from specialist companies as noted, and will be able to reach true pilot plant scales shortly.

Whether an application justifies the higher installation and running costs of microwave chemistry on scale will depend on the benefit it brings. This will almost certainly need to be determined on a case-by-case basis. What seems certain is that some beneficial applications should be expected in process chemistry for which microwave heating offers a distinct advantage and is economically viable; and they will be viable only if issues of sustainability have been addressed. The advantages of small-scale microwave heating in process chemistry for increasing reaction understanding should also not be overlooked. However, for now, a limited scale-up capability is possible in process chemistry, with the potential for much more.

REFERENCES

1. Metaxas, A. C.; Meredith, R. J. Industrial Applications and Economics. In *Industrial Microwaves Heating*; Peter Peregrinus Ltd for the Institute of Electrical Engineers: London, 1983; Chap. 11; pp 296–321 (reprinted 1993).
2. (a) Repic, O. *Principles of Process Research and Chemical Development in the Pharmaceutical Industry*; Wiley: New York, 1997. (b) Atherton, J. H.; Carpenter, K. J. *Process Development: Physicochemical Concepts*; Oxford University Press: Oxford, 1999. (c) Anderson, N. G. *Practical Process Research and Development*; Academic Press: San Diego, 2000.
3. Bogdal, D.; Prociak, A. *Chimica Oggi/Chemistry Today* 2007, *25(3)*, 30–33.
4. Loupy, A. Ed. *Microwaves in Organic Synthesis,* 2nd ed.; Wiley-VCH: Weinheim, 2006 (2 volumes).
5. Kappe, O.; Stadler, A. *Microwaves in Organic and Medicinal Chemistry*; Wiley-VCH: Weinheim, 2005.
6. Hayes, B. L. *Microwave Synthesis: Chemistry at the Speed of Light*; CEM: Matthews, NC, 2002.
7. (a) Kappe, C. O. *Angew. Chem. Int. Ed.* 2004, *43*, 6250–6284. (b) Hayes, B. L. *Aldrichimica Acta* 2004, *37*, 66–76. (c) Lidström, P.; Tierney, J.; Wathey, B.; Westman, J. *Tetrahedron* 2001, *57*, 9225–9283. (d) Perreux, L.; Loupy, A. *Tetrahedron* 2001, *57*, 9199–9223.
8. Varma, R. S.; Ju, Y. Organic Synthesis Using Microwaves and Supported Reagents. In *Microwaves in Organic Synthesis,* 2nd ed.; Loupy, A. Ed.; Wiley-VCH: Weinheim, Germany, 2006; Vol 1; Chap 8; pp 362–415.
9. Kappe, O.; Stadler, A. *Microwaves in Organic and Medicinal Chemistry*; Wiley-VCH: Weinheim, Germany, 2005; p 30.
10. Raner, K. D.; Strauss, C. R.; Trainor, R. W.; Thorn, J. S. *J. Org. Chem.* 1995, *60*, 2456–2460.
11. Cablewski, T.; Faux, A. F.; Strauss, C. R. *J. Org. Chem.* 1994, *59*, 3408–3412.
12. Roberts, B. A.; Strauss, C. R. *Acc. Chem. Res.* 2005, *38*, 653–661.
13. Kremsner, J. M.; Stadler, A.; Kappe, C. O. *Top. Curr. Chem.* 2006, *266*, 233–278.
14. Ondruschka, B.; Bonrath, W.; Stuerga, D. Development and Design of Laboratory and Pilot Scale Reactors for Microwave-assisted Chemistry. In *Microwaves in Organic Synthesis,* 2nd ed.; Loupy, A. Ed.; Wiley-VCH: Weinheim, 2006; Vol 1; Chap 2; pp 62–107.
15. Howarth, P.; Lockwood, M. *The Chemical Engineer* 2004, 29–31.

16. Wolkenberg, S. E.; Shipe, W. D.; Lindsley, C. W.; Guare, J. P.; Pawluczyk, J. M. *Current Opinion in Drug Discovery and Development*, 2005, *8*, 701–708.
17. Moseley, J. D. *Chimica Oggi/Chemistry Today* 2009, *27*, 6–10.
18. Kappe, O.; Stadler, A. *Microwaves in Organic and Medicinal Chemistry*; Wiley-VCH: Weinheim, Germany, 2005; p 21.
19. Kremsner, J. M.; Kappe, C. O. *Eur. J. Org. Chem.* 2005, 3672–3679.
20. Leadbeater N. E.; Schmink, J. R. *Tetrahedron* 2007, *63*, 6764–6773.
21. Alcazar, J.; Diels, G.; Schoentjes, B. *QSAR Comb. Sci.* 2004, *23*, 906–910.
22. Stadler, A.; Yousefi, B. H.; Dallinger, D.; Walla, P.; Van der Eycken, E.; Kaval, N.; Kappe, C. O. *Org. Process Res. Dev.* 2003, *7*, 707–716.
23. Ferguson, J. D. *Mol. Diversity*, 2003, *7*, 281–286.
24. Lehmann, F.; Pilotti, Å.; Luthman, K. *Mol. Diversity* 2003, *7*, 145–152.
25. Nüchter, M.; Ondruschka, B.; Bonrath, W.; Gum, A. *Green Chem.*, 2004, *6*, 128–141.
26. Baghurst, D. R.; Mingos, D. M. P. *Chem. Soc. Rev.* 1991, *20*, 1–47.
27. (a) www.anton-paar.com. (b) www.biotage.com. (c) www.cem.com. (d) www.mile-stonesrl.com.
28. Moseley, J. D.; Lenden, P.; Lockwood, M.; Ruda, K.; Sherlock, J.-P.; Thomson, A. D.; Gilday, J. P. *Org. Process Res. Dev.* 2008, *12*, 30–40.
29. Bowman, M. D.; Holcomb, J. L.; Kormos, C. M.; Leadbeater, N. E.; Williams, V. A. *Org. Process Res. Dev.* 2008, *12*, 41–57.
30. Kappe, O.; Stadler, A. Equipment Review. *Microwaves in Organic and Medicinal Chemistry*; Wiley-VCH: Weinheim, Germany, 2005; Chap 3; pp 29–56.
31. Personal communications to the author.
32. Tilstam, U. *Org. Process Res. Dev.* 2004, *8*, 421.
33. Laird, T. *Org. Process Res. Dev.* 2008, *12*, 904 and following themed-issue papers.
34. For examples see citation 8 in Reference 28.
35. For example: (a) www.alfalaval.com. (b) www.NiTechSolutions.co.uk. (c) www.syrris.com. (d) www.thalesnano.com. (e) www.uniqsis.com. (f) www.vapourtec.co.uk.
36. Barnard, T. M.; Leadbeater, N. E.; Boucher, M. B.; Stencel, L. M.; Wilhite, B. A. *Energy Fuels* 2007, *21*, 1777–1781.
37. Razzaq, T.; Kappe, C. O.; *Tetrahedron Lett.* 2007, *48*, 2513–2517.
38. Carey, J. S.; Laffan, D.; Thomson, C.; Williams, M. T. *Org. Biomol. Chem.* 2006, *4*, 2337–2347.
39. Heller, E.; Lautenschläger, W.; Holzgrabe, U. *Tetrahedron Lett.* 2005, *46*, 1247–1249.
40. Quai, M.; Repetto, C.; Barbaglia, W.; Cereda, E. *Tetrahedron Lett.* 2007, *48*, 1241–1245.
41. Vanier, G. S. *Synlett* 2007, 131–135.
42. (a) Larhed, M.; Moberg, C.; Hallberg, A. *Acc. Chem. Rev.* 2002, *35*, 717–727. (b) Appukkuttan, P.; Van der Eycken, *Eur. J. Org. Chem.* 2008, 1133–1155.
43. Loones, K. T. J.; Maes, B. U. W.; Rombouts, G.; Hostyn, S.; Diels, G. *Tetrahedron*, 2005, *61*, 10338–10348.
44. Lehmann, H.; LaVecchia, L. *JALA*, 2005, *10*, 412–417.
45. Chemat, F.; Esveld, E. *Chem. Eng. Technol.* 2001, *24*, 735–744.
46. Moseley, J. D.; Lenden, P.; Thomson, A. D.; Gilday, J. P. *Tetrahedron Lett.* 2007, *48*, 6084–6087.
47. Herrero, M. A.; Kremsner, J. M.; Kappe, C. O. *J. Org. Chem.* 2008, *73*, 36–47.
48. Loupy, A.; Petit, A.; Bogdal, D. Microwaves and Phase-transfer Catalysis. In *Microwaves in Organic Synthesis*, 2nd ed.; Loupy, A. Ed.; Wiley-VCH: Weinheim, Germany, 2006; Vol 1; Chap 6; pp 278–326.
49. Kappe, O.; Stadler, A. Limitations and Safety Aspects. In *Microwaves in Organic and Medicinal Chemistry*; Wiley-VCH: Weinheim, 2005; Section 5.4; pp 103–105.
50. Moseley, J. D. Unpublished data.

51. Nüchter, M.; Ondruschka, B. *Mol. Diversity*, 2003, *7*, 253–264.
52. Amore, K. M.; Leadbeater, N. E.; Miller, T. A.; Schmink, J. R. *Tetrahedron Lett.* 2006, *47*, 8583.
53. Bowman, M. D.; Schmink, J. R.; McGowan, C. M.; Kormos, C. M.; Leadbeater, N. E. *Org. Process Res. Dev.* 2008, *12*, 1078–1088.
54. Slade, J.; Parker, D.; Girgis, M.; Wu, R.; Joseph, S.; Repic, O. *Org. Process Res. Dev.* 2007, *11*, 721–725.
55. Iannelli, M.; Bergamelli, F.; Kormos, C. M.; Paravisi, S.; Leadbeater, N. E. *Org. Process Res. Dev.* 2009, *13*, 634–637.
56. Kaval, N.; Dehaen, W.; Kappe, C. O.; Van der Eycken, E. *Org. Biomol. Chem.* 2004, *2*, 154–156.
57. (a) Kormos, C. M.; Leadbeater, N. E. *Org. Biomol. Chem.* 2007, *5*, 65–68. (b) Kormos, C. M.; Leadbeater, N. E. *Synlett*, 2006, 1663–1666.
58. Barnard, T. M.; Vanier, G. S.; Collins, M. J. *Org. Process Res. Dev.* 2006, *10*, 1223–1237.
59. Leadbeater, N. E.; Williams, V. A.; Barnard, T. M.; Collins, M. J. *Org. Process Res. Dev.* 2006, *10*, 833–837.
60. Leadbeater, N. E.; Williams, V. A.; Barnard, T. M.; Collins, M. J. *Synlett*, 2006, 2953–2958.
61. Savin, K. A.; Robertson, M.; Genert, D.; Green, S.; Hembre, E. J.; Bishop, J. *Mol. Diversity*, 2003, *7*, 171–174.
62. Moseley, J. D.; Woodman, E. K. *Org. Process Res. Dev.* 2008, *12*, 967–981.
63. Arvela, R. K.; Leadbeater, N. E.; Collins, M. J. *Tetrahedron*, 2005, *61*, 9349–9355.
64. Pitts, M. R.; McCormack, P.; Whittall, J. *Tetrahedron*, 2006, *62*, 4705–4708.
65. Ioannidis, P.; Lundin, R. at www.biotage.com/DynPage.aspx?id=56907/Applications/ Poster (accessed July 14, 2009).
66. (a) Shieh, W.-C.; Dell, S.; Repic, O. *Tetrahedron Lett.* 2002, *43*, 5607–5609. (b) Shieh, W.-C.; Lozanov, M.; Repic, O. *Tetrahedron Lett.* 2003, *44*, 6943–6945.
67. Moseley, J. D.; Lawton, S. J. *Chimica Oggi/Chemistry Today* 2007, *25*, 16–19.
68. Tye, H.; Whittaker, M. *Org. Biomol. Chem.* 2004, *2*, 813–815
69. Glasnov, T. M.; Tye, H.; Kappe, C. O. *Tetrahedron* 2008, *64*, 2035–41.
70. Strauss, C. R. *Angew. Chem. Int. Ed.* 2002, *41*, 3589–3590.
71. Gilday, J. P.; Lenden, P.; Moseley, J. D.; Cox, B. G. *J. Org. Chem.* 2008, *73*, 3130–3134.
72. Lombard, C. K.; Myers, K. L.; Platt, Z. H.; Holland, A. W. *Organometallics*, 2009, *28*, 3303–3306.
73. Pivonka, D. E.; Empfield, J. R. *Appl. Spectrosc.* 2004, *58*, 41–46.
74. Leadbeater, N. E.; Smith, R. J. *Org. Lett.* 2006, *8*, 4589–4591
75. Schmink, J. R.; Holcomb, J. L.; Leadbeater, N. E. *Chem. Eur. J.* 2008, *14*, 9943–9950.
76. Schmink, J. R.; Leadbeater, N. E. *Org. Biomol. Chem.* 2009, 3842–3846
77. Yates, A. *The Complete Microwave Book*; Clarion: Tadworth, Surrey, 1995.
78. Nüchter, M.; Müller, U.; Ondruschka, B.; Tied, A.; Lautenschläger, W. *Chem. Eng. Technol.* 2003, *26*, 1207.
79. Rosen, M. A.; Dincer, I. *Int. J. Energy. Res.* 2004, *28*, 917–930.
80. Gronnow, M. J.; White, R. J.; Clark, J.; Macquarrie, D. *Org. Process Res. Dev.* 2005, *9*, 516–518 (corrigendum 2007, *11*, 293).
81. Hoogenboom, R.; Wilms, T. F. A.; Erdmenger, T.; Schubert, U. S. *Aust. J. Chem.* 2009, *62*, 236–243.
82. Moseley, J. D.; Woodman, E. K. *Energy Fuels*, 2009, *23*, 5438–5447.
83. Razzaq, T.; Kappe, C. O. *ChemSusChem.* 2008, *1*, 123–132.
84. www.cambrex.com.
85. www.ctechinnovation.com.

86. Muir, J. E. *Microwaves Come of Age*, Microwave and Flow Chemistry Conference, Antigua, January 28–31, 2009.
87. http://www.accelbeamsynthesis.com.
88. Schmink, J. R.; Kormos, C. M.; Devine, W. G.; Leadbeater, N. E. *Org. Process Res. Dev.* 2010, *14*, 205–214.
89. http://www.sairem.com.
90. Metaxas, A. C.; Meredith, R. J. Dielectric Properties. In *Industrial Microwaves Heating*; Peter Peregrinus Ltd for the Institute of Electrical Engineers: London, 1983; Chap 3; pp 26–69 (reprinted 1993).
91. Gabriel, C.; Gabriel, S.; Grant, E. H.; Halstead, B. S. J.; Mingos, D. M. P. *Chem. Soc. Rev.*, 1998, *27*, 213–223.

6 Microwave Heating as a Tool for the Undergraduate Organic Chemistry Laboratory

Cynthia B. McGowan and Nicholas E. Leadbeater

CONTENTS

6.1 INTRODUCTION

Laboratory instruction has been an integral part of science-based courses in institutes of higher education since the seventeenth century. However, up until the start of the twentieth century, it was conducted almost entirely in demonstration format by the lecturer.[1,2] Today, individual laboratory work for students is commonplace, and the concept of laboratory-based learning has become firmly entrenched in the teaching of science.[3] Looking at a standard organic chemistry laboratory manual, a range of reactions are covered, but these represent only a few of the many possibilities. In addition, the equipment used in undergraduate organic chemistry laboratories has not changed significantly in the last 40 years. This is very different from industrial facilities and most academic research laboratories, where state-of-the-art equipment is often found. One class of reactions of which students often do not gain experience is those that require extended heating. This is because they often cannot be fitted into an average laboratory period of approximately 3 h. This rules out a number of synthetic transformations, in particular, many metal-catalyzed reactions. Using microwave heating as a tool, these time constraints can often be overcome. Also, if a reaction is complete within a few minutes of heating, students may be able to repeat an experiment in the case of a mistake being made in preparing the reaction mixture or, more excitingly, they may be able to perform additional self-designed experiments based around the general reaction being investigated in the lab period. In this chapter, the use of microwave heating as a tool to facilitate student learning and broaden their laboratory experience will be discussed. Particular attention will be paid to the organic chemistry curriculum, the broad range of reactions that can be performed, and the incorporation of green chemistry principles.

6.2 MOTIVATION FOR INCORPORATION OF MICROWAVE HEATING AS A TOOL IN THE UNDERGRADUATE LABORATORY

Microwave heating as a tool for organic synthesis has made significant inroads in academic and industrial chemical laboratories.[4,5] This is mainly due to the shorter reaction times, increased product yields, and minimization of secondary reactions that are possible. In addition, the operational simplicity of microwave apparatus is one of its greatest attributes, as researchers can run dozens of screening reactions in a single day. Another advantage of microwave-assisted chemistry is the precise control over reaction parameters such as temperature, pressure, time, and microwave power input. Thus, finding the optimal conditions for a specific reaction can be greatly simplified. As the successful applications of this heating technology continue to expand, the need to train future scientists in microwave chemistry theory and practice becomes a compelling need in the undergraduate chemistry curriculum.

The short reaction times possible when using microwave heating open avenues for incorporation of a wider range of experiments into the organic chemistry laboratory curriculum, the first published report appearing in 1988.[6] Students gain more time for activities involved in reaction planning, setup, workup, purification, and identification and spend less time watching a reaction reflux. Microwave reactions can be run in either open-vessel or closed-vessel mode using scientific microwave apparatus. In both cases, the manufacturers provide built-in safety features. For each reaction, the equipment can be programmed with a heating profile that limits the temperature and pressure that the reaction mixture can reach. This is followed by a cooling period so that the reaction vessel is at a temperature safe for the student to handle. In the rare case of a reaction vessel leaking or failing during the course of a run, the microwave unit contains the contents in the microwave cavity and vents any volatiles directly into a fume hood.

An advantage of sealed-vessel processing is the flexibility possible when choosing a solvent in which to perform the reaction. It is not necessary to depend on boiling point to obtain the desired reaction temperature. Reactions can be run in less expensive and lower-boiling solvents that are easier to remove at the end. It must be noted, however, that not all solvents are amenable to microwave heating. Low-microwave-absorbing solvents such as hexane, toluene, and dichloromethane tend not to be good solvent choices, since they are difficult to heat.

Depending on the microwave system employed, pressurized closed vessels can hold between 0.25 and 10 mL of a reaction mixture, allowing for reactions to be run on micromole-to-millimole scale. The reaction vessels generally consist of a heavy-wall glass tube with a fitted cap. Although the glass tubes may be a little more costly than a conventional round-bottom flask, overall, the cost of the glassware is comparable since there is no need for reflux condensers, hoses, and clamps.

When considering incorporation of microwave heating into an undergraduate laboratory class, the initial cost of the apparatus is often considered to be a major obstacle. The cheapest option is to use a domestic microwave oven. However, there are significant drawbacks in doing this, the most important of which are safety-related. Since domestic microwave ovens are not designed for performing synthetic chemistry, they are not built to withstand a vessel failure. As a result, there is risk of injury when performing reactions in sealed vessels. Even when using open vessels, there is the possibility of toxic vapors escaping from the microwave cavity and, if flammable, there is the potential for a fire. Another problem when using domestic microwave ovens is the lack of direct temperature control. As a result, reactions have to be controlled by setting microwave power levels. This, together with the inhomogeneity of the microwave field in a domestic oven and the potential for overheating a reaction mixture, create both safety and reproducibility issues.

Although more costly, scientific microwave apparatus has many advantages. It makes performing reactions safer, controllable, more easily monitored, and reproducible. Depending on the size of the laboratory, either a monomode (smaller class) or multimode (larger class) unit may be more appropriate. There are actually two cost-saving avenues available: academic discounts that microwave manufacturers are often willing to offer and the fact that some industrial chemistry companies may be interested in donating surplus equipment. The faculty at Merrimack College in Massachusetts has tested three multimode units from three major manufacturers for use in undergraduate organic chemistry laboratories. The CEM MARS and Milestone MicroSYNTH units can each hold up to 40 sealed reaction vessels. The Anton Paar Synthos 3000 can hold up to 96 microscale reactions in a 4 × 24 well-plate format, each vessel being a standard Wheaton vial sealed with a septum and screw cap. Since these multimode units can hold multiple reactions vessels, often only a single unit is needed at any one time. The less expensive monomode systems process only one reaction at a time, which typically necessitates a dedicated unit for approximately every four students per lab. At Merrimack College, the CEM Discover and the Biotage Initiator have both been used. They can be purchased with a robotic automation attachment for running multiple reactions sequentially over time. The monomode systems are especially useful in smaller advanced laboratory settings where students are often involved in individual or small group mini-research projects, so the demand for the microwave apparatus is often staggered. Regardless, the short reaction times possible when using microwave heating allows for maximum use of the limited laboratory time. Student productivity at Merrimack College has increased significantly since the incorporation of microwave heating into the laboratory curriculum, and students routinely use the apparatus for any reaction that requires heating as opposed to the conventional tools also available to them.

At the 229th National Meeting of the American Chemical Society in March 2005, a symposium was dedicated to the incorporation of green chemistry into the undergraduate curriculum and laboratory experience. The report summarizing the contributions stated that "the green chemistry community is expanding efforts to develop educational materials for students."[7] In addition, the article claims that "green chemistry is a useful tool to increase awareness and teach sophisticated problem solving skills in the chemistry context." One of the principles of green chemistry is to use safer solvents and auxiliaries.[8] One way to achieve this is either to eliminate organic solvents from a reaction, therefore performing it neat, or to use a more environmentally benign solvent. When using microwave heating and sealed-vessel processing, reactions can often be performed solvent free or, alternatively, water can be used as a solvent if the reagents are slightly soluble at elevated temperatures. At the completion of the reaction, products often precipitate out of the water and can be easily collected by vacuum filtration. Another principle of green chemistry is to minimize waste by using catalytic reactions. Reactions using metal catalysts are highly amenable to microwave heating. In many cases, catalyst loadings can be reduced substantially. This, in addition to the short reaction times possible, allows students to get exposure to this class of industrially very useful reactions.

6.3 EXAMPLES OF PUBLISHED TEACHING MATERIALS INCORPORATING MICROWAVE HEATING INTO THE UNDERGRADUATE LABORATORY

While undergraduate laboratories provide the opportunity for students to understand science through a hands-on experience, time is often restricted to approximately 3 h per session. Many colleges and universities are converting experiments that use conventional heating into microwave-promoted analogs to better manage laboratory time. Reactions that once required more than a laboratory period for completion can now be added to the curriculum. Additionally, smaller reaction volumes and the use of more benign solvents are helping deliver a more sustainable approach to the undergraduate laboratory experience. What follows is a sample of published experiments using microwave heating as a tool in the undergraduate organic chemistry laboratory as well as, for completeness, a brief overview of applications to other chemistry disciplines.

6.3.1 SOLVENT-FREE REACTIONS

The use of solvent-free conditions streamlines synthetic protocol in the undergraduate laboratory. Solvents can be costly to purchase and dispose of. In addition, adding and removing solvent can be time-consuming. A number of protocols for the Diels–Alder reaction have demonstrated these advantages. The reaction of 2,3-dimethyl-1,3-butadiene with either maleimide or maleic anhydride in a 2:1 molar ratio,

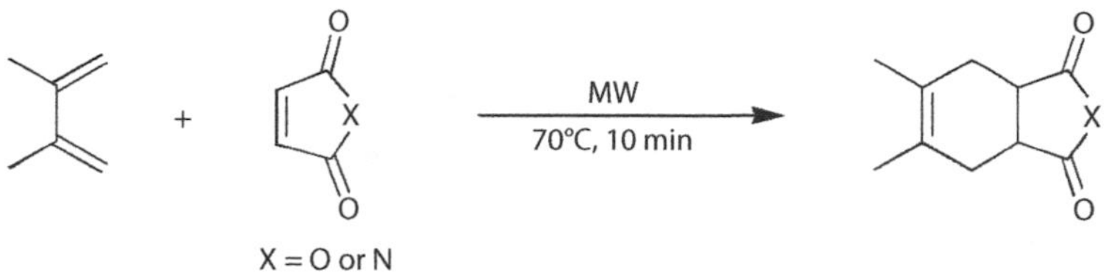

X = O or N

SCHEME 6.1

respectively, can be performed solvent free and in 10 min at 70 °C in sealed reaction vessels (Scheme 6.1).[9,10] In both cases, the products are isolated by evaporation of the excess 2,3-dimethyl-1,3 butadiene. The products may be recrystallized, but this is not required to obtain acceptable melting points and for spectral analysis. Both protocols have been used by classes of students. In the case of the reaction with maleimide, the reaction has been performed using both monomode and multimode microwave units. As a note of comparison, conventional methods for the reaction with maleimide require a 90 min reflux in DMF and the purification procedure includes column chromatography. In an interesting extension of the Diels–Alder chemistry, the reaction of 2,3-dimethyl-1,3-butadiene with dimethyl fumarate has also been reported for use in an undergraduate laboratory.[9] The reagents are not sufficiently microwave absorbant to perform the reaction solvent-free when using a multimode microwave unit. Instead, by using a small quantity of water as a microwave absorber, the chemistry can be performed in 10 min at 140 °C in a sealed vessel. The reagents essentially float on the water and, at the end of the reaction, the product can be isolated by means of an extraction.

Other examples of carbon–carbon bond-forming reactions using microwave heating in conjunction with solvent-free conditions include the Wittig reaction[11] and Knoevenagel condensation.[9,12] The Wittig reaction is used widely in synthetic chemistry but generally requires the use of strong, often pyrophoric bases such as butyllithium or sodium amide, and can take time to perform and reach completion. While the use of commercially available "instant ylide" (methyltriphenylphosphonium bromide-sodium amide) has opened avenues for performing the reaction in an undergraduate setting, it still proves challenging. Variants have been developed that involve phase-transfer conditions and sodium hydroxide as base.[13] In addition, a solvent-free approach has been reported, which involves grinding the reagents in a mortar and pestle.[14] When using microwave heating, the reaction can be completed in 5–11 min, depending on the substrates and scale used. The experiment was designed to be part of a discovery-based sequence for students in a second-semester organic laboratory course. Each student was given a general reaction procedure and a unique aldehyde starting material (Scheme 6.2). They were guided through the proper use of the microwave oven and, when their reaction was complete, the students were allowed to design their own purification procedure and were asked to prove the identity of their products using spectroscopic analysis. Reactions were run on the ~1 g scale, and a domestic microwave oven was used. First, potassium carbonate and (ethyl)triphenylphosphonium iodide were combined and ground into a fine powder before transferring them to an Erlenmeyer flask. The aldehyde was added, mixed

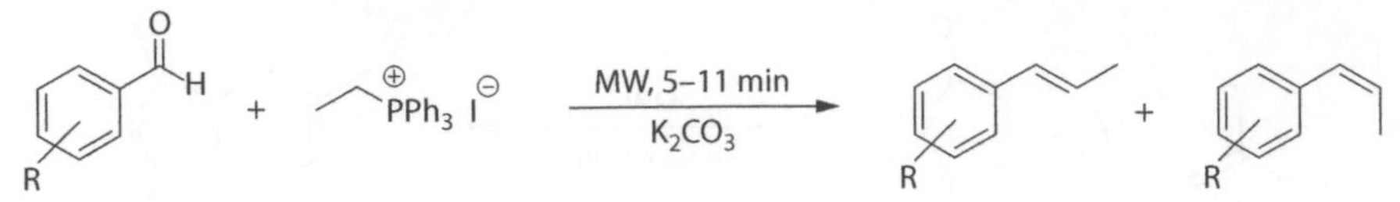

SCHEME 6.2

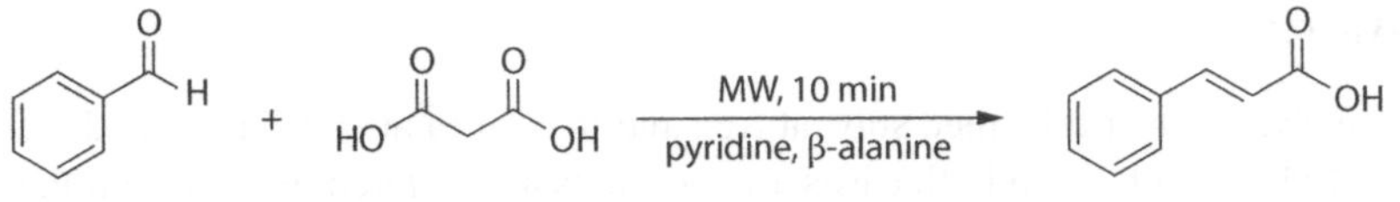

SCHEME 6.3

into the contents of the flask, and then a watch glass placed over the top before placing the flask into the microwave oven. A beaker of ice was also placed in the oven, the objective being to modulate the heating of the reaction mixture. The oven was then programmed to heat at full power for the set time. The students were able to synthesize a variety of β-methylstyrene derivatives in good yields. Since the reactions could be completed in short times, the class had ample time to prepare and analyze their products. The experiment also provided the instructor with a single experiment that could generate discussions on topics ranging from organophosphorus chemistry, preparation of alkenes, isomers, and spectroscopic analysis, as well as the novel areas of microwave synthesis and solvent-free/green chemistry.

Cinnamic acid has been prepared in an undergraduate laboratory setting using a solvent-free protocol.[9] Benzaldehyde and malonic acid were used as substrates, and the product was formed via a Verley–Doebner-modified Knoevenagel condensation using pyridine and β-alanine as cocatalysts (Scheme 6.3). Since the reaction involves a decarboxylative step, it was performed in an open reaction vessel. Because only one open-vessel reaction could be performed at a time using a microwave unit, it was suggested that the experiment be run as a group exercise. Enough material could be made for 12 students in one run. When using a multimode microwave unit, this was the minimum scale recommended. For monomode apparatus, it could be scaled down so that each student could perform the reaction individually on the ~1 g level. The experiment formed part of a multistep synthesis project in a second-semester organic chemistry laboratory. A variant of the reaction using ethanol as a solvent has also been reported.[15] Also, by means of a solvent-free Knoevenagel condensation, a very simple synthesis of coumarins has been developed.[12] It could be completed in under 10 min using microwave heating and involves the condensation of salicylaldehyde derivatives with ethylacetate derivatives in the presence of piperidine as a base catalyst. The reaction was again performed in an open vessel. The products were isolated by crystallization from the appropriate solvent upon cooling.

A microwave-promoted solvent-free Knoevenagel condensation has been used as a tool for exposing students to new preparative chemistry techniques, identification of a product using analytical methods, and use of the Karplus rule for identification of stereoisomers by ^{1}H-NMR spectroscopy.[16] The experiment was designed for and tested in a second-year organic chemistry laboratory. It involved the reaction of methyl

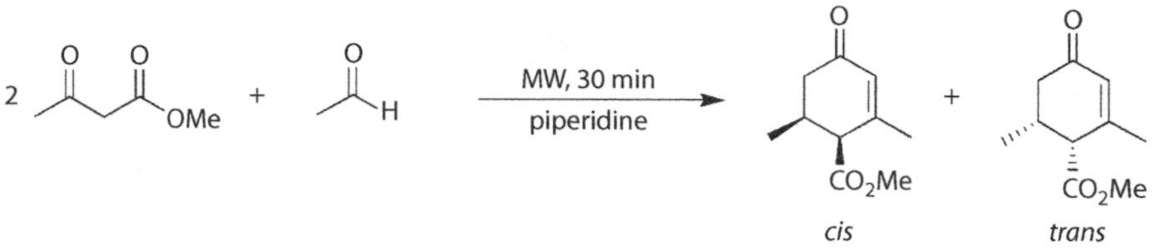

SCHEME 6.4

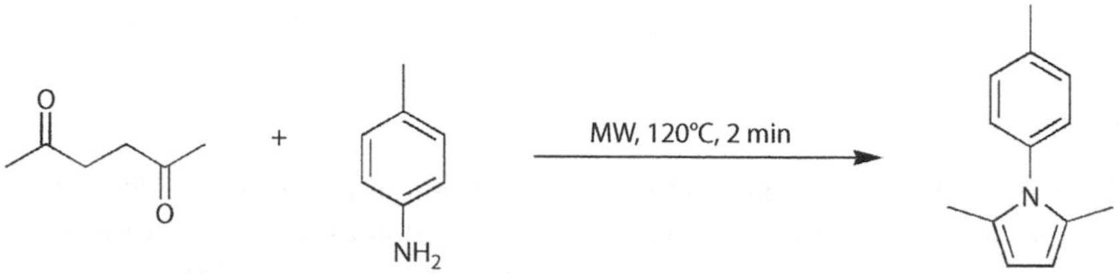

SCHEME 6.5

acetoacetate and acetaldehyde using piperidine as a catalyst from which two stereoisomers of the product were produced (Scheme 6.4). A sealed-vessel approach was taken, heating the reaction mixture to 85 °C over a period of 5 min and then holding at this temperature for a further 25 min. The ¹H-NMR of the product was recorded and, using the Karplus rule, the *cis* and *trans* isomers differentiated and their relative proportions determined. Again with an emphasis on stereochemistry, a *cis–trans* mixture of dioxolanones has been prepared in an undergraduate laboratory setting using microwave heating and then analyzed using NOESY 2-D NMR spectroscopy.[17]

Procedures for preparing a number of heterocycles have been developed for use in the undergraduate laboratory, all being performed solvent-free using a domestic microwave oven.[18] Each entails loading a beaker with the reactants, covering it with a watch glass, and then placing it into the oven and heating it at a fixed microwave power for a fixed time. To overcome issues associated with use of volatile or temperature-sensitive reagents, pulses of microwave power were used. While making some interesting compounds and offering potential for pre- and postlab discussion based around heterocyclic chemistry and key organic transformations, the safety and reproducibility of these protocols is a concern. A better approach involves the use of scientific microwave apparatus, and a procedure for the solvent-free Paal–Knorr pyrrole synthesis has been developed and incorporated into an organic chemistry laboratory course (Scheme 6.5). Heating at 120 °C for 5 min in a monomode microwave unit was sufficient to obtain a 75% yield of the product.

A solvent-free procedure for the preparation of methylenedioxyprecocene (MDP), a natural insecticide, has been developed for use in the undergraduate organic chemistry laboratory; a section of 15 students is being used as a test group.[19] The synthesis involves, in tandem, electrophilic addition of 3-methyl-2-butenal to sesamol, followed by dehydration and an intramolecular hetero-Diels–Alder cyclization (Scheme 6.6). Montmorillonite-K10, a commercially available basic clay, is used as a support

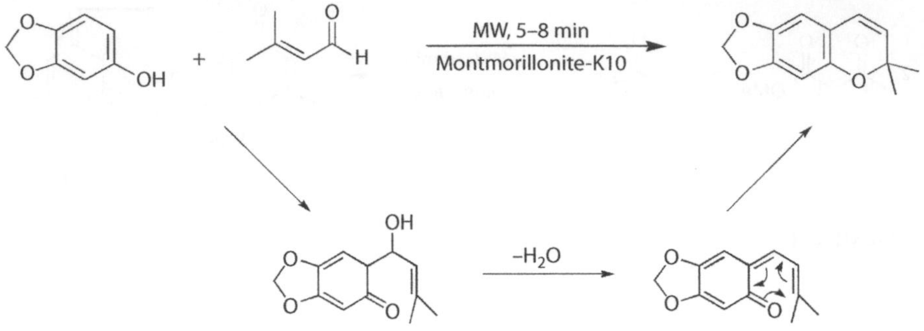

SCHEME 6.6

and catalyst. The reaction is carried out by combining the sesamol, aldehyde, and clay in a glass vial and heating for 5–8 min in a domestic microwave oven. The product is then isolated by washing the clay with ethyl acetate, collecting the filtrate, and removing the organic solvent. The green chemistry aspects of the protocol can be used as teaching tools. They can be further emphasized by the fact that both the clay and the ethyl acetate may be recycled. The student's interest can be piqued by the fact that they are preparing a physiologically active compound.

Also using K-10 clay as a support, an experiment illustrating the Fries rearrangement has been reported for use in an undergraduate laboratory, in which the β-napthylacetate and the steroid estrone are used as substrates (Scheme 6.7).[20] Drawbacks, however, are the use of a domestic microwave, open beakers as reaction vessels, and the vague experimental details provided. The protocol calls for heating the reaction at power level 9 for 10 min. In the absence of temperature data, it would be hard to reproduce the experiment without reoptimization.

The silica-supported synthesis of tetraphenylporphyrin has been modified for use in a green organic chemistry curriculum (Scheme 6.8).[21] The aim was to adapt a report in the scientific literature[22] to fit one 3 h laboratory period. Although the literature protocol was successfully carried out by some students with very little modification, reoptimization was required before it could be reliably incorporated into the undergraduate laboratory. The majority of the refinements came in the purification step; new chromatography conditions being required to separate the porphyrin product successfully. The modified protocol required mixing pyrrole and benzaldehyde in an Erlenmeyer flask before adding silica and agitating until it was evenly covered in reagents. Heating in a domestic microwave oven for 10 min in 2 min intervals at 1000 W followed by chromatography using a hexane–ethylacetate mixture was found to be optimal. Since the porphyrin product is colored purple, it is easy to track on the column. In the event that a student makes an error in preparing the reaction mixture, the likely outcome is that the first product eluting from the column is green rather than purple. This allows the student to make a quick assessment of the success of the experiment and, if needed, repeat it within the same laboratory period. Interesting extensions to the protocol have also been discussed, these including metalation of the porphyrin using zinc acetate or synthesis and characterization of *ortho*-substituted tetraphenylporpyrins. In the case of the former, the metalated

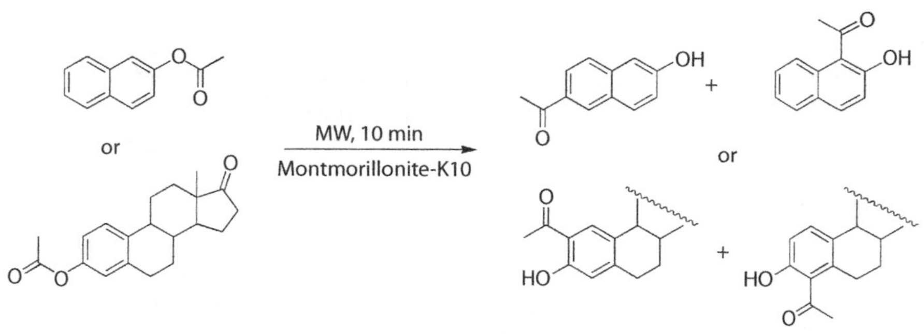

SCHEME 6.7

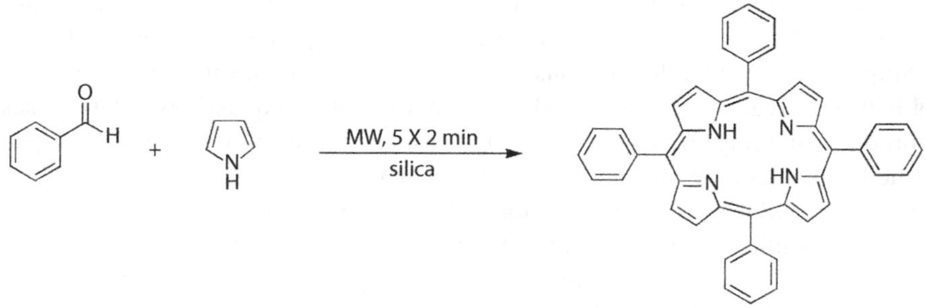

SCHEME 6.8

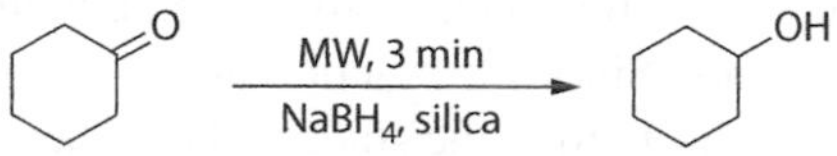

SCHEME 6.9

product could easily be prepared at room temperature and the reaction monitored using visible spectroscopy. In the case of the synthesis of ortho-substituted analogs, the atropoisomerism exhibited by the products proves an interesting challenge for students to study using NMR spectroscopy, introducing concepts such as temperature dependence and determination of interconversion rates using line-broadening methods. In addition, the products formed could be used to prepare a solar cell using another published laboratory procedure.[23]

With silica as a support and sodium borohydride as a reducing agent, cyclohexanol has been prepared from cyclohexanone in less than 3 min using a scientific microwave unit (Scheme 6.9).[24] The reduction of carbonyl compounds by sodium borohydride is traditionally time consuming. It can take 2–4 h to reach completion, this equating to a whole laboratory period, and usually another lab period is required for product isolation and characterization. Using the microwave procedure, the entire experiment can be completed in one 2.5 h session. The reaction was performed in closed tubes

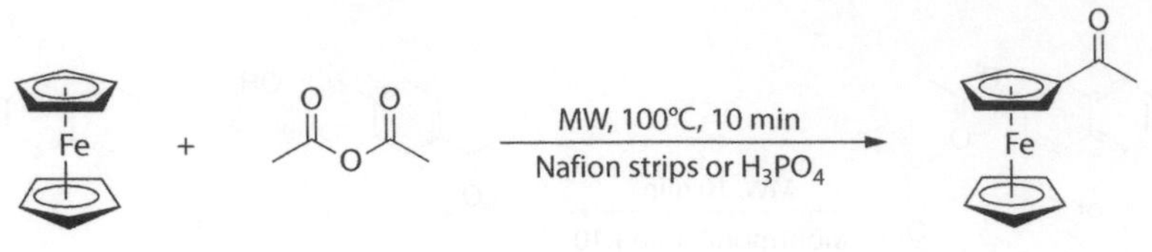

SCHEME 6.10

although the tops were not firmly screwed down to avoid overpressurization since the vessels used did not have any direct pressure measurement device attached.

The acid-catalyzed acyclation of ferrocene with acetic anhydride has been converted into a greener reaction using microwave heating (Scheme 6.10).[25] Concentrated mineral acid was replaced by polymeric sulfonic acid strips (Nafion), and the reaction carried out solvent free at 100 °C for 10 min in a sealed vessel. The product was purified by column chromatography, which, although not a "green" method, is an important technique for students to learn. However, an advantage of the protocol is that the reaction proceeds with substantially reduced quantities of tarry black byproducts, making the purification stage easier. The reaction could also be performed with a molar equivalent of phosphoric acid in place of Nafion strips should they not be available or if the instructor prefers not to use them. The reaction conditions were identical, and a similar yield of the product was obtained.

A number of acetate esters can be prepared by performing the reaction solvent-free in a sealed vessel using the alcohol substrate as the limiting reagent, sulfuric acid as the catalyst, and silica beads as water scavengers (Scheme 6.11).[9] Heating to 120 °C for 5 min is sufficient to take the reaction to completion. Students can select from a number of alcohols, reacting them with acetic acid, and the resultant esters have a range of distinct odors. Since the reaction and workup (sodium bicarbonate wash) can be performed in short order, the students have the potential to make more than one ester in a laboratory period.

Microwave heating has been used in the preparation of aspirin from salicylic acid and acetic anhydride.[26] Four acids and four bases are used as potential catalysts for the reaction, each student being assigned one. For comparison, the reaction is also

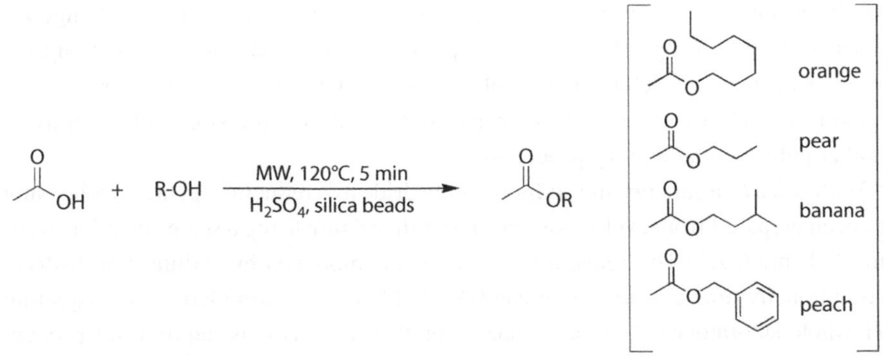

SCHEME 6.11

performed in the absence of a catalyst. This report is concerning from a number of perspectives. The first is that the reaction performed catalyst free is reported to yield no product when heated conventionally but gives an 80% yield after 10–13 min in a domestic microwave oven. This difference is most likely due to the fact that, in the case of the reaction performed using microwave heating, some of the acetic anhydride was hydrolyzed to acetic acid, this serving as an acid catalyst for the reaction.[27] Second, there is extensive discussion of nonthermal microwave effects,[28] suggesting that microwave irradiation can stabilize polar transition states and change the activation energy of a reaction. Along these lines, a mechanism for the purported catalyst-free reaction is suggested. Importantly, nonthermal effects have been proved to be spurious.[29] Undergraduates should not be taught that the principles of organic or physical chemistry are any different when using microwave as opposed to conventional heating. Aspirin can be prepared very rapidly conventionally by using sulfuric acid or phosphoric acid as a catalyst in the undergraduate or high school laboratory. Heating in a water bath at 85 °C for 5–15 min is sufficient.[30]

6.3.2   REACTIONS PERFORMED IN GREENER SOLVENTS

When performing a reaction under solvent-free conditions is not practical, the use of less expensive and less toxic solvents still demonstrates the benefits of a greener approach to synthesis. The use of microwave heating in the undergraduate teaching laboratory has also provided a unique opportunity to use water as the solvent in organic reactions. Due to the high temperatures and pressures that can be obtained in a sealed reaction vessel, organic compounds can react cleanly either on the surface of the water or in solution due to increased solubility under these conditions. Addition of phase-transfer agents or a cosolvent such as ethanol can also facilitate reactions. The preparation of 3-methyl-2-cyclopenten-1-one from hexane-2,5-dione has been performed via an intramolecular aldol condensation in basic water (Scheme 6.12). The reaction can be used as a tool for teaching students the key concepts of carbonyl chemistry as well as differences between inter- and intramolecular reactions. The reaction is complete in sealed vessels after 5 min at 175 °C in a multimode microwave unit (up to 40 reactions being performed at a time) and 5 min at 150 °C in a monomode system.

Hydrolysis of a nitrile[9] and an amide[15] have both been converted for use in undergraduate laboratory classes (Scheme 6.13). The base-catalyzed hydrolysis of o-tolunitrile can be performed using either a monomode (150 °C for 10 min) or multimode unit (150 °C for 6 min) in sealed-vessel format. The shorter time required at the target temperature when using a multimode unit as opposed to the monomode system is at first counterintuitive. However, the time taken to reach 150 °C is significantly

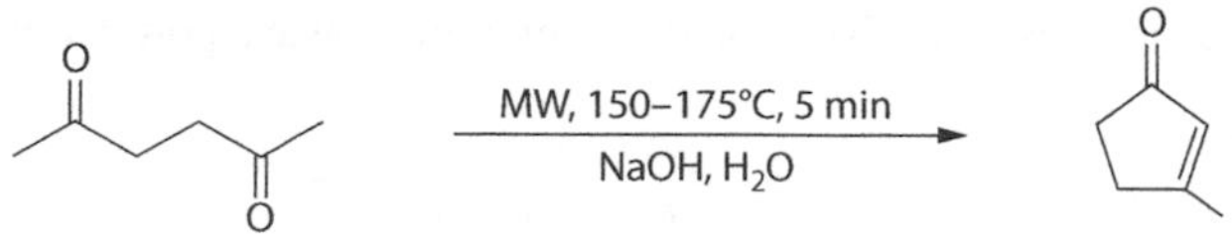

$$\underset{\text{NaOH, H}_2\text{O}}{\xrightarrow{\text{MW, 150–175°C, 5 min}}}$$

SCHEME 6.12

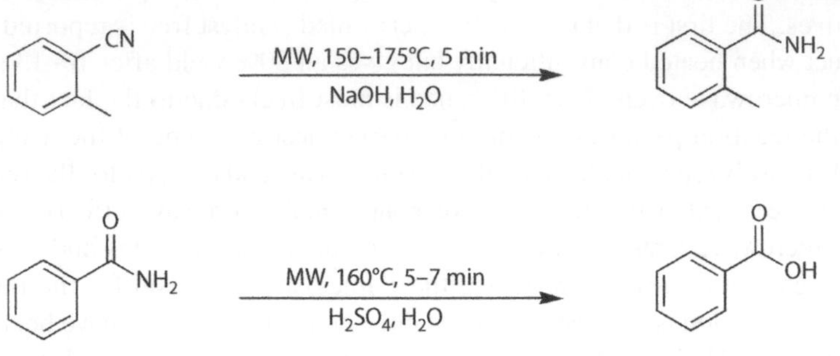

SCHEME 6.13

longer for the former, meaning that the total reaction time is longer. The fast heat-up and cool-down times when using a monomode microwave system necessitate a longer hold time in order for the reaction to reach completion. The acid-catalyzed hydrolysis of benzamide to benzoic acid was performed in a monomode microwave unit by heating to 160 °C for 5–7 min.

Three analgesic drugs (acetanilide, acetaminophen, and phenacetin) have been prepared by reaction of aniline, *p*-aminophenol, and *p*-phenetidin with acetic acid in water. The reactions were developed using a microwave oven, 5 min at 30% power. In the absence of any temperature data, reproducibility is a concern. In addition, these reactions can be performed at room temperature in short times. Indeed, it is likely that they are near to completion by the time they are put into the microwave oven. When using microwave heating and domestic equipment, another concern is safety. The acetic acid is generated in situ by hydrolysis of acetic anhydride, and the reactions are performed in beakers simply covered with a watch glass. The authors specifically mention concern over splashing that can occur due to inhomogeneous heating because stirring is not possible in a domestic oven. They do state that this can be moderated by using an oven equipped with a turntable, but it may be better to perform the reactions at room temperature, even if the time required is a little longer than the 5 min when using microwave heating.

The Williamson etherification of *p*-iodophenol (*p*-cresol) with 1-iodobutane can be performed in 10 min at 100 °C using water as the solvent, sodium hydroxide as base, and tetrabutylammonium bromide as a phase-transfer agent (Scheme 6.14).[9] The experiment was initially developed for use by one section of students as part of an organic laboratory course. When performing the procedure with much larger class loads, a problem arose with the odor of *p*-cresol, this being the net effect of each student weighing out and transferring the chemical.[31] The problem was overcome by using 4-nitrophenol in place of *p*-cresol, with slight modifications to the procedure. The temperature was increased to 150 °C, and the input microwave power was reduced.[32]

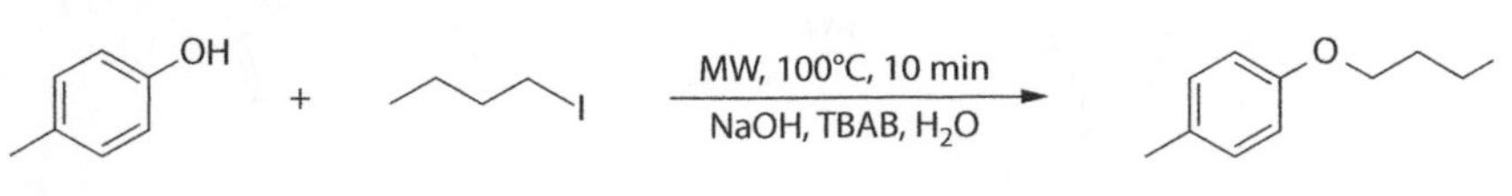

SCHEME 6.14

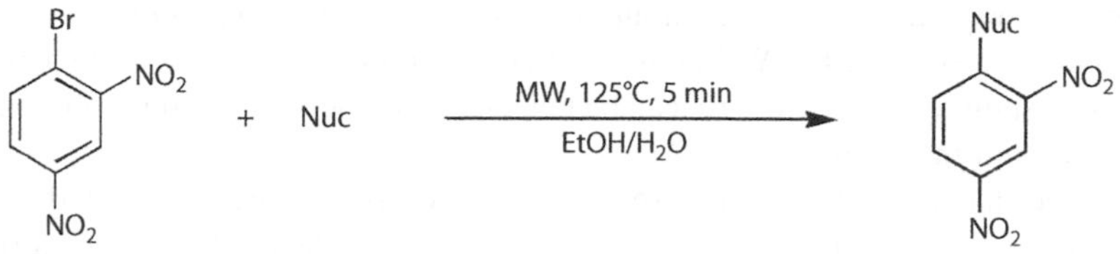

SCHEME 6.15

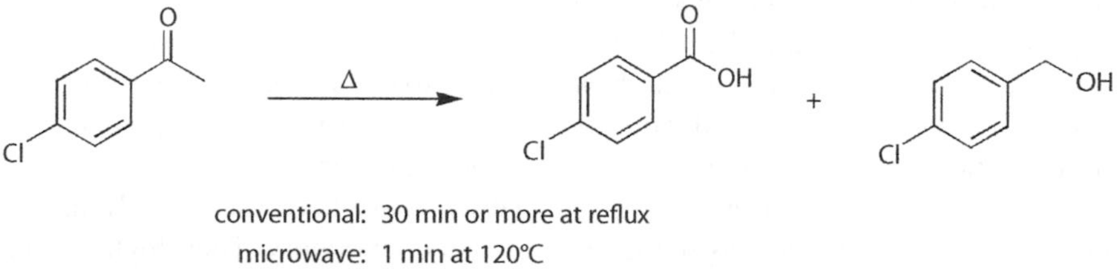

conventional: 30 min or more at reflux
microwave: 1 min at 120°C

SCHEME 6.16

Alcohols and water–alcohol mixtures also feature highly in procedures developed for undergraduate teaching laboratories using microwave heating. Nucleophilic aromatic substitution is an organic reaction that is covered in most undergraduate classes. A traditional microscale approach involves a two-phase reaction mixture (toluene and water).[33] A phase-transfer reagent is added to carry the nucleophilic anion into the toluene phase, and the reaction is complete after 2 h at reflux. By performing the reaction in a sealed vessel using microwave heating, the reaction can be performed in aqueous ethanol (Scheme 6.15).[9] It is complete in 5 min at 125 °C, and the product precipitates from the reaction medium upon cooling, thus eliminating the conventional need for extractions, solvent removal, and recrystallization. A range of nucleophiles can be used, offering the students choice.

The Cannizzaro reaction can be performed using methanol as the solvent (Scheme 6.16). The reaction demonstrates the simultaneous oxidation and reduction of an aromatic nonenolizable carbonyl compound to form the corresponding benzoic acid and benzyl alcohol. It allows instructors the opportunity to discuss the chemistry of the carbonyl group and also gives students exposure to the techniques required for separation and quantification of the two products. In the conventional microscale approach, 4-chlorobenzaldehyde is dissolved in methanolic sodium hydroxide and the mixture heated to reflux.[34] After 30 min, students start to monitor the progress of the reaction using TLC. In the microwave approach, the reaction mixture is heated to 120 °C in a sealed vessel and held at this temperature for 1 min, after which it is cooled (Scheme 6.16).[32] When using a monomode microwave unit, since only one reaction mixture can be processed at any one time, the conventional approach may be faster if the number of microwave units is limited and the class size is medium or large. When using a multimode unit, the microwave approach can have time saving associated with it. Even factoring in the time taken for multiple students to load their vessels and place them into a carousel and the time taken to ramp to the target temperature, the hold time of just 1 min means that the students can move on to the workup stage in short order. In another example of short reaction times using an alcohol as a solvent, the Williamson etherification of *m*-cresol with iodobutane using potassium hydroxide as base can be performed in methanol using a monomode microwave unit in 1 min at 80 °C.[10]

Methanol has proved a good solvent for the one-pot preparation of methyl salicylate (oil of wintergreen) from aspirin (Scheme 6.17).[35] The tandem transesterification–Fischer esterification process has been performed both conventionally (90 min at reflux) or using microwave heating (5 min at 120 °C in a sealed vessel). Commercial aspirin tablets were added to methanol and the insoluble binder removed by filtration, leaving the acetylsalicylic acid in solution. To this, concentrated sulfuric acid was added and the reaction mixture heated. Upon work-up, student yields of up to 70% were on occasion achieved. However, typical student yields were around 25% for conventional heating and 48% when using microwave heating.

Ethylene glycol and polyethylene glycols (PEGs) have been used as solvents for microwave-promoted reactions in the undergraduate laboratory. Their high boiling points and microwave absorbtivities make them useful for performing reactions at elevated temperatures in open vessels. Other advantages of glycols include the fact that they are benign, biodegradable, and soluble in water. Ethylene glycol has found application in the Wolff–Kishner reduction. This reaction, involving the full reduction of a carbonyl moiety to an alkane using hydrazine, is traditionally performed at temperatures around 200 °C in the presence of a strong base. The Huang–Minlon modification[36] is a convenient variation of the reaction and involves heating the carbonyl compound, potassium hydroxide, and hydrazine hydrate in ethylene glycol in a one-pot reaction, taking in the region of 1 h at reflux to reach completion.[37] Using microwave heating and a two-step approach, the reaction has been incorporated into a teaching laboratory (Scheme 6.18).[38] Isatin was used as the substrate for reduction. In the first step, this was combined with hydrazine hydrate and a small volume of ethylene glycol in a beaker, which was then placed inside a domestic microwave and heated for 30 s at medium power. After cooling in an ice bath, the isatin 3-hydrazone product was isolated. In the second step, potassium hydroxide was dissolved in

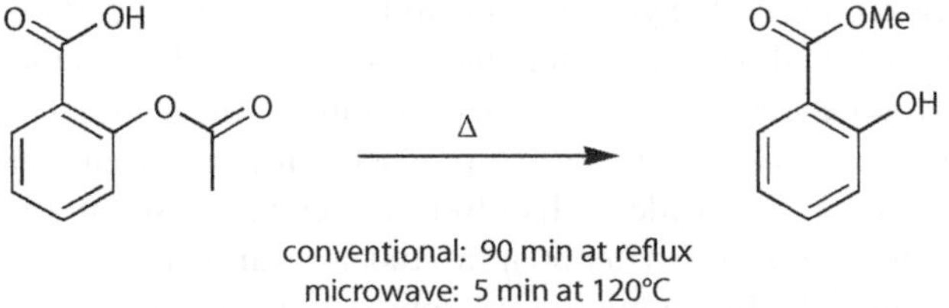

conventional: 90 min at reflux
microwave: 5 min at 120°C

SCHEME 6.17

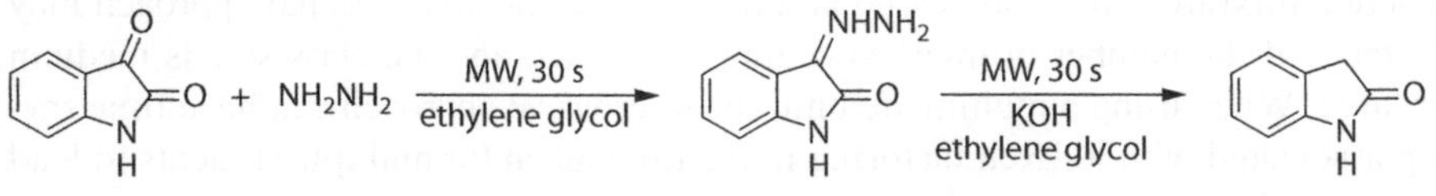

SCHEME 6.18

ethylene glycol by heating the two in the microwave oven for 10 s before adding the hydrazone and heating for a further 10 s. Workup comprised acidification, extraction with diethyl ether, and recrystallization from water. The oxindole product is reported in 32% yield. No temperature measurement was made for either of the heating steps. Given the short time exposed to microwave irradiation, the small scale on which the reaction was performed (1.7 mmol), and the fact that a domestic oven was used, it is unlikely that a significant temperature rise was observed. However, the second stage should be performed cautiously when working on gram scales since reports suggest that, when the temperature becomes sufficient for dissolving of KOH in glycols to start, it continues exothermically with evolution of heat sufficient to drive material out of the top of the reaction vessel if the source of external heat is not removed immediately.[37] The microwave procedure offers significant advantages over a previous report of an undergraduate protocol for the preparation of oxindole from Isatin, also using a Wolff–Kishner reduction but requiring anhydrous solvents and the preparation of fresh sodium ethoxide by addition of sodium metal to ethanol.[39]

By using ethylene glycol as a solvent, Raney nickel or palladium on carbon as a catalyst, and ammonium formate as a hydrogen donor, a procedure for the microwave-promoted transfer hydrogenation of β-lactams has been reported.[40] The reaction was carried out successfully by students at a college and at high schools. Using a domestic microwave oven, the reaction mixture was heated to 110–120 °C and held at this temperature for 3–4 min. Given that both Raney nickel and Pd/C can be pyrophoric and that the reaction is performed in an open flask in air, care is needed when preparing the reaction mixture and, during the heating stage, a beaker of water is used as a moderator.

Polyethylene glycol has been used as solvent for the Diels–Alder reaction of 2,3-dimethylbutadiene with maleic anhydride.[41] Using conventional heating, the reaction was complete after 1 h at 50–70 °C in an open flask equipped with a reflux condenser. In a sealed vessel, the reaction was complete after 100 s of microwave heating at 150 W. Although a scientific microwave unit containing a carousel of 16 sealed tubes was used, no temperature data was noted.

6.3.3 METAL-CATALYZED REACTIONS

The use of transition-metal catalysts allows synthetic chemists to perform a wide range of reactions easily and efficiently.[42] As a result, they are used very extensively in both academia and industry. However, since reaction times can be long and the metal complexes can be costly, introduction into the undergraduate laboratory curriculum has been somewhat limited. However, some innovative experiments have been reported in the more recent chemical education literature.[43] The Suzuki reaction, the palladium-catalyzed cross-coupling of aryl halides or triflates with boronic acids, is one example. This reaction is one of the most versatile and utilized reactions for the selective construction of carbon–carbon bonds, in particular, for the formation of biaryls.[44] The first report of the Suzuki coupling in the chemical education literature came in 2001.[45] It used a phosphine-ligated palladium catalyst, formed in situ from palladium acetate and triphenylphosphine. The coupling was performed in a water–propanol solvent mix at reflux under a nitrogen atmosphere for 1 h. The

Suzuki coupling has also been used to show how automated synthesis robots can be used for screening of reaction conditions.[46] The coupling of 2-bromofluorene and phenylboronic acid using $Pd(OAc)_2$ or $Pd(PPh_3)_4$ as a catalyst was initially performed by the students. These substrates were chosen because the 2-phenylfluorene product is fluorescent, providing a visual indication of reaction progress. The procedure required the use of THF as solvent, exclusion of air, and a reaction time of 16 h at 60 °C. Following this, a robotic system was used to perform the same coupling in parallel, with different solvents and bases being screened. In another approach to the reaction, the coupling of 4-iodophenol and phenylboronic acid has been performed in water as the solvent using 10 mol% palladium on carbon as the catalyst.[47] The coupling is complete after 30 min at room temperature; this is not surprising given that the substrates are water soluble, aryl iodides are highly reactive, and a very high catalyst loading is used. The protocol is presented as a greener approach to the reaction. However, as the authors acknowledge, while this may in a sense be true, aryl iodides are not "green" substrates.

With microwave heating, the Suzuki coupling reaction can be performed using low loadings of simple palladium salts as the catalyst and either water/tetrabutylammonium bromide (TBAB) or water/ethanol as the solvent.[48,49] The reaction is complete within 5 min at 150 °C in a sealed tube or 20 min in an open vessel at reflux.[50] The coupling of bromoarenes with phenylboronic acid has been taken into the undergraduate laboratory (Scheme 6.19).[9] Students could select from one of three aryl bromides, namely, 4-bromoanisole, 4-bromotoluene, or 4-bromoacetophenone. As catalyst, 0.4 mol% $Pd(OAc)_2$ was used, and water/TBAB was employed as the reaction medium. Both monomode and multimode scientific microwave apparatus were used, with results being identical. A protocol using significantly lower catalyst loadings has been used with two groups of high school students.[51] An ICP palladium standard was used as a stock solution for delivery of 0.0004 mol% of palladium chloride into the reaction vessel containing water and ethanol as cosolvents, 4-bromoanisole and phenylboronic acid as substrates, and potassium carbonate as base. The experiment was used to show how the principles of green chemistry can be applied to an industrially relevant reaction. In addition, since such a small volume of the palladium ICP standard was used in each reaction, the catalyst cost was minimal. The boronic acid proved to be the most expensive reagent.

The Heck coupling is an example of another key metal-catalyzed transformation. Again, this has been incorporated into the undergraduate laboratory curriculum in conjunction with the use of microwave heating (Scheme 6.20).[52,53] Similar conditions to those developed for the Suzuki reaction were used with either palladium acetate (0.4 mol%) or palladium chloride (ICP standard, 0.004 mol%) as catalyst,

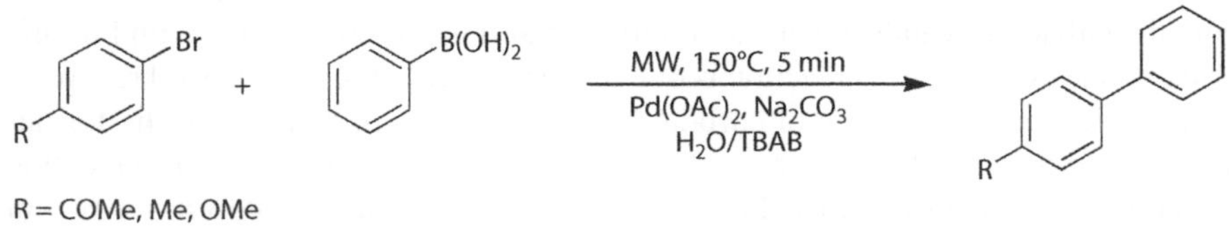

R = COMe, Me, OMe

SCHEME 6.19

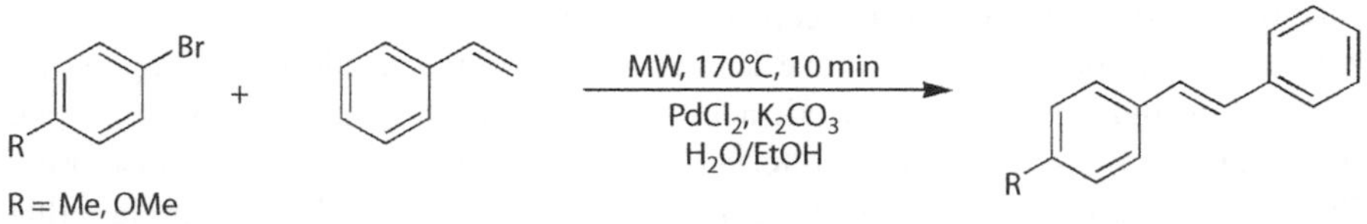

R = Me, OMe

SCHEME 6.20

and water/ethanol as solvent. Styrene was coupled with one of two aryl bromide substrates (4-bromoanisole or 4-bromotoluene). The reaction was performed in a sealed vessel at 170 °C for 10 min in either a monomode or multimode microwave unit. To put this in context, the first report of the Heck coupling in the chemical education literature came in 1988, this using conventional heating.[54] The palladium acetate-mediated protocol utilized iodobenzene as a substrate and tributylamine as base. The reaction was complete after heating for 1–2 h, depending on the alkene coupling partner used. The coupling of bromoiodobenzene with acrylic acid has been performed on the microscale in acetonitrile as solvent using 0.225–0.375 mol% palladium acetate as the catalyst and is complete within 1 h at 90–100 °C.[55] The reaction is chemoselective, bromocinnamic acid being the only product obtained. Iodoacetophenone has been coupled with acrylic acid in water as a solvent, using Pd/C as the catalyst and refluxing for 60–75 min.[56] A palladacyclic complex has been prepared and used as a catalyst for the Heck reaction within an undergraduate laboratory setting, the objective being to illustrate the concepts of organometallic chemistry and catalysis.[57] The steps to make the catalyst are not trivial, and the Heck coupling cited (bromobenzene and styrene as substrates) takes 6 h to reach completion at 130 °C in *N,N*-dimethylacetamide as solvent.

6.3.4 ADDITIONAL REPORTS IN THE LITERATURE

The microwave-promoted synthesis of 2,4,5-triphenyl-1*H*-imidazole (lophine) has been used as a tool not only for teaching a new preparative technique but also for providing an opportunity for students to employ the principles of carbonyl chemistry in devising a mechanism to explain the formation of the product.[58] The reaction involves heating benzaldehyde, benzil, and ammonium acetate in a sealed tube, using glacial acetic acid as solvent (Scheme 6.21). The reaction mixture was heated from room temperature to 120 °C over the period of 10 min and then to 125 °C over a further 5 min in a scientific multimode unit, with 16 reaction vessels being processed simultaneously in a carousel. Although not explicitly stated, the reason for the slow temperature ramp is probably because the vessels used for the reaction

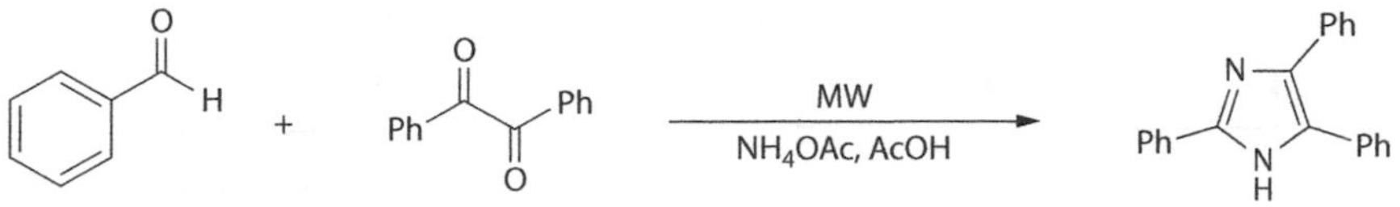

SCHEME 6.21

were fitted with pressure-release caps that vent vapors from the tube in the event that the internal pressure rises above 1.5 bar. Faster heating could lead to a pressure increase and thus the release of acetic acid vapors. Alongside the protocol developed for a scientific microwave unit, an alternative for a domestic oven was also disclosed. The reagents were placed into a beaker that was then topped with a watch glass that held a piece of dry ice. Heating at 30% power for 10 min resulted in similar product yields to those obtained in the scientific unit. Once prepared, the ^{1}H and ^{13}C-NMR spectra of recrystallized lophine were recorded, and students used these, in conjunction with their knowledge of organic chemistry and results obtained using molecular modeling, to determine the mechanism of the reaction. In an interesting extension to the laboratory experience, students have performed either both the conventional and microwave-promoted synthesis of lophine or else the conventional preparation of another heterocycle (2,5-dimethyl-1-phenylpyrrole) and the microwave-promoted synthesis of lophine. Since the conventional heating route to either product takes some time, students can set this up, move on to the microwave protocol, and have that finished by the time the conventional synthesis reaches completion. All this can be fitted into one laboratory period, and exposes students to a range of chemistry while also allowing them to gain a fuller appreciation of the advantages of microwave heating as a tool for organic synthesis.

Incorporation of multistep synthesis into the undergraduate laboratory comes much closer to real-life experiences students will have in graduate school or industry since the preparation of most target molecules generally requires more than one step. Such projects also require a higher level of skill from students than do single-step reactions. A number of articles offering multistep syntheses have appeared in the chemical education literature.[59] Augmenting these is a report of a four-step sequence, all using microwave heating.[9] The procedure first introduced the concept of retrosynthetic analysis and used this to show how *trans*-stilbene can be prepared from benzaldehyde. The sequence was devised such that it could be successfully performed in one laboratory period per step. It introduced students to a range of chemistry, specifically, a Knovenagel condensation, bromination, elimination, and Suzuki coupling (Scheme 6.22). In addition, since each step starts with a commercially available compound, if a student does not obtain the requisite material required to move to the next stage, the instructor can provide additional reagent. Another advantage of this is that each step could also be performed as a stand-alone experiment. In the first step, benzaldehyde was converted to *trans*-cinnamic acid by a base-catalyzed Knoevenagel condensation with malonic acid. Since it involved a decarboxylation step, it was performed in an open-reaction vessel. This was followed in step 2 by bromination using

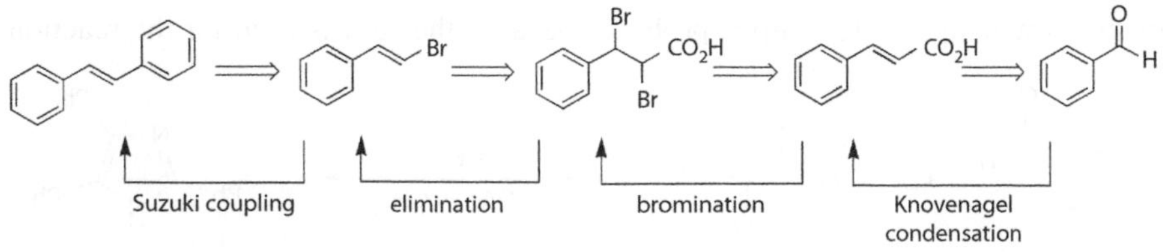

SCHEME 6.22

pyridinium tribromide as a solid, easier-to-handle source of bromine. Heating to 120 °C for 5 min yielded 2,3-dibromo-3-phenylpropanoic acid. In step three, the dibromo compound was converted to 1-bromo-2-phenylethene (β-bromostyrene) via a base-mediated elimination reaction. Two procedures are given, one for the formation of the *trans* product (water as solvent, 100–150 °C for 5 min) and one for the *cis* product (acetone as solvent, 115–125 °C for 5 min). The experiment allows the instructor to discuss the differences between E1 and E2 mechanisms and the effect of solvent on the outcome of the reaction. In the final step of the sequence, *trans* β-bromostyrene is coupled with phenylboronic acid in a palladium-catalyzed Suzuki reaction to yield *trans*-stilbene. The coupling is performed using water as the solvent, TBAB as a phase-transfer agent, and palladium acetate as the catalyst. Heating for 5 min at 130 °C is sufficient.

The preparation of biodiesel has been used as a tool for teaching students the concepts of carbonyl chemistry, equilibrium, and NMR spectroscopy, as well as the differences between conventional and microwave heating.[60] Biodiesel is generally prepared by an acid- or base-catalyzed transesterification between vegetable oil and methanol (Scheme 6.23). In the laboratory setting, engineering students in a one-semester organic chemistry course participated in a discovery-based experiment. The students were polled before and after the lab class on their knowledge in the areas of biofuels and microwave heating as a tool for preparative chemistry. From experience in the lab, students demonstrated increased learning regarding the theory behind microwave heating and its practical use, the concepts behind the preparation of biodiesel by a transesterification reaction, and the application of NMR spectroscopy as a tool for analysis of organic molecules. Each student was given a specific set of reaction conditions based on the molar ratio of vegetable oil to methanol (1:3 or 1:6), catalyst loading (1 wt% or 3 wt% KOH), and the heating method used (conventional or microwave). A minimum of three students performed any given set of conditions to afford statistical comparisons. Students then submitted a sample for ^{1}H-NMR analysis. At the conclusion of the laboratory exercise, all students calculated their product conversion by comparing integrations of signals due to product and starting materials in their NMR spectrum. In the next laboratory period, students with similar reaction conditions were then grouped together to compare and report results to the entire class. This was the first time all the students in the class were able to see the overall outcome of the experiment. The design of the experiment was to simulate a "research-like" environment in which it was not known to the students which set of reaction conditions would garner the highest yield of biodiesel product. Hence, the opportunity arose for students to compare directly conventional and microwave heating

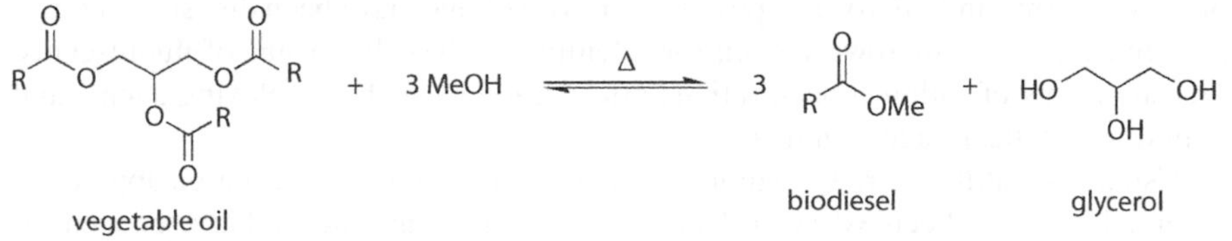

SCHEME 6.23

under various reaction conditions in a discovery-based undergraduate organic chemistry laboratory setting. At the start of the laboratory exercise, the students predicted unanimously that the optimal conditions would comprise the highest oil:methanol ratio, the highest catalyst loading, and use of microwave heating. While the highest oil:methanol ratio and catalyst loading did result in higher yields, conventional heating proved more successful than microwave heating. The lower yields obtained by the students using microwave heating may be symptomatic of insufficient mixing of the reactants by a small stirbar in the reaction vessel, as opposed to swirling and glass-rod agitation with the conventional heating procedure.

6.3.5 BEYOND ORGANIC CHEMISTRY

The use of microwave heating in the undergraduate laboratory goes beyond just organic chemistry. Examples can be found in fields as diverse as sample digestion, extraction of essential oils, synthesis of organometallic complexes, and preparation of nanomaterials. To highlight these areas, selected examples are given.

6.3.5.1 Analytical Chemistry

Sample preparation forms an important part of analytical chemistry, but students get little exposure to the methods involved. This is often because it can be time consuming and limits what can be performed in a single laboratory period. Microwave heating is used very extensively for digestion and extraction in academic and industrial settings; indeed, much of the early literature involving microwave heating in chemistry was on the application to sample preparation.[61,62] Respondents to surveys of employers in the analytical field specifically commented on their desire for students to have experience using microwave apparatus.[63] In addition to this, by introducing microwave-assisted digestion into the undergraduate laboratory, it is possible for students to prepare and analyze their samples in one class period.[64] As an example, snack foods such as potato chips have been decomposed in acid solution using sealed-vessel microwave heating.[65] While the heating step was under way, students could familiarize themselves with the analytical technique to be used, prepare a run a series of standards, and then be ready to process their sample.

A domestic microwave oven has been used for extraction of iron from an ore sample.[66] While the heating was performed in a test tube placed inside a plastic bottle with a vent to the outside of the oven, safety is still an issue, especially as the procedure requires heating the ore in concentrated acid with no temperature measurement device in place. Using this method, students were able to probe the efficiency of extraction using different acids as well as study the effect of varying processing time and microwave power. Also, Brazil nuts have been digested in nitric acid in a domestic microwave oven for selenium analysis by means of fluorescence measurements of cation concentrations after treatment with complexing agents and a modified standard addition method.[67]

Using a scientific microwave unit equipped with a modified distillation apparatus, essential oils have been extracted from plant material, and their relative concentrations determined. Fresh orange peel was used as the sample, and the students gained exposure to extraction, chromatographic, and spectroscopic techniques.[68]

To show an application of the method of linear least squares to data collected in a laboratory, a procedure has been developed in which a beaker containing water was heated in a domestic microwave oven, and the water temperature measured as a function of time and power.[69] Students obtained a regression line for each power level screened and determined the intercept and slope of the line. They then compared and contrasted the values obtained for initial temperature and power input using the linear regression with those set experimentally, outlining sources of error.

6.3.5.2   Inorganic and Organometallic Chemistry

The number of reports describing the use of microwave heating for the preparation of inorganic and organometallic compounds is small compared with those related to the synthesis of organic compounds. Similarly, there are far fewer reports of microwave heating as a tool in the undergraduate inorganic chemistry laboratory as compared to organic chemistry. However, the advantages of speed and ease of operation are being seen. The synthesis of the metalated macrocycle, 6,8,15,17-tetraphenyl-dibenzo[*b,i*][1,4,8,11] tetraazacyclotetradecinato nickel (II) is an example (Scheme 6.24).[70] The published synthesis of this compound requires the reagents to be heated at 250–260 °C for 90–120 min under a stream of argon.[71] By performing the reaction in a domestic microwave oven, comparable (albeit low) yields were obtained after 20 min. On the 10–20 mmol scale, a typical experiment yielded 50–100 mg of the macrocycle, this being sufficient for students to perform IR, NMR, UV-Vis, or electrochemical studies. Microwave heating has also been used for the dehydration of several transition-metal complexes, the objective being to illustrate the principles behind gravimetric analysis and highlight the concept of coordinated water.[72]

In the area of organometallic chemistry, a modified domestic microwave oven has been used for the preparation of molybdenum carbonyl complexes.[73] Issues associated with performing transition-metal carbonyl chemistry in the undergraduate laboratory are that binary metal carbonyl compounds either are very reactive and require the use of strictly anaerobic conditions or else are inert, and extended times at elevated temperatures are required to perform ligand substitution reactions. Using microwave heating, the disubstituted molybdenum complex *cis*-Mo(CO)$_4$(piperidine)$_2$ could be prepared in 40 min by combining Mo(CO)$_6$ and an excess of piperidine in an open flask and using a diglyme–THF mix as solvent (Scheme 6.25). This compares favorably to the 4 h required for the conventional method in *n*-heptane.[74] To control the heating, a glass U-tube connected to a cold water supply was used as a moderator. The tube entered and exited the microwave oven by means of holes drilled in the side. The reaction vessel sat in the middle of the microwave cavity and

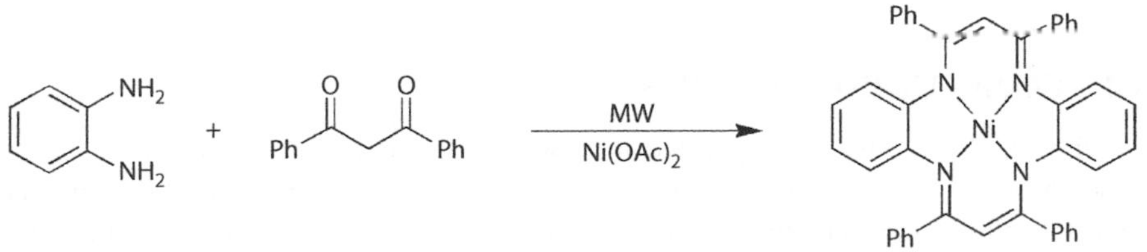

SCHEME 6.24

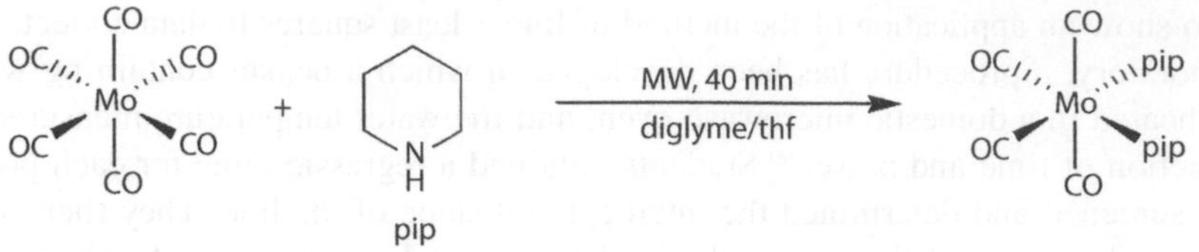

SCHEME 6.25

was connected to a water condenser outside by means of a glass tube passing through a hole drilled in the top of the microwave oven. Once the contents of the reaction vessel started to boil, the flow of water into the U-tube was adjusted so as to keep the reaction mixture at reflux. The $Mo(CO)_4(piperidine)_2$ complex was then used in a ligand-substitution reaction with triphenylphosphine, the kinetically favored *cis*-$Mo(CO)_4(PPh_3)_2$ being formed initially and, upon heating, being converted into the thermodynamically more stable *trans* isomer. Using infrared spectroscopy, these products can all be characterized and students introduced to the use of group theory for structural prediction, the *cis* isomer (C_{2v} symmetry) showing four bands in the CO stretching region of the IR spectrum ($2a_1 + b_1 + b_2$) compared to a single band (*eu*) in the case of the *trans* isomer (D_{4h} symmetry). The dimeric iridium complex $[Ir(COD)Cl]_2$ has also been prepared using a similarly modified domestic microwave oven.[75] Using $(NH_4)_2IrCl_6$ and 1,5-cyclooctadiene as starting materials and a distilled water–2-propanol mix as solvent, 40 min at reflux was sufficient to give a good yield of the desired product. Preparing the dimer in-house offered a significant cost savings over directly purchasing it while at the same time allowing its rich chemistry to be explored in an undergraduate laboratory setting.

6.3.5.3 Materials Chemistry

The application of microwave heating to materials chemistry in the undergraduate laboratory is exemplified by the preparation and characterization of silver nanoparticles.[76] They were formed from an aqueous solution containing silver nitrate and D-glucose, which was heated in a domestic microwave oven for 8 s on high power. The temperature at the end of the heating step was around 80 °C. Neither a capping reagent nor an oxygen-free environment was necessary. Having prepared the nanoparticles, the rest of the laboratory period could be spent analyzing them, looking at the size and shape of the nanoparticles. In addition, the effects of reactant concentration or heating time on the physical properties of the resultant nanoparticles could be probed. Overall, the experience enhanced students' understanding of nanoscience as well as analytical techniques.

6.4 CONCLUDING REMARKS

Merrimack College has incorporated microwave chemistry into the teaching laboratory over the last 5 years. Currently, 80% of the organic laboratory experiments undertaken by students involve microwave heating. The technology has provided the opportunity to incorporate experiments that would otherwise have been beyond the scope of a single laboratory period. To show the differences between conventional and microwave heating, students perform both. The class runs a microscale reaction such as the radical bromination of anthracene, which requires 1 h refluxing in a sand bath. This allows them the opportunity to set up a reaction with a reflux condenser, and there is more than enough time during the waiting period to prepare for the experiment workup and take any spectra needed of the starting materials. The following week they run a reaction using scientific microwave apparatus. They quickly realize the ease with which equipment can be set up. Instead of reflux condenser, water hoses, clamps, and sand baths, they have one simple reaction vessel with a cap. Once the class has placed their reactions in the microwave oven, there is enough time for them to set up the required equipment for the workup procedure, in most cases, in less than 10 min. Students appreciate the advantages that a short reaction time brings them, namely ample opportunity for the isolation, purification, and identification of the product. They express a strong interest in using this new technology in further laboratory classes as opposed to conventional heating.

When a reaction fails to produce the desired product, students can become very discouraged. Because when using microwave heating the reaction can be complete in the matter of a few minutes, if a problem occurs during the product isolation steps, students often request to repeat the procedure during the same laboratory period. At other times, students have asked if they could try to optimize a reaction by rerunning it with changes sanctioned by the instructor. As a result, more time is spent in problem solving as opposed to waiting for reactions to reach completion. In addition, students appreciate the opportunity to improve their work by having time to repeat a procedure and not having to write up a failed experiment.

Overall, faculty have seen a 25% increase in productivity in the laboratories. The incorporation of microwave heating has expanded the range of chemical concepts that can be introduced through hands-on learning. There is less crowding at the end of the laboratory time around the NMR and IR spectrometers and melting point apparatus, since the time for collecting data has nearly doubled.

Students are also very environmentally aware. Adding the microwave component has provided the opportunity to introduce many of the concepts of green chemistry. The role of solvents in a reaction can be exemplified by having students compare changes in reactions that under traditional microscale procedure used toluene or DMF as solvents but are now run in ethanol and water. The issues of solubility and boiling point properties are compared to environmental impact and waste management. For example, water need not necessarily be the optimal solvent for a reaction.[77] Not only does the reaction step itself have to be considered but also the environmental impact of product workup and waste disposal in order to make an informed decision on the benefits of a particular solvent.

Chemistry majors who continue in the higher-level courses are quick to utilize the microwave apparatus when designing their own research projects in the advanced organic/inorganic laboratory and senior research. In some cases, after performing a reaction by a published method, students will convert it to a microwave procedure. By doing this, they have often been able in a single laboratory period to optimize a reaction to give a much higher yield of product than noted in the published work. Students are very accepting of the new technology since there are no preconceived notions on how reactions are typically run.

In conclusion, with the increased productivity, ability to run a wider range of experiments, and the positive, enthusiastic student participation that results, microwave heating is clearly proving its place as a valuable tool in the undergraduate teaching laboratory.

REFERENCES

1. Leicester, H. M. *The Historical Background of Chemistry*; Wiley: New York, 1956.
2. Blick, D. J. *J. Chem Educ.* 1955, *32*, 264–266.
3. Domin, D. S. *J. Chem Educ.* 1999, *76*, 543–547.
4. For books see: (a) Loupy, A., Ed. *Microwaves in Organic Synthesis,* 2nd ed.; Wiley-VCH: Weinheim, Germany, 2006. (b) Tierney, J. P.; Lidstrom, P., Eds. *Microwave Assisted Organic Synthesis;* Blackwell Scientific: Boca Raton, FL, 2005. (c) Kappe, C. O.; Stadler, A. *Microwaves in Organic and Medicinal Chemistry;* Wiley-VCH: Weinheim, Germany, 2005. (d) Hayes, B. L. *Microwave Synthesis: Chemistry at the Speed of Light;* CEM Publishing: Matthews, NC, 2002.
5. For reviews see: (a) Lidström, P.; Tierney, J.; Wathey, B.; Westman, J. *Tetrahedron,* 2001, *57*, 9225–9283. (b) Kappe, C. O. *Angew. Chem., Int. Ed.* 2004, *43*, 6250–6284. (c) Caddick, S.; Fitzmaurice, R. *Tetrahedron,* 2009, *65*, 3325–3355. (d) Kappe, C. O.; Dallinger, D. *Mol. Divers.* 2009, *13*, 71–193.
6. Gedye, R.; Smith, F.; Westaway, K. *Educ. Chem.* 1988, *25*, 55–56.
7. Haack, J.; Hutchinson, J.; Kirchhoff, M.; Levy, I. *J. Chem. Educ.* 2005, *82*, 974–976.
8. Anastas, P. T.; Warner, J. C. *Green Chemistry: Theory and Practice*; Oxford University Press: New York, 1998.
9. Leadbeater, N. E.; McGowan, C. B. *Clean, Fast Organic Chemistry: Microwave-Assisted Laboratory Experiments*, CEM Publishing, Matthews, NC, 2006.
10. Katritzky, A.; Cai, C.; Collins, M. D.; Scriven, E. F. V.; Singh, S. K.; Barnhardt, E. K. *J. Chem Educ.* 2006, *83*, 634–636.
11. Martin, E.; Kellen-Yuen, C. *J. Chem. Educ.* 2007, *84*, 2004–2006.
12. Bogdal, D. *J. Chem. Res.(S)*, 1998, 468–469.
13. (a) Warner, J. C.; Anastas, P. T.; Anselme, J.-P. *J. Chem. Educ.* 1985, *62*, 346–346. (b) Broos, R.; Tavernier, D.; Anteunis, M. *J. Chem. Educ.* 1978, *55*, 813–813.
14. Leung, S. H.; Angel, S. A. *J. Chem. Educ.* 2004, *81*, 1492–1493.
15. Kappe, C. O.; Murphree, S. *J. Chem. Educ.* 2009, *86*, 227–229.
16. Cook, A. G. *J. Chem. Educ.* 2007, *84*, 1477–1479.
17. Friebe, T. L. *Chem. Educator* 2003, *8*, 33–36.
18. Musiol, R.; Tyman–Szram, B.; Polanski, J. *J. Chem. Educ.* 2007, *84*, 1477–1479.
19. Dintzner, M.; Wucka, P.; Lyons, T. W. *J. Chem. Educ.* 2006, *83* 270–272.
20. Trehan, I. R.; Brar, J. S.; Arora, A. K.; Kad, G. L. *J. Chem. Educ.* 1997, *74*, 324–324.
21. Warner, M. G.; Succaw, G. L.; Hutchison, J. E. *Green Chem.* 2001, *3*, 267–270.
22. Petit, A.; Loupy, A.; Maillard, P.; Momenteau, M. *Synth. Commun.* 1992, *22*, 1137–1140.
23. Durantini, E. N.; Otero, L. *Chem. Educator* 1999, *4*, 144–146.
24. White, L. L.; Kittredge, K. W. *J. Chem. Educ.* 2005, *82*, 1055–1056.
25. Birdwhistell, K.; Nguyen. A.; Ramos, E.; Kobelja, R. *J. Chem. Educ.* 2008, *85*, 261–262.
26. Montes, I.; Sanabria, D.; García, M.; Castro, J.; Fajardo. J. *J. Chem. Educ.* 2006, *83*, 628–631.
27. Acetic acid is used as a catalyst for the reaction in an undergraduate laboratory setting at University College Dublin, Ireland: http://www.ucd.ie/chem/Lab%20manual%201.pdf.

28. For reviews discussing non-thermal microwave effects, see: (a) Perreux, L.; Loupy, A. In *Microwaves in Organic Synthesis* 2nd Ed.; Loupy, A., Ed.; Wiley-VCH: Weinheim, Germany, 2006 Ch. 4, pp 134–218. (b) De La Hoz, A.; Diaz-Ortiz, A.; Moreno, A. *Chem. Soc. Rev.* 2005, *34*, 164–178.

29. See, for example: (a) Schmink, J. R.; Leadbeater, N. E. *Org. Biomol. Chem.* 2009, 3842–3846. (b) Obermayer, D.; Gutmann, B.; Kappe, C. O. *Angew. Chem. Int. Ed.* 2009, *48*, 8321–8324.

30. For examples, see: (a) Barry, E.; Borer, L. L. *J. Chem. Educ.* 2000, *77*, 354–355. (b) Olmsted, J. *J. Chem. Educ.* 1998, *75*, 1261–1263. (c) Brown, D. B.; Friedman, L. B. *J. Chem. Educ.* 1973, *50*, 214–215.

31. This was brought to the author's attention by the late Dr. Ahamindra Jain, Director of Undergraduate Laboratories and Lecturer at Harvard University.

32. Leadbeater, N. E.; McGowan, C. B. In *Microscale Organic Laboratory: With Multistep and Multiscale Syntheses,* 5th ed.; Mayo, D. W.; Pike, R. M.; Forbes, D. C. Eds.; Wiley: New York, 2010.

33. Mayo, D. W.; Pike, R. M.; Trumper, P. K. *Microscale Organic Laboratory: With Multistep and Multiscale Syntheses,* 4th ed.; Wiley: New York, 2000.

34. Mayo, D. W.; Pike, R. M.; Forbes, D. C. *Microscale Organic Laboratory: With Multistep and Multiscale Syntheses,* 5th ed.; Wiley: New York, 2010.

35. Hanna, J. M.; Hartel, A. M. *J. Chem. Educ.* 2009, *86*, 475–476.

36. Huang-Minlon *J. Am. Chem. Soc.* 1946, *68*, 2487–2488.

37. Durham, L. J.; McLeod, D. J.; Cason, J. *Org. Synth.* 1958, *38*, 34–37.

38. Parquet, E.; Lin, Q. *J. Chem. Educ.* 1997, *74*, 1225–1225.

39. Soriano, D. S. *J. Chem. Educ.* 1993, *70*, 332–332.

40. Banik, B. K.; Barakat, K. J.; Wagle, D. R.; Manhas, M. S.; Bose, A. K. *J. Org. Chem.* 1999, *64*, 5746–5753.

41. McKenzie L. C.; Huffman, L. M.; Hutchison, J. E.; Rogers, C. E.; Goodwin, T. E.; Spessard, G. O. *J. Chem. Educ.* 2009, *86*, 488–493.

42. For books, see: (a) Crabtree, R. H. *The Organometallic Chemistry of the Transition Metals*; Wiley: New York, 2009. (b) Caprio, V. Williams, J. M. *Catalysis in Asymmetric Synthesis*; J. Wiley-Blackwell: Chichester, 2008. (c) Beller, M.; Bolm, C., Eds. *Transition Metals for Organic Synthesis: Building Blocks and Fine Chemicals;* Wiley-VCH: Weinheim, Germany, 2004.

43. For examples, see: (a) Moorhead, E. J.; Wenzel, A. G. *J. Chem. Educ.* 2009, *86*, 973–975. (b) Feng, Z. V.; Lyon, J. L.; Croley, J. S.; Crooks, R. M.; Vanden Bout, D. A.; Stevenson, K. J. *J. Chem. Educ.* 2009, *86*, 368–372. (c) Kirk, S. R.; Silverstein, T. P.; Holman, K. L. M. *J. Chem. Educ.* 2008, *85*, 676–677. (d) Pappenfus, T. M.; Hermanson, D. L.; Ekerholm, D. P.; Lilliquist, S. L.; Mekoli, M. L. *J. Chem. Educ.* 2007, *84*, 1998–2000. (e) Mucientes, A. E.; de la Pena, M. A. *J. Chem. Educ.* 2006, *83*, 1643–1644. (f) Sharpless, W. D.; Wu, P.; Hansen, T. V.; Lindberg, J. G. *J. Chem. Educ.* 2005, *82*, 1833–1836. (g) Howard, D. L.; Tinoco, A. D.; Brudvig, G. W.; Vrettos, J. S.; Allen, B. C. *J. Chem. Educ.* 2005, *82*, 791–794. (h) Seen, A. J. *J. Chem. Educ.* 2004, *81*, 383–384.

44. For reviews, see: (a) Brennfuhrer, A.; Neumann, H.; Beller, M. *Angew. Chem. Int. Ed.* 2009, *48*, 4114–4133. (b) Alonso, F.; Beletskaya, I. P.; Yus, M. *Tetrahedron,* 2008, *64*, 3047–3101. (c) Hong, B. C.; Nimje, R. Y. *Curr. Org. Chem.* 2006, *10*, 2191–2225. (d) Bellina, F.; Carpita, A.; Rossi, R. *Synthesis*, 2004, 2419.

45. Callam, C. S.; Lowary T. L. *J. Chem. Educ.* 2001, *78*, 947–948.

46. Hoogenboom, R.; Meier, M. A. R.; Schubert, U. S. *J. Chem. Educ.* 2005, *82*, 1693–1696.

47. Aktoudianakis, E.; Chan, E.; Edward, A. R.; Jarosz, I.; Lee, V.; Mui, L.; Thatipamala, S. S.; Dicks, A. P. *J. Chem. Educ.* 2008, *85*, 555–557.

48. For a review: Leadbeater, N. E. *Chem. Commun.* 2005, 2881–2902.

49. (a) Arvela, R. K.; Leadbeater, N. E.; Sangi, M. S.; Williams, V. A.; Granados, P.; Singer, R. D. *J. Org. Chem.* 2005, *70*, 161–168. (b) Leadbeater, N. E.; Marco, M. *J. Org. Chem.* 2003, *68*, 5660–5667. (c) Leadbeater, N. E.; Marco, M. *Org. Lett.* 2002, *4*, 2973–2976.

50. Leadbeater, N. E.; Williams, V. A.; Barnard, T. M.; Collins, M. J. *Org. Process Res. Dev.* 2006, *10*, 833–837.

51. The experiment formed part of a six-day outreach program at the University of Connecticut.

52. McGowan, C. B.; Leadbeater, N. E. In *Experiments in Green and Sustainable Chemistry*, Roesky, H. W.; Kennepohl, D. Eds. Wiley-VCH: Weinheim, Germany, 2009, Ch. 17, pp 97–107.

53. Procedure based on: Arvela, R. K.; Leadbeater, N. E. *J. Org. Chem.* 2005, *70*, 1786–1790.

54. Lauron, H.; Mallet, J.-M.; Mestdagh, H.; Ville, G. *J. Chem. Educ.* 1988, *65*, 632–632.

55. Martin, W. B.; Kateley, L. J. *J. Chem. Educ.* 2000, *77*, 757–759.

56. Cheung, L. L. W.; Aktoudianakis, E.; Chan, E.; Edward, A. R.; Jarosz, I.; Lee, V.; Mui, L.; Thatipamala, S. S.; Dicks, A. P. *Chem. Educator* 2007, *12*, 77–79.

57. Herrmann, W. A.; Böhm, V. P. W.; Reisinger, C.-P. *J. Chem. Educ.* 2000, *77*, 92–95.

58. Crouch, R. D.; Howard, J. L.; Zile, J. L.; Baker, K. H. *J. Chem. Educ.* 2006, *83*, 1658–1660.

59. For examples, see: (a) Betush, M. P.; Murphree, S. S. *J. Chem. Educ.* 2009, *86*, 91–93. (b) Smith, T. E.; Richardson, D. P.; Truran, G. A.; Belecki, K.; Onishi, M. *J. Chem. Educ.* 2008, *85*, 695–697. (c) Ji, C.; Peters, D. G. *J. Chem. Educ.* 2006, *83*, 290–291. (d) Stocksdale, M. G.; Fletcher, S. E. S.; Henry, I.; Ogren P. J.; Berg, M. A. G.; Pointer, R. D.; Benson, B. W. *J. Chem. Educ.* 2004, *81*, 388–390. (e) Graham, K. J.; Schaller, C. P.; Johnson, B. J.; Klassen, J. B. *Chem. Educator* 2002, *7*, 376–378. (f) Williams, B. D.; Williams, B.; Rodino, L. *J. Chem. Educ.* 2000, *77*, 357–359. (g) Howell, J.; deLannoy, P. *J. Chem. Educ.* 1997, *74*, 990–991.

60. Miller, T. M.; Leadbeater, N. E. *Chem. Educator*, 2009, *14*, 98–104.

61. Kingston, H. M.; Haswell, S. J., Eds. *Microwave-Enhanced Chemistry: Fundamentals, Sample Preparation, and Applications,* American Chemical Society: Washington, DC, 1997.

62. For reviews see: (a) MacKenzie, K.; Dunens, O.; Harris, A. T. *Sep. Purif. Technol.* 2009, *66*, 209–222. (b) Halko, R.; Hutta, M. *Chem. Listy* 2007, *101*, 649–656. (b) Eskilsson, C. S.; Bjorklund, E. *J. Chromatogr. A* 2000, *902*, 227–250.

63. Fahey, A. M.; Tyson, J. F. *Chem. Educator* 2006, *11*, 445–450.

64. For the first published report see: Smith, F. E.; Cousins, B. G.; Maillet, J.-Y. *Educ. Chem.* 1987, *24*, 13.

65. Freeman, R. G.; McCurdy, D. L. *J. Chem. Educ.* 1998, *75*, 1033–1034.

66. Goltz, D. M.; Hall, T.; Grant, A.; Segstro, E. *J. Chem. Educ.* 2000, *77*, 1486–1488.

67. Sheffield, M.-C.; Nahir, T. M. *J. Chem. Educ.* 2002, *79*, 1345–1347.

68. Ferhat, M. A.; Meklati, B. Y.; Visinoni, F.; Vian, M. A.; Chemat, F. *Chim. Oggi.* 2008, *26*, 48–50.

69. Feitosa, C. M.; Neto, J. M. M.; Moita, G. C. *Chem. Educator* 1999, *4*, 16–18.

70. Hayes, J. W.; Taylor, C. J.; Hotz, R. P. *J. Chem. Educ.* 1996, *73*, 991–992.

71. Hotz, R. P.; Purrington, S. T.; Hochgesang, P. J.; Singh, P.; Bereman, R. D. *Synth. React. Inorg. Met. Org. Chem.* 1991, *21*, 253–262.

72. Yoshikawa, N.; Takashima, H. *Chem. Educator* 2002, *2*, 354–355.

73. Ardon, M.; Hayes, P. D.; Hogarth, G. *J. Chem. Educ.* 2002, *79*, 1249–1251.

74. Darensberg, D. J.; Kump, R. L. *Inorg. Chem.* 1978, *17*, 2680–2682.

75. Cooke, J. *Chem. Educator* 2008, *13*, 353–357.

76. Dong, Z.; Richardson, D.; Pelham, C.; Islam, M. R. *Chem. Educator*, 2008, 13, 240–243.

77. Blackmond, D. G.; Armstrong, A.; Coombe, V.; Wells, A. *Angew. Chem. Int. Ed.* 2007, *46*, 3798–3800.

7 Microwave Heating as a Tool for Inorganic and Organometallic Synthesis

Gregory L. Powell

CONTENTS

7.1   INTRODUCTION

One of the two publications in 1986 reporting the first organic transformation performed using microwave heating could also be considered as the first report of a microwave-promoted inorganic reaction. The permanganate oxidation of toluene to benzoic acid was performed, and the redox reaction that occurred resulted in the conversion of $KMnO_4$ into MnO_2.[1] Of course, this particular inorganic reaction is so common that it easily goes unnoticed despite the fact that it addresses the concerns some scientists had about the safety of heating metal compounds in a microwave appliance. This initial report and ensuing studies by other researchers have clearly shown that, under the proper conditions, a variety of metal-containing chemicals and even powdered metals themselves may be used as reagents and catalysts in microwave-promoted reactions. The number of journal articles describing the use of microwave heating for the preparation of inorganic and organometallic compounds is, however, a small fraction of the number of articles concerned with the microwave synthesis of organic compounds. As a result, ample opportunity exists for further research in this area of chemistry.

The goal of this chapter is to present an overview of reports in the scientific literature that involve the use of microwave heating in the preparation of inorganic and organometallic compounds. For practical purposes, no attempt has been made to cover compounds of all elements. The scope has been limited to coordination compounds and organometallic complexes containing transition metals. The focus is on molecular compounds and not on materials that could be classified as nanoparticles, polymers, supported catalysts, metal-organic frameworks, or solid-state materials.

Pioneering work in microwave-promoted organometallic synthesis was carried out just a few years after the first articles describing microwave-assisted organic reactions. In 1989, the use of microwave heating for the synthesis of $[M(diene)Cl]_2$ complexes from $RhCl_3$ and $IrCl_3$, as well as $[Cp_2Rh]PF_6$ from $RhCl_3$, using methanol or ethanol as solvent was reported.[2] That same year, microwave heating was used to prepare the compounds $PhBiCl_2$ and $RPhHgCl$ in ethanol.[3] Both these procedures used sealed Teflon reaction vessels placed inside domestic microwave ovens. In 1990, further examples were reported using the same experimental setup. Coordination compounds of the substitutionally inert transition metal ions Cr(III), Ru(II), Ir(I), Ir(III), Pt(II), and Au(III) were prepared, reaction times being dramatically reduced compared to traditional methods of synthesis.[4] In at least one case, a ligand substitution reaction that could not be performed by traditional methods was observed, the complex $[Ru(9aneS3)_2](PF_6)_2$ being prepared directly from $RuCl_3 \cdot xH_2O$. The conventional synthesis of this compound requires the initial conversion of the ruthenium(III) chloride to a labile complex such as $Ru(SO_3CF_3)_3$. Moreover, the yield obtained using microwave heating was more than twice that of the conventional method. These early experiments suggested that microwave heating might be effective in the syntheses of a variety of other metal-containing compounds using simpler procedures and/or fewer steps than previously reported. Subsequent studies have shown that this is often the case.

7.1.1 GENERAL CONSIDERATIONS

The majority of the publications describing the microwave-promoted synthesis of transition-metal compounds mention using domestic microwave ovens rather than scientific microwave units. Dedicated laboratory reactors provide many features for safe and reliable work, such as the ability to monitor temperature and pressure accurately, to adjust microwave power in response to temperature or pressure changes, and to isolate the reaction vessel in an explosion-proof enclosure. Domestic microwave ovens are not designed to handle the corrosive atmospheres and potentially high pressures associated with many chemical reactions. Dedicated reactors also provide built-in stirring and a more homogeneous microwave field. This makes for more reproducible procedures. Results of experiments conducted with domestic ovens are still presented in this chapter in order to provide a more comprehensive view of the field. Some of these experiments make use of special containers such as digestion vessels that have been designed for safe heating in domestic appliances, but most could presumably be carried out in dedicated reactors with few modifications.

Since microwave heating allows for easy access to reaction temperatures that are often higher than it is practical to achieve using conventional heating, it comes as no surprise that reaction times are often dramatically shorter. When sealed reaction vessels are used, solvents may reach temperatures 30–50% higher than their normal boiling points. For almost every procedure presented in this chapter, reactions have been reported to be 5 to 2500 times faster than those carried out by conventional thermal methods. Rate enhancement is assumed and is therefore seldom mentioned.

In a handful of cases, microwave heating and conventional heating have been directly compared.[5–8] For example, in the preparation of bis(aryl)zinc(II) complexes, it was found that when the zinc starting material and the arenes C_6H_4R (R = CO_2Et or $CONEt_2$), were combined with THF and heated in an oil bath at 120 °C for 5 h, yields of the arylzinc complexes were in the 10–20% range.[5] After microwave heating at 120 °C for 5 h, the desired products were obtained in greater than 90% yield. When attempting to prepare $Cp_2Zr(-C=CR)Cl$ complexes, hydrozirconation of alkynes at 100 °C in toluene using microwave heating resulted in good yields of alkenylchlorozirconocenes in 5–15 min, whereas conventional heating at 100 °C in toluene resulted in significant (25–30%) side-product formation.[6] Another example is the preparation of a polynuclear La_3Ni_6 complex ion in water.[7] Using microwave heating, the complex ion was obtained in 40% yield after 30 min at 100 °C, while under conventional hydrothermal conditions the yield was only 28.5% after 3 days at 100 °C.

The origins of the differences in outcome between microwave and conventional heating have been the subject of debate.[9] Some authors attribute them to nonthermal microwave effects being at play. It is more likely that the differences can be explained simply as the result of a rate acceleration in a microwave field that cannot be achieved or duplicated by conventional heating but is still temperature related in origin. An example is the ability of microwave energy to be preferentially absorbed by highly polar reagents.

Yield is often a more important consideration for inorganic and organometallic reactions than it is for organic reactions due to the significant cost associated with many metal-containing starting materials. Sometimes, the use of microwave heating results in higher yields than conventional approaches but this is certainly not always the case. It should be pointed out again that in almost every instance reactions performed using microwave heating are deliberately performed at elevated temperatures, whereas conventional procedures may not be. While greater thermal energy leads to faster reaction rates, it also increases the likelihood that side reactions or degradation of products will occur. In some cases, therefore, conventional heating may be preferable. In order to assist the reader in making such decisions, a yield comparison is regularly included in this chapter in the form of a yield improvement factor (YIF) that is calculated by dividing the yield obtained with microwave heating by the yield obtained with conventional heating. YIF values greater than one indicate that the yield is higher for the microwave method. For example, the microwave yield is 20% greater for a reaction with YIF = 1.2.

In contrast to conventional syntheses, many procedures using microwave heating have not been optimized. Indeed, the difficulty of monitoring a reaction in a scientific microwave reactor or a domestic microwave oven makes optimization especially challenging. In-situ Raman spectroscopy has been used as a tool for monitoring microwave-promoted ligand-substitution reactions of $Mo(CO)_6$.[10] The v_{CO} bands in the carbonyl complexes are easily observed and can be monitored at intervals of 5–6 s. Ligand substitution reactions of $Ru_3(CO)_{12}$ have been monitored in real time by means of a digital camera interfaced with a microwave reactor.[11] Such methods may prove to be especially helpful in deciding when to stop a reaction in order to prevent unwanted side reactions or product decomposition.

Solvent choice can sometimes affect the efficiency and outcome of a reaction performed using microwave heating. Polar solvents are normally chosen because they are directly heated efficiently via interaction with microwave energy. Some microwave syntheses have been carried out with excellent results under solvent-free conditions. Other synthetic procedures require a nonpolar reaction medium. In such cases, graphite is often used as a microwave absorber that in turn heats the reaction mixture. The metal complex itself, however, is often capable of absorbing microwave irradiation. An example from the literature is the observation of rapid heating of a toluene solution containing zirconocene hydrochloride under microwave irradiation.[6]

Interestingly, microwave heating can be used for physical as well as chemical transformations. For example, the dehydration of tetrakis(acetylacetonato)zirconium octahydrate to yield anhydrous $Zr(acac)_4$ was accomplished by heating a powdered sample for about 4 min.[12] In addition, some groups have used a microwave-promoted physical change such as the melting of a solid reagent as an initial step in preparative procedures that otherwise involve conventional heating.

7.2 MICROWAVE-PROMOTED COORDINATION CHEMISTRY

7.2.1 LIGAND SUBSTITUTION

The majority of reports concerning the microwave-promoted synthesis of metal complexes describe ligand substitution reactions. A variety of ligands containing different numbers of donor atoms have been used, but ligands with nitrogen donor atoms are by far the most common. A range of polyatomic ligands are referred to throughout this chapter and many of these are illustrated in Figure 7.1. A list of abbreviations for ligands appears in Table 7.1, along with a list of most of the coordination compounds that have been prepared using microwave heating. This table also provides a comparison of product yields obtained using microwave heating with yields from conventional methods of heating.

Ruthenium is the most frequently used transition metal in microwave-promoted synthesis. The heavier member of Group 9, iridium, is a distant second. This could be due, at least in part, to the fact that ruthenium complexes such as [Ru(bpy)3]$^{2+}$ are of great interest because of potential applications in catalysis, photovoltaics, biodetection, and photosensitization.[59] While polar compounds with relatively high boiling points such as ethylene glycol, water, or dimethylformamide are the most commonly used solvents, simple alcohols with lower boiling points have also found application. In some instances, researchers simply use the solvent that was used in a previous conventional procedure. At other times, a particular reaction medium is chosen to take advantage of rapid microwave heating. The volume of solvent used

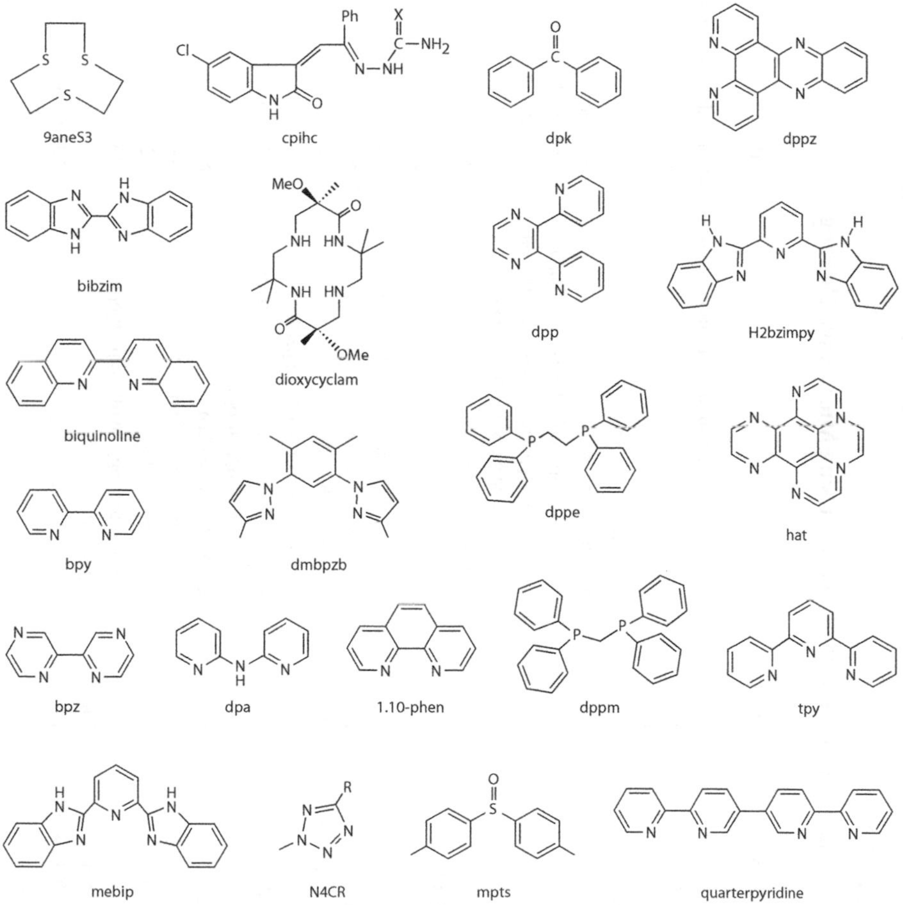

FIGURE 7.1 Selected ligands.

TABLE 7.1
Coordination Compounds Prepared Using Microwave Heating

Complex	Metal	Ligand[a]	YIf[b]	Solvent Used	Microwave Unit Used[c]	Reference
		Compounds with monodentate and/or bidentate ligands				
MCl$_2$L$_3$	Ru	PPh$_3$	1.1	MeOH	domestic	13
MOCl$_3$L$_2$	Re	PPh$_3$	0.99	EtOH/H$_2$O	domestic	13
M(NH$_3$)$_4$L$_2$	Co	NO$_2$N4C	1.3	H$_2$O	monomode*	14,15
[ML$_3$]$^{2+}$	Ru	bpy, dbbpy, phen	0.92 (bpy)	MeOH	domestic	16,17
[ML$_3$]$^{2+}$	Ru (a few Os)	bpy, bibzim, bpz, dpk, dpkox, dpa, dmphen, dpphen, dmdpphen, en, mphen, pdptz, phen, pphen, Cl-phen, NO$_2$-phen, dmbpy	0.95 (bpy), 3.0 (dpa)	EG^d	domestic	18,19
[ML$_3$]$^{2+}$	Ru	pydppn, pydppx	NA	EG	domestic	20
[ML$_3$]$^{2+}$	Ru	dpphen	12	DMF^e	monomode	21
[ML$_3$]$^{2+}$	Ru	pbpy, ppbpy	NA^f	EG	monomode	22
[ML$_3$]$^{3+}$	Ir	bpy, phen, bqn, dmbpy, dpphen, dpbpy	NA	EG	domestic	23
[ML$_3$]$^{3+}$	Eu	bqn, bpy, bpyca, dmbpy, pyre	NA	EG	monomode*	24
[ML$_2$L']$^{2+}$	Ru	L = bpy, dmbpy, dbbpy L' = phen, bpm, tmbiH2, dppzR2	NA	DMF/H$_2$O	monomode*	25,26
[ML$_2$L']$^{2+}$	Ru	L = bpy, phen, bpz, dpp, tmbpy, dtfbpy, ttfbpy L' = hat, bpz, dpp, ttfbpy	0.90–1.1	EG	domestic	27–29
[ML$_2$L'$_2$]$^{2+}$	Ru	L = MeCN, L' = phen	NA	EG	monomode*	30
[M(bpy)$_2$L]$^+$	Ru	acac, dbm, F-quol, dpm	NA	EG	domestic	19
ML$_2$Cl$_2$	Ru	bpy, dmbpy, dbbpy, dcmb	1.2	DMF	monomode*	25,31
ML$_2$Cl$_2$	Ru	pyrtbpy, pyrt$_2$bpy, pyrdbpy, pyrd$_2$bpy	NA	DMF	monomode	32
MCl(bpy)$_2$L	Ru	mpts, epts	1.0	EG	monomode*	30
[ML$_2$X$_2$]$^+$, X = Cl, Br, I	Rh	bpy	0.92–0.96	EG/H$_2$O	domestic	33

Compound	Metal	Ligands	Ratio	Solvent	Mode	Ref.
$[ML_2Cl_2]^+$	Ir	bpy, phen, bqn, dcmb, tmphen, dpphen, dmdpphen, dpbpy, pdmphen	NA	EG	domestic	23,34,35
$M(PR_3)_2L_2$	Pt	N4CR	0.90–1.1	MeCN or DMF	monomode	36–40
MCl_2L_2	Pd, Pt	NCNOC	1.0	RCN	monomode	41–43
ML_2	Cu	phhy	NA	melamine	domestic	44
ML_2	Mn	cpihc, cpihct	1.0, 1.1	MeOH	domestic	45
$M(L)Cl(H_2O)$	Mn	cpihc, cpihct	1.1, 1.1	MeOH	domestic	45
$MOCl_2L$	V	cpihc, cpihct	1.0, 1.1	MeOH	domestic	45
MOL_2Cl	V	cpihc, cpihct	1.2, 1.1	MeOH	domestic	45
MO_2L_2	Mo	cpihc, cpihct	1.1, 1.1	MeOH	domestic	45
$[\{ML_2\}_2(\mu\text{-}L')]^{4+}$	Ru	L = bpy, phen, dmbpy, dmphen; L′ = hat, bpm	NA	EG	domestic	27, 46
$[M_2(\mu\text{-}Cl)_2L_2]^+$	Os	PEt_2Ph	0.67	EtOH	monomode	47
Compounds with tridentate ligands						
$[MLCl]^+$	Pt, Au	tpy	NA	H_2O	domestic	48
MCl_3L	Ir	9aneS3	0.91	MeOH	domestic	48
$[ML_2]^{2+}$	Ru	9aneS3	2.2	MeOH	domestic	17
$[ML_2]^{2+}$	Ru	ppty, pptpy, dqpR	NA	EG	monomode	22, 49–51
$[ML_2]^{2+}$	Ru	tpy, pydppz, phenq	1.1 (tpy)	EG	domestic	18, 52, 53
$[ML_2]^{2+}$	Co	bzimpy	1.1	EG	domestic	54
$[ML_3]^{3+}$	Eu	tptz, tpy	NA	EG	monomode*	24
ML(ppy)X, X = Cl, Br, I, CN	Ir	Mebib, Phbib, Mebip, bpzb, dmbpzb, dfbpzb	NA	glycerol	multimode	55, 56
Compounds with tetradentate ligands						
ML	Mn	cpe, cpp, cpp-2, cppp	1.3–1.8	H_2O	domestic	57
$[M_2L_3]^{4+}$	Ru	di-mbpy	NA	EG	domestic	58

(continued)

TABLE 7.1 (CONTINUED)
Coordination Compounds Prepared Using Microwave Heating

a Ligand abbreviations: **acac** = acetylacetonate, **9aneS3** = 1,4,7-trithiacyclononane, **bibzim** = 2,2′-bibenzimidazole, **bpm** = 2,2′-bipyrimidine, **bqn** = 2,2′-biquinoline, **bpy** = 2,2′-bipyridine, **bpz** = 2,2′-bipyrazine, **bpzb** = 1,3-bis(3-methylpyrazolyl)benzene, **bpyca** = bipyridine-4,4′-dicarboxylic acid, **bzimpy** = 2,6-bis(benzimidazol-2-yl)pyridine, **Cl-phen** = 5-chloro-1,10-phenanthroline, **cpe** = *N,N*-bis-(3-carboxy-1-oxopropanyl)-1,2-ethylenediamine, **cpihc** = 5-chloro-1,3-dihydro-3-[2-(phenyl)-ethylidene]-2H-indol-2-one-hydrazinecarbothioamide, **cpp** = *N,N*-bis-(3-carboxy-1-oxo-propanyl)-1,2-phenylenediamine, **cpihct** = 5-chloro-1,3-dihydro-3-[2-(phenyl)-ethylidene]-2H-indol-2-one-hydrazinecarbothioamide, **cpp** = *N,N*-bis-(3-carboxy-1-oxopropanyl)-1,2-phenylenediamine, **cpp-2** = *N,N*-bis-(3-carboxy-1-oxoprop-2-enyl)-1,2-phenylenediamine, **cppp** = *N,N*-bis-(2-carboxy-1-oxophenelenyl)-1,2-phenylenediamine, **dbm** = 1,3-diphenylpropane-1,3-dionate, **dcmb** = 4,4′-dimethoxycarbonyl-2,2′-bipyridine, **dfbpzb** = 1,5-difluoro-2,4-bis(3-methylpyrazolyl)benzene, **di-mby** = 5,5″-dimethyl-2,2′:5′,5″:2″,2‴-quaterpyridine, **dmbpy** = 4,4′-dimethyl-2,2′-bipyridine, **dmbpzb** = 1,5-dimethyl-2,4-bis(3-methylpyrazolyl)benzene, **dmdpphen** = 2,9-dimethyl-4,7-diphenyl-1,10-phenanthroline, **dmphen** = 2,9-dimethyl-1,10-phenanthroline, **dpa** = di-2-pyridylamine, **dpbpy** = 4,4′-diphenyl-2,2′-bipyridine, **dpk** = di-2-pyridylketone, **dpkox** = di-2-pyridylketone oxime, **dpm** = 2,2,6,6-tetramethyl-3,5-heptadionate, **dpp** = 2,3-bis(2-pyridyl)pyrazine, **dppe** = 1,2-bis(diphenylphosphino)ethane, **dpphen** = 4,7-diphenyl-1,10-phenanthroline, **dppm** = bis(diphenylphosphino)methane, **dppzR2** = 11,12-substituted dipyridophenazine, **dqpR** = 2,6-di(quinolin-8-yl)pyridine (R=H, 4-NH₂, 4-OMe, 4-PhMe), **dtfbpy** = 4,4′-bis(trifluoromethyl)-2,2′-bipyridine, **hat** = 1,4,5,8,9,11,4,5,8,9,12-hexaazatriphenylene, **epts** = (R)-(+)- or (S)-(−)-ethyl p-tolyl sulfoxide, **F-quol** = 5-fluoro-8-quinolinolate, **Mebib** = bis(*N*-methylbenzimidazolyl)benzene, **Mebip** = bis(*N*-methylbenzimidazolyl)pyridine, **mphen** = 5-methyl-1,10-phenanthroline, **mpts** = (R)-(+)- or (S)-(−)-methyl p-tolyl sulfoxide, **NO₂-phen** = 5-nitro-1,10-phenanthroline, **N4CR** = 5-R-tetrazolate (R = Me, Et, Pr, Ph, etc.), **NCNOC** = cyclic oxadiazoline, **NO₂N4C** = 5-nitro-tetrazolate, **pbpy** = 2-(4-(pyrrolidin-1-yl)pyridin-2-yl)pyridine, **pdmphen** = 5-phenyl-2,9-dimethyl-1,10-phenanthroline, **pdptz** = 3-(2-pyridyl)-5,6-diphenyl-1,2,4-triazine, **Phbib** = bis(*N*-phenylbenzimidazolyl)benzene, **phen** = 1,10-phenanthroline, **phenq** = 8-(1′,10′-phenanthrolin-2′-yl)quinoline, **phhy** = anion of 5-(2-pyridylmethylene)hydantoin, **ppbpy** = 4-(Pyrrolidin-1-yl)-2-(4-(pyrrolidin-1-yl)pyridin-2-yl)pyridine, **pptpy** = 4-(pyrrolidin-1-yl)-2-(6-(4-(pyrrolidin-1-yl)pyridin-2-yl)pyridin-2-yl)pyridine, **pphen** = 5-phenyl-1,10-phenanthroline, **pptpy** = 3-(2-pyridyl)-1,10-phenanthroline, **pydppm** = 3-(pyrid-2′-yl)-4,5,9,16-tetraaza-dibenzo[*a,c*]naphthacene, **pydppx** = 3-(pyrid-2′-yl)-11,12-dimethyl-dipyrido[3,2-*a*:2′,3′-*c*]phenazine, **pydppz** = 3-(pyrid-2′-yl)dipyrido[3,2-*a*:2′,3′-*c*]phenazine, **pyre** = bispyridinoethylene, **pyrdbpy** = 4-pyrrolidinyl-2,2′-bipyridine, **pyrd₂bpy** = 4,4′-dipyrrolidinyl-2,2′-bipyridine, **pyrrbpy** = 4-pyrrolyl-2,2′-bipyridine, **pyrr₂bpy** = 4,4′-dipyrrolyl-2,2′-bipyridine, **pzpy** = ethyl[3-(2-pyridyl)-1-pyrazolyl]acetate, **dbbpy** = 4,4′-di-*tert*-butyl-2,2′-bipyridine, **tmbiH2** = 5,5′,6,6′-tetramethyl-2,2′-bibenzimidazole, **tmbpy** = 4,4′,5,5′-tetramethyl-2,2′-bipyridine, **tmphen** = 3,4,7,8-tetramethyl-1,10-phenanthroline, **tptz** = 2,4,6-tripyridil-s-triazin, **tpy** = terpyridine, **ttfbpy** = 4,4′,5,5′-tetrakis(trifluoromethyl)-2,2′-bipyridine.

b YIF = Yield Improvement Factor as defined in the Introduction.

c An asterisk (*) indicates that the nature of the microwave reactor was not explicitly stated.

d EG = ethylene glycol.

e DMF = dimethylformamide.

f NA indicates either (i) not applicable since the compounds are new or (ii) not available because the published report does not give all of the needed data.

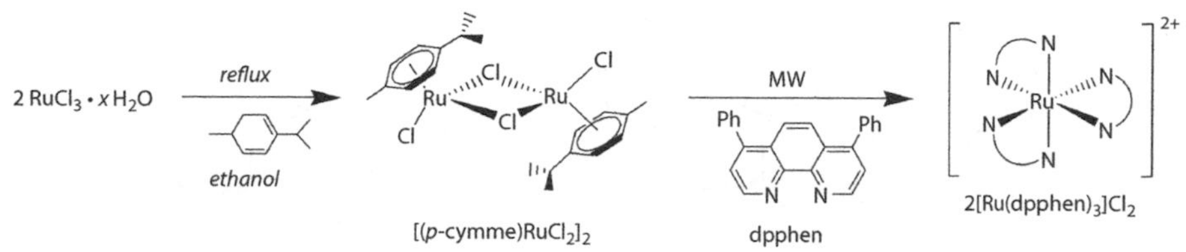

SCHEME 7.1

is also a consideration. In general, significantly less solvent is used for procedures involving microwave heating as compared to conventional analogs.

A number of the reactions represented in Table 7.1 merit special attention. The bidentate ligand 4,7-diphenyl-1,10-phenanthroline (dpphen) is difficult to complex with ruthenium(II) using conventional heating. The reaction of $RuCl_3$ with dpphen in ethanol gives a 6.3% yield of [Ru(dpphen)$_3$]Cl$_2$ after refluxing for 48 h.[60] Using microwave heating, this complex was prepared in 77% yield after only 5 min in a scientific microwave unit.[21] An extra 4-h step of preparing [RuCl$_2$(p-cymene)]$_2$ from $RuCl_3$ was required before the microwave procedure, but this still represents a much more efficacious overall synthesis (Scheme 7.1). In fact, the extra step could most likely have also been accomplished with microwave heating in a short time since an analogous reaction yielded [RuCl$_2$(benzene)]$_2$ from $RuCl_3$ in less than 1 min.[2] Similarly, the Ru^{II} complex [Ru(9aneS3)$_2$](PF$_6$)$_2$ containing two tridentate cyclic thioether ligands was prepared in over twice the yield of the conventional method and in one *less* step.[17]

A number of manganese(II) complexes with diamido ligands bearing two pendant carboxyoxogroupssuchas N,N'-bis-(3-carboxy-1-oxo-propanyl)-1,2-phenylenediamine complexes have been prepared in high yield using microwave heating (Scheme 7.2).[57] Yield improvement factors of 1.3 to 1.8 over conventional heating were observed. IR spectroscopy revealed that the ligands cpe, cpp, cpp-2, and cppp were coordinated through the amide oxygen and carboxylate oxygen atoms, resulting in a tetrahedral environment around the Mn atom.

Complexes bearing porphyrins and phthalocyanine ligands are not included in Table 7.1. However, examples of their synthesis using microwave heating are numerous (Figure 7.2). Metalloporphyrins are studied because they mimic biological systems such as cytochrome P450 and have potential use in solar cells, light-emitting diodes, molecular thermometers, chemical sensors, or chemotherapeutic agents. Most of the reports relating to use of microwave heating for their preparation describe

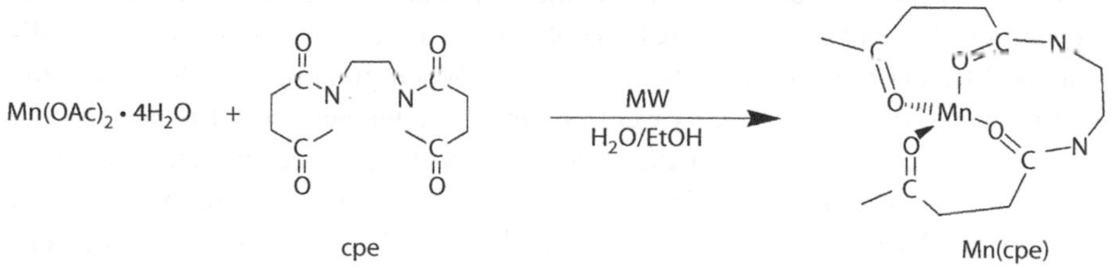

SCHEME 7.2

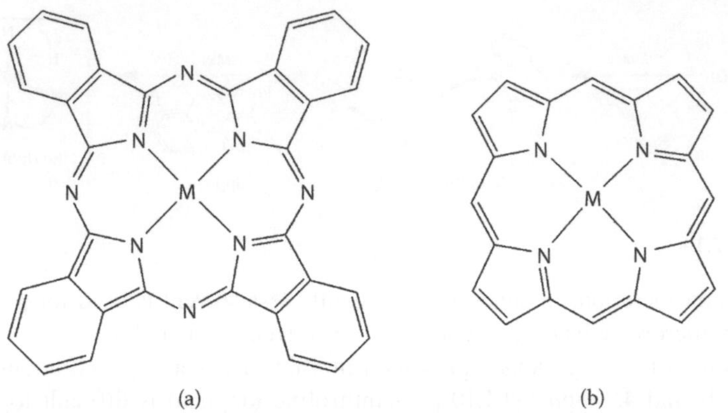

(a)   (b)

FIGURE 7.2   Core portions of (a) a metallophthalocyanine and (b) a metalloporphyrin.

reactions performed using domestic ovens.[61–70] There is one report on the application of a scientific microwave unit for the insertion of Ni, Pd, and Pt into several different porphyrin ligands.[71] The reactions gave good yields in most cases but of note is that when heat-sensitive porphyrin ligands were used, several byproducts were also observed.

Similar to porphyrins, phthalocyanines serve as models for biologically important systems and as excellent dyes and pigments. Metallophthalocyanines have potential applications in display devices, photodynamic reagents, optical read/write discs, and liquid crystal displays. A number of these compounds have been prepared using microwave heating, many in higher yield than previously reported for conventional heating. Again, the vast majority of the work has been performed using domestic microwave ovens,[61–66,72–98] but a few reports using scientific microwave apparatus have recently appeared.[7,8,99–103] New Co, Ni, and Zn phthalocyanines bearing bulky phenoxy substituents have been prepared in good yields by heating a mixture of the metal acetate and the phthalonitrile in the presence of DBU and using hexanol as solvent.[102]

The synthesis of copper(II) phthalocyanines at various temperatures in a scientific microwave unit has been compared to results obtained using an oil bath. Microwave heating always resulted in a higher product yield (YIF = 1.01–2.23).[7] An extension of this work showed that the best conventional yield (85.1%) is obtained at 180 °C for 4 h, while the best microwave yield is obtained at 170 °C for 4 h.[101] An open-vessel approach has also been used to compare microwave and conventional heating for the preparation of copper(II) and cobalt(II) phthalocyanines under solvent-free conditions at different temperatures (Scheme 7.3).[104] When phthalic anhydride was the starting material, the best results were obtained at a temperature of about 150 °C. When phthalonitrile was the starting reagent, the best yield was obtained at approximately 210 °C. Microwave heating was shown to give products of higher purity in better yields (YIF = 2–4 for phthalic anhydride and YIF = 7–15 for phthalonitrile).

A series of metalloporphyrazines (M = Co, Ni, Cu, Zn) have been prepared directly from substituted maleonitriles using both microwave and conventional

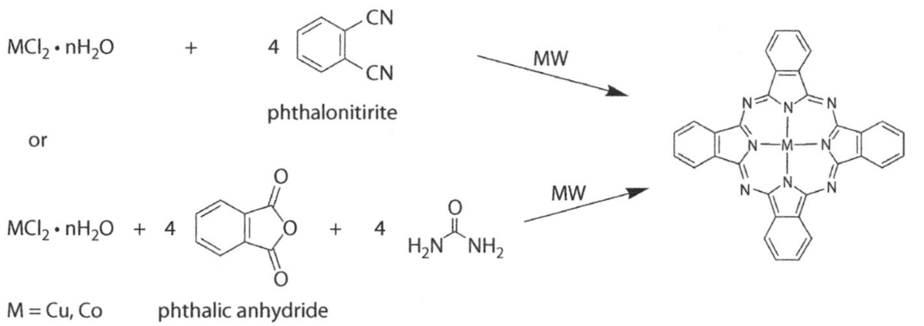

M = Cu, Co   phthalic anhydride

SCHEME 7.3

SCHEME 7.4

methods (Scheme 7.4).[105] In every instance, the products were obtained in higher yield using microwave heating.

7.2.2 LIGAND TRANSFORMATION

Ligands bound to transition metals have been modified by means of a range of organic synthetic transformations using microwave heating. A number of studies have been carried out involving organic reactions of ligands coordinated to platinum. The conversion of Pt(II) nitrile complexes to Pt(II) imine complexes occurred with YIF of 1.5–2.7 (Scheme 7.5).[106,107] This work was extended to produce chiral unsymmetrical imine complexes (YIF = 1.1–1.3) (Scheme 7.6).[108] Pt(II) and Pt(IV) nitrile complexes were converted into oxadiazoline complexes via cycloaddition of nitrones with yields similar to those observed using conventional methods (Scheme 7.7).[41,43] Lastly, the conversion of Pt(II) diazido complexes to bis(tetrazolato)Pt(II) analogs by cycloaddition reactions with organonitriles took place with YIF of 0.90–1.1 (Scheme 7.8).[36–39]

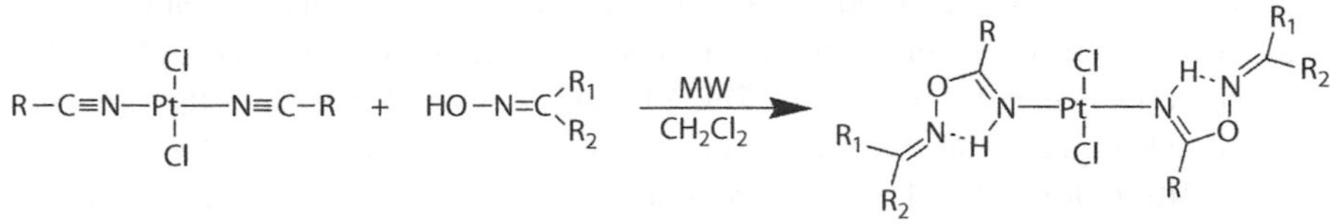

SCHEME 7.5

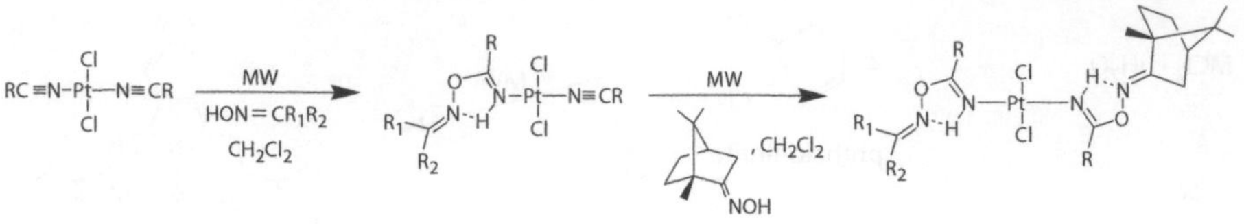

SCHEME 7.6

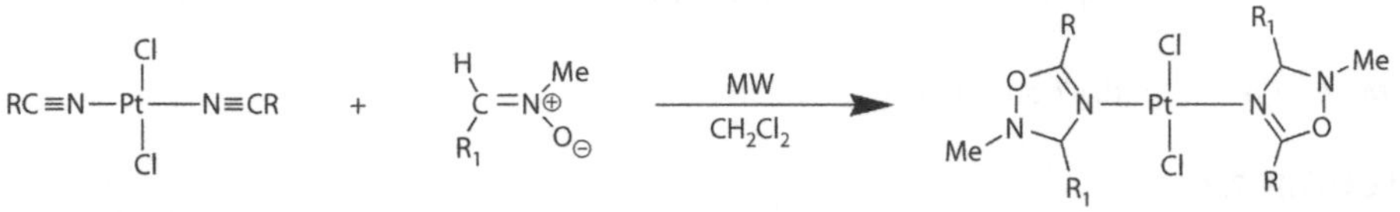

SCHEME 7.7

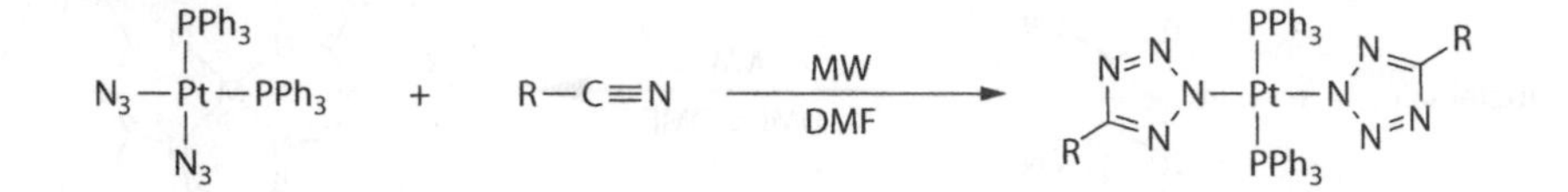

SCHEME 7.8

A different ligand transformation occurred when an ethylene glycol mixture of 4,4'-diethoxycarbonyl-2,2'-bipyridine and $RuCl_3 \cdot xH_2O$ was heated using microwave irradiation for 5 min at 225 °C in a sealed vessel.[109] The unexpected decarboxylation of the functionalized bpy ligand resulted in the formation of the $[Ru(bpy)_3]^{2+}$ complex ion. The pressure in the reaction vessel exceeded 20 bar during the course of the reaction, ethanol and CO_2 being confirmed as byproducts. No such reaction occurred in a control experiment in the absence of the ruthenium(III) chloride. Likewise, no decarboxylation was observed during experiments conducted with other Group 8 metal chlorides.

7.2.3 SYNERGISTIC COMBINATION OF MICROWAVE AND CONVENTIONAL HEATING METHODS

In some cases, microwave heating has been used to initiate a reaction before conventional methods of heating are used. This methodology can be important when trying to anneal compounds and cool very slowly over the period of days from high temperature to room temperature. Microwave heating can be leveraged for initial preparation of the metal complexes and then conventional heating used for prolonged annealing and slow cooling. As an example, a number of rare earth metal complexes have been prepared in this way. Pulses of 5 s of microwave irradiation were used to superheat elemental metal particles (Ce, Nd, Sm, Ho, Er, Tm, Yb, and Sc) mixed with 2,2'-dipyridylamine in a sealed quartz tube.[110] The reaction mixture was then heated to 190 °C for 6 h and to 210 °C for 168 h. The reaction mixture was cooled to 90 °C over the period of 400 h and to room temperature in another 6 h. These solvent-free reactions yielded M_2(dipyridylamide)$_6$ complexes. For M = Ho, Er, and Yb, microwave activation was compared to two other forms of activation: formation of an mercury amalgam and dissolution in liquid ammonia. Yields for the microwave method are somewhat higher in the case of Ho and Yb, but are significantly lower in the case of Er.

7.2.4 SYNTHESIS OF CLUSTER AND POLYNUCLEAR METAL COMPLEXES

A number of complexes containing multiple metal atoms have been prepared by microwave heating. Those with significant metal–metal bonding (clusters) and those with three or more metal atoms without M-M bonds (polynuclear complexes) are discussed in this section. The binuclear complexes listed in Table 7.1 are not mentioned here, because they do not contain M-M bonds. Clusters appear to be particularly attractive targets for microwave synthesis since they are often prepared conventionally by heating reaction mixtures at high temperatures for extended periods of time.[111] Metal carbonyl clusters are addressed in a later section discussing the use of microwave heating for preparing organometallic compounds.

7.2.4.1 Polymetallic Complexes Prepared Using Domestic Microwave Ovens

The synthesis of metal chloride clusters (Figure 7.3) is possible in minutes instead of days using metal powders and no solvent in a sealed ampoule inside a domestic microwave oven.[112] Starting with MCl_5, $AlCl_3$, NaCl, and Al, clusters of the type $[M_6Cl_8]Cl_4$ with M = Mo or W have been prepared in 56% and 48% yields, respectively. The clusters $[M_6Cl_{12}]Cl_2$ (M = Nb or Ta) were synthesized in 35–40% yield from MCl_5 and Al. A higher yield (45%) of the niobium cluster was produced when the starting materials were $NbCl_5$, Nb, and NaCl.

Two trinuclear ruthenium(II) complexes of the type $[\{Ru(N–N)_2\}_3(\mu\text{-hat})]^{6+}$ have been prepared, where N–N is bpy or phen and hat is a bridging ligand with six donor N atoms capable of chelating three metal atoms.[27] Ethylene glycol mixtures of hat and $Ru(N–N)_2Cl_2$ were heated for 6 min (Scheme 7.9). The bpy complex had been synthesized previously by conventional means, and the yield was essentially the same when using microwave heating. In a follow-up study, the stereoselective synthesis of two diastereoisomers of $[\{Ru(bpy)_2\}\{Ru(phen)_2\}\{Ru(dmbpy)_2\}(\mu\text{-hat})]^{6+}$ was performed by heating a mixture of $[\{Ru(bpy)_2\}(\mu\text{-hat})\{Ru(phen)_2\}][PF_6]_4$ and $Ru(dmbpy)_2Cl_2$ in ethylene glycol/water for 4 min.[113]

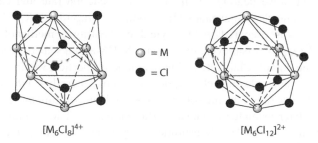

$[M_6Cl_8]^{4+}$ 　　　　　　　　　　　$\bigcirc$ = M 　$\bullet$ = Cl 　　　　　　　　　　$[M_6Cl_{12}]^{2+}$

FIGURE 7.3   Structures of common metal chloride clusters.

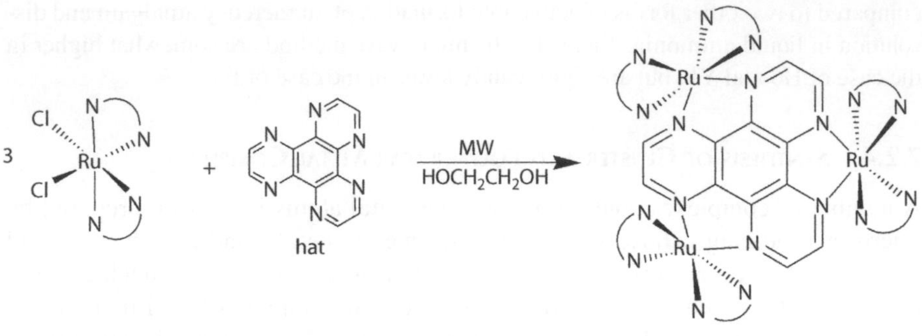

SCHEME 7.9

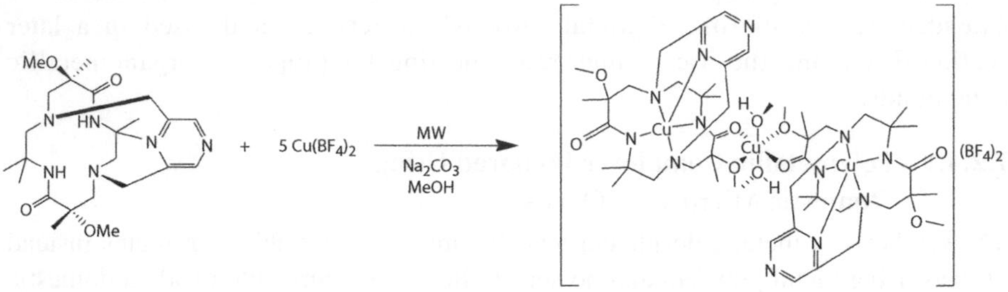

SCHEME 7.10

A pyrazine-capped 5,12-dioxocyclam ligand has been serendipitously used to prepare a trinuclear copper(II) compound having a central octahedral Cu atom coordinated to two pyrazine/cyclam-bound Cu units through the oxygen atoms of the amide and methoxy groups of the cyclam moiety.[114] A methanol solution of Cu(BF$_4$)$_2$•6H$_2$O, K$_2$CO$_3$ and the pyrazine–cyclam ligand was heated for 1 min (Scheme 7.10). Conventional thermal heating produced the expected mononuclear copper(II) complex with a pentadentate pyrazine–cyclam ligand and no trace of the trinuclear compound.

7.2.4.2 Polymetallic Complexes Prepared Using Scientific Microwave Apparatus

The use of microwave heating has improved the reaction rate and enhanced the yield of the hexanuclear manganese(III) complex Mn$_6$O$_2$(sao)$_6$(O$_2$CH)$_2$(MeOH)$_4$ (Figure 7.4). It was prepared in 80% yield by heating a methanol solution of Mn(ClO$_4$)$_2$•4H$_2$O, saoH$_2$ (salicylaldoxime) and NaOMe in a sealed vessel at 110 °C for 5 min.[115] The complex had interesting magnetic properties, which is noteworthy given that single-molecule magnet compounds are not generally synthesized in good yield in the presence of a reducing solvent at elevated temperature and pressure. This chemistry was later extended to include different carboxylate ligands (O$_2$CR, R = Me, Et, Ph) in place of O$_2$CH.[116] Substitution of pyridine for methanol in the reaction mixture led to the formation of the triangular complexes Mn$_3$O(sao)$_3$(O$_2$CR)(H$_2$O)-(py)$_3$ (Figure 7.5) (R = Me or Ph).

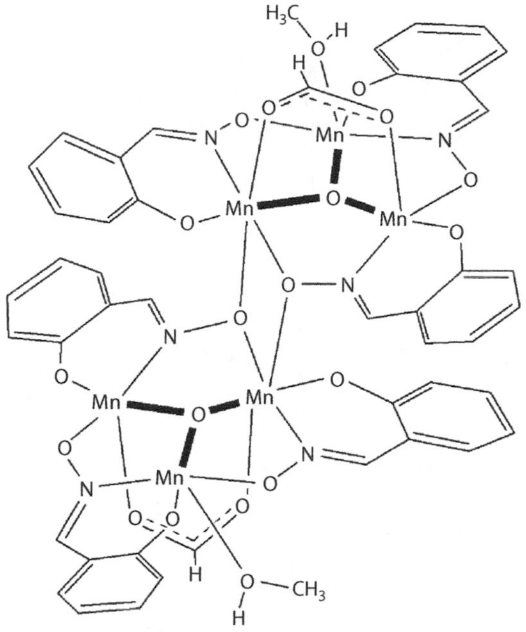

FIGURE 7.4   Structure of $Mn_6O_2(sao)_6(O_2CH)_2(MeOH)_4$.

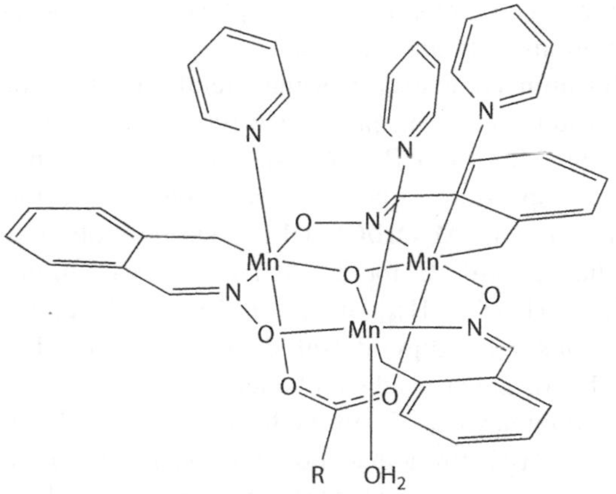

FIGURE 7.5   Structure of $Mn_3O(sao)_3(O_2CR)(H_2O)(py)_3$.

Along the same lines, an octametallic iron(III) complex $Fe_8O_4(sao)_8(py)_4$ was prepared in 70% yield by heating a mixture of $Fe(OAc)_2$, $saoH_2$, and pyridine in a sealed vessel at 120 °C for 2 min.[117] The iron atom arrangement is best described as a cube inside a tetrahedron. Heating $M(OAc)_2 \cdot 4H_2O$ (M = Ni or Co) and 1 equiv of NaN_3 in pyridine/MeOH as solvent at 120 °C for 4 min led to the formation of the trinuclear complexes $M_3(N_3)_3(OAc)_3(py)_5$ (Figure 7.6) in 55% yield for Ni and 40% yield for

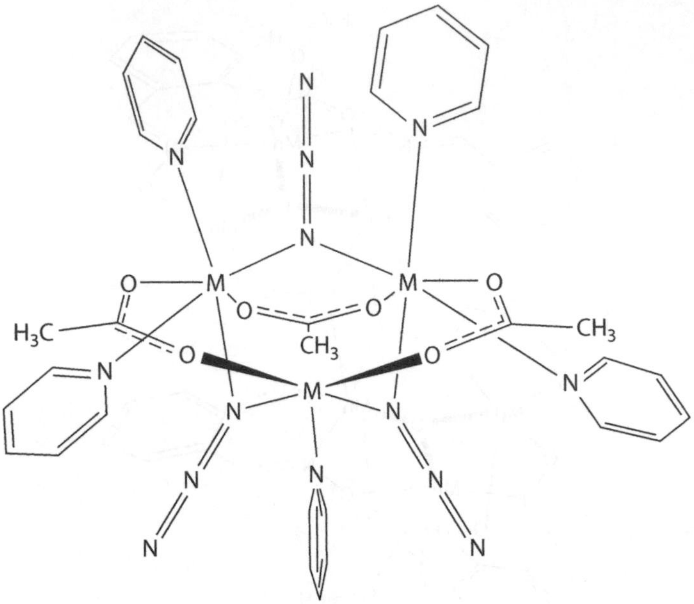

FIGURE 7.6 Structure of $M_3(N_3)_3(OAc)_3(py)_5$, M = Co, Ni.

Co.[118] Numerous attempts to prepare these complexes using conventional heating or under ambient conditions were unsuccessful.

Dimeric molybdenum complexes bearing metal–metal quadruple bonds have been prepared in better yield than previously reported conventionally. The tetra-carboxylate complexes $Mo_2(\mu-O_2CR)_4$ (R = Me, Et, Ph) were synthesized by heating $Mo(CO)_6$ and the corresponding carboxylic acid at 180-200 °C for 15–32 min.[119]

The treatment of a $MeOH/CH_2Cl_2$ solution of $Cu(BF_4)_2 \cdot H_2O$ and 2,2′-dipyridyldisulfide at 140 °C for 2 h led to the formation of the polymeric Cu^I complex $\{[Cu_9(C_5H_5NS)_8(SH)_8](BF_4)\}n$ in 55% yield and the Cu^{II} complex $[\{Cu(dps)_2\}_2(\mu-S)]^{2+}$, dps = 2,2′-dipyridylsulfide, in 4% yield.[120] In this unusual reaction, C–S and S–S bonds are cleaved simultaneously.

The nonanuclear complex $[La_3Ni_6(iminodiacetate)_6(OH)_6(H_2O)_{12}]^+$ has been prepared from $Ni(NO_3)_2 \cdot 6H_2O$, iminodiacetic acid and $La(NO_3)_3 \cdot 6H_2O$ at 100 °C in water using microwave heating or under standard reflux conditions.[7] The complex ion was obtained in 40% yield after 30 min when using microwave heating, but under conventional hydrothermal conditions the yield was only 28.5% after 3 days.

The mixed-valent manganese phosphonate complex $Mn^{II}_8Mn^{III}_2O_2(O_3PC_{10}H_7)_4$-$(O_2CPh)_{10}(py)_4(H_2O)_2$ could be prepared in 82% yield by sealing an acetonitrile solution of $Mn_3O(O_2CPh)_6(py)_2(H_2O)$ and 1-naphthyl phosphonic acid in a glass tube and heating it at 120 °C for 5 min.[121] A different complex, $Mn^{II}_4Mn^{III}_9O_4(OH)_2(O_3PC_{10}H_7)_{10}$-$(O_2CPh)_5(py)_8(H_2O)_6$, was prepared in low yield by placing the same reaction mixture in a water bath at 45 °C.

TABLE 7.2
Selected Transition Metal Carbonyl Complexes Prepared by Microwave Heating

Complex	Metal	Ligand	YIF	Solvent Used	Microwave Unit Used	Reference
MCl(CO)L$_2$	Ir	PPh$_3$	0.79	DMF	domestic	48
[MCl(CO)L$_2$]$^+$	Ru	bpy	NA	EtOH/H$_2$O	domestic	48
MCl$_2$(CO)$_2$L	Os	bpy, dmbpy, phen,	6.1 (bpy)	EtOH	domestic	122
M(CO)$_4$L	Cr, Mo, W	en, bpy, phen, dppm, dppe	0.67–5.2	toluene/ diglyme	domestic	123, 125
M(CO)$_5$L	Mo	PPh$_3$	NA	diglyme/ THF	domestic	125
M(CO)$_4$L$_2$	Mo, W	PPh$_3$, pip, py	NA	diglyme/ THF	domestic	125
M(CO)$_4$L$_2$	Mo, W	pzpy	0.97	diglyme/ toluene	multimode	126

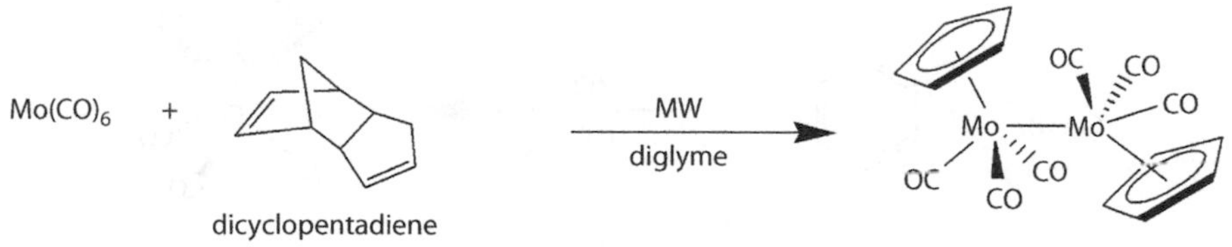

SCHEME 7.11

7.3 MICROWAVE-PROMOTED ORGANOMETALLIC CHEMISTRY

7.3.1 METAL CARBONYL COMPLEXES

7.3.1.1 Synthesis Using Domestic Microwave Ovens

The preparation of metal carbonyl complexes using domestic microwave ovens has been the subject of several reports, the results being summarized in Table 7.2. The earliest reported studies focused on the synthesis of IrCl(CO)(PPh$_3$)$_2$ (Vaska's compound) and the complex ion [RuCl(CO)(bpy)$_2$]$^+$, both of which could be significantly accelerated under microwave heating.[48] Following this, several OsCl$_2$(CO)$_2$(N–N) complexes have been prepared in high yield.[122]

A series of ligand substitution reactions of Group 6 metal carbonyls have been performed in a Teflon-lined digestion vessel equipped with a temperature probe and designed for safe use in a domestic microwave oven.[123] Significant rate enhancements and yield improvements compared to conventional heating were observed in most cases. A modified household oven equipped with a reflux condenser and a water-cooling loop were used to explore additional microwave-assisted substitution reactions of Group 6 metal carbonyls.[124,125] In one case, the complex [CpMo(CO)$_3$]$_2$ was prepared in 94% yield by heating a diglyme solution of Mo(CO)$_6$ and dicyclopentadiene for 1 h (Scheme 7.11). The conventional synthesis of this complex requires at least two steps that must be carried out under an inert atmosphere.

7.3.1.2 Synthesis Using Scientific Microwave Apparatus

The formation of (η^6-arene)Cr(CO)$_3$ complexes from Cr(CO)$_6$ and arenes has been reported.[127] As with conventional heating, solvent selection proves to be very important. For example, in the case of mesitylene as the arene reagent, when diglyme was used as solvent no Cr(CO)$_3$(mesitylene) was formed, but a 79% yield of this complex was obtained upon changing to THF. Optimal conditions for preparing a range of the arene complexes were found to be heating to 160 °C for 1 h (Scheme 7.12). Of interest is that Cr(CO)$_3$(anisole) was prepared in 77% yield by this method, whereas when using a domestic microwave oven, the same compound was prepared in only 45% yield after 4 h.[125]

Complexes bearing the bidentate nitrogen ligand pzpy have been prepared in good yield directly from M(CO)$_6$ (M = Mo, W; pzpy =

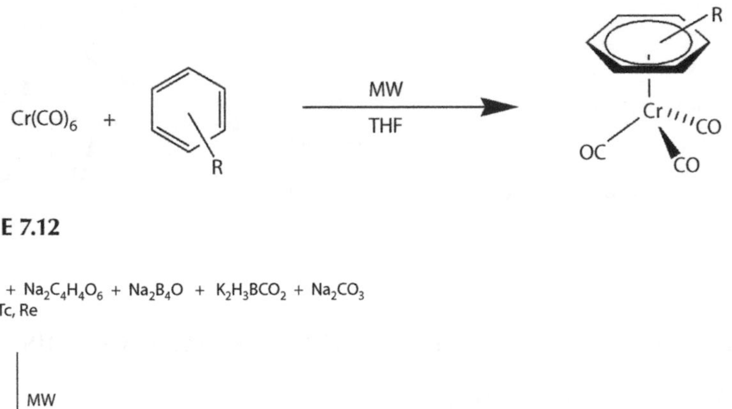

SCHEME 7.12

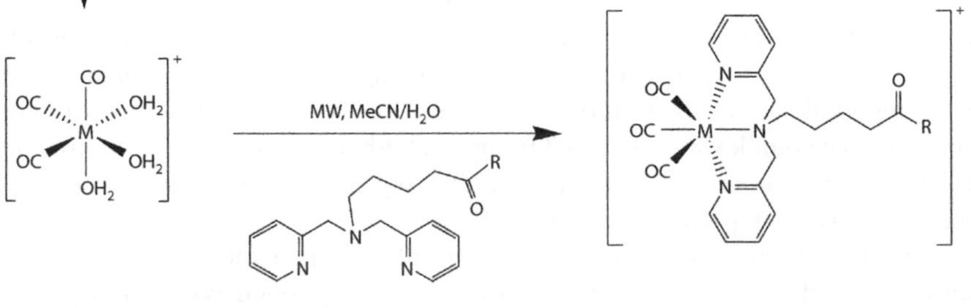

SCHEME 7.13

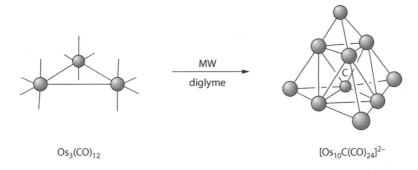

$Os_3(CO)_{12}$ $[Os_{10}C(CO)_{24}]^{2-}$

SCHEME 7.14

ethyl[3-(2-pyridyl)-1-pyrazolyl]acetate).[126] Reactions were carried out using diglyme as solvent and took just 30 s in the case of Mo and 15 min for W. Conventional methods for preparing $M(CO)_4(pzpy)$ typically require the additional step of preparing more labile adducts such as $Mo(CO)_4(pip)_2$ and $W(CO)_5(THF)$.

Several technetium(I) and rhenium(I) tricarbonyl compounds bearing a tridentate dipicolyl ligand have been prepared in a multistep protocol that is significantly faster than all previous methods (Scheme 7.13).[128] The route, starting from pertechnate or perrhenate, could have a significant impact on the preparation of gamma-emitting radiopharmaceuticals.

The versatile trinuclear complex $Os_3(CO)_{11}(NCMe)$ has been prepared in high yield by microwave heating of $Os_3(CO)_{12}$ in acetonitrile at 150 °C for 5 min.[129] This complex is used widely as a starting material for other substituted trinuclear complexes as well as higher nuclearity clusters. The trinuclear ruthenium complexes $Ru_3(CO)_{12-n}(PPh_3)_n$ ($n = 1–3$) have been generated in high yield by heating a 1,2-dichloroethane (DCE) solution of $Ru_3(CO)_{12}$ and PPh_3 at 110 °C for 1 min, the reaction being followed in situ using a digital camera interfaced with the microwave unit.[11] Using phenylacetylene instead of PPh_3 and heating for 5 min, the acetylide cluster $Ru_3(H)(CO)_9(C\equiv CPh)$ was afforded in excellent yield with none of the side products produced by the traditional preparation. The microwave-accelerated reactions of $Ru_3(CO)_{12}$ and $Os_3(CO)_{12}$ with H_2 in DCE produced $H_4Ru_4(CO)_{12}$ and $H_2Os_3(CO)_{10}$ in near quantitative yield in only 15 min.

Metal carbonyl clusters have traditionally been prepared by somewhat lengthy pyrolysis or thermolysis reactions, often in an autoclave or pressurized reaction vessel.[130] Two recent reports have described far more efficient synthetic methods by using microwave heating. The binary cluster $Ru_3(CO)_{12}$ has been prepared in 90% yield by heating a methanol solution of $RuCl_3$ and Cs_2CO_3, under an atmosphere of CO (3.5 bar) at 110 °C for 10 min.[11] The standard synthesis of this compound generally requires 8 h of heating at 125 °C under a high pressure (66 bar) of CO gas. It is noteworthy, however, that attempts to prepare the hexanuclear cluster $Ru_6C(CO)_{17}$ using microwave heating were unsuccessful. The tetra-capped octahedral osmium cluster $[Os_{10}C(CO)_{24}]^{2-}$ has been synthesized 57% yield by heating a diglyme solution of $Os_3(CO)_{12}$ at 220–230 °C for 1 h (Scheme 7.14).[118] An almost identical yield

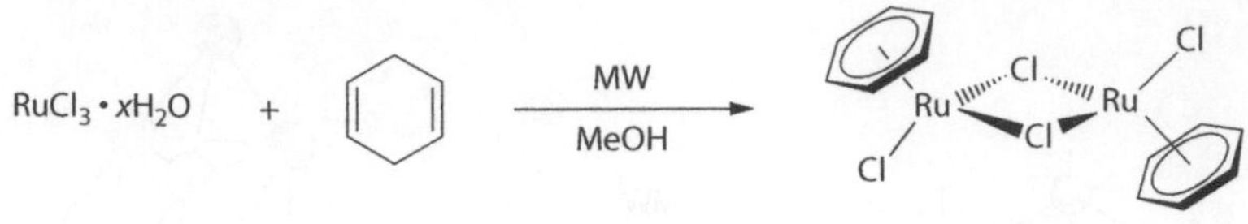

SCHEME 7.15

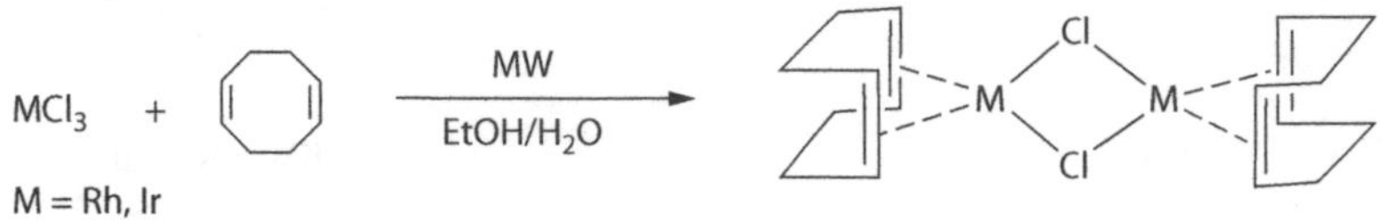

M = Rh, Ir

SCHEME 7.16

is obtained through a conventional synthetic method, but it requires an additional preparative step followed by 64 h of heating.

7.3.2 SANDWICH COMPOUNDS

7.3.2.1 Synthesis Using Domestic Microwave Ovens

One of the first reports of microwave heating as a tool for organometallic synthesis featured a sandwich compound. The [Cp$_2$Rh]$^+$ complex ion was formed directly from RhCl$_3$•xH$_2$O and cyclopentadiene in only 30 s.[2] The starting materials were dissolved in methanol and sealed in a Teflon vessel before being placed in the microwave oven. This reaction represented a chemical shortcut over the conventional approach, which takes over 24 h and requires the use of a Grignard reagent (CpMgBr) although the yield is 30% greater than that obtained in the microwave approach. The high-yield synthesis of [Ru(C$_6$H$_6$)Cl$_2$]$_2$ from RuCl$_3$•xH$_2$O and 1,3-cyclohexadiene as well as that of [M(diene)Cl]$_2$ complexes (M = Rh or Ir; diene = cyclooctadiene or norbornadiene) were also reported (Schemes 7.15 and 7.16).

A microwave oven fitted with a novel dry ice condenser was used to prepare over 20 complexes of the type [CpFe(η^6-arene)]$^+$ in high yield (YIF = 1.2–7.3).[131,132] Since it is transparent to microwaves, solid CO$_2$ was placed directly in the microwave oven in a beaker positioned above another beaker containing the reaction mixture. In this early work, no solvent was used other than the arene to be complexed, but more recently these researchers have employed a similar experimental arrangement using 1,2,4-trichlorobenzene or DMF as solvent to synthesize many additional [CpFe(η^6-

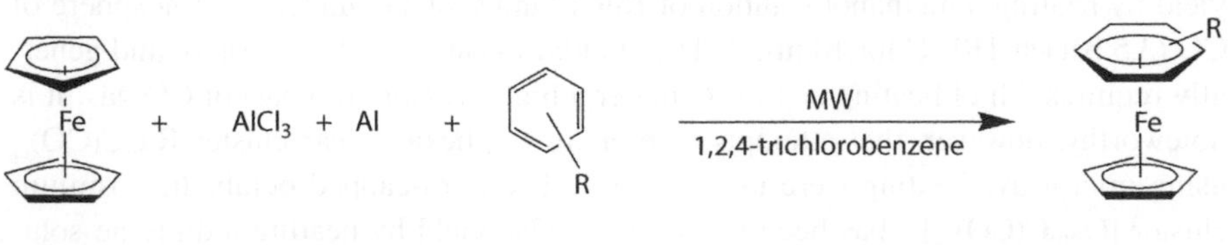

SCHEME 7.17

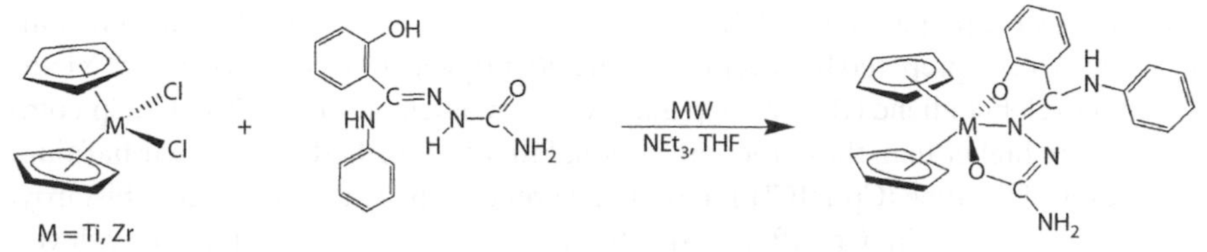

SCHEME 7.18

arene)]$^+$ complexes with halogen, oxygen, carbonyl, or amino substituents on the arenes (Scheme 7.17).[133–135]

Titanium(IV) and zirconium(IV) complexes of the types Cp$_2$MCl(HPO), Cp$_2$M(HPO)$_2$, Cp$_2$M(O–N–O), and Cp$_2$M(O–N–S) have been prepared by reaction of titanocene dichloride or zirconocene dichloride with the appropriate ligand using THF as the solvent (Scheme 7.18). The reactions are complete in 4–7 min as opposed to 14–18 h conventionally. Yields were always higher for the microwave approach (YIF = 1.1–1.2).

The bis(arene)chromium complex ion [Cr(η^6-C$_6$H$_5$CH$_3$)$_2$]$^+$ has been prepared from metal powders using two methods.[137] The starting materials in the first approach were Cr metal, AlCl$_3$, and toluene, while for the second method Al metal, AlCl$_3$, CrCl$_3$, and toluene were used. Reaction times were 20 and 30 min, respectively, and the product yields were comparable to those obtained using conventional heating (YIF = 1.0 and 0.92). The second method was also used to prepare the [Cr(η^6-C$_6$H$_6$)$_2$]$^+$ complex ion in 45 min. Extreme caution must be exercised when carrying out such procedures since the susceptibility of metal powders to arcing in the presence of microwave radiation poses a significant hazard, especially when flammable solvents such as toluene are involved.

7.3.2.2 Synthesis Using Scientific Microwave Apparatus

Ruthenium(II) complexes of the type CpRu(dppm)SR (R = Ph, CH$_2$CH$_2$Ph, CH$_2$(2-furyl), CH$_2$CO$_2$Et, CH$_2$CH(NHAc)CO$_2$H) have been synthesized by both conventional and microwave heating and the product yields found to be quite similar.[138] Several planar chiral cobalt metallocenes have been prepared via diastereoselective reactions between ether- or ester-linked chiral diynes and CpCo(CO)$_2$ (Scheme 7.19).[139] The five-membered-ring ether complexes were prepared in similar yields to conventional approaches (YIF = 1.0–1.1), while the seven-membered-ring ester

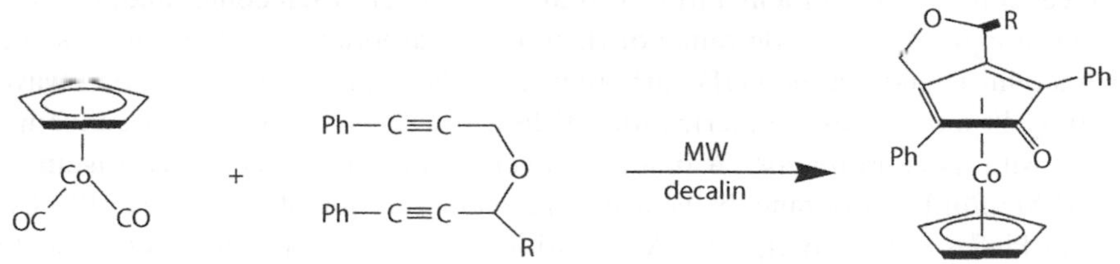

SCHEME 7.19

complexes were prepared with YIF = 1.4–1.7. In related work, cobalt sandwich complexes have been prepared by reaction of $CpCo(CO)_2$ with diaryl acetylenes.[140] Yields for both cyclobutadiene $(R_4C_4)CoCp$ and cyclopentadienone $(R_4C_4C=O)CoCp$ complexes were higher than those for conventional thermal methods. Dinuclear hafnium complexes of the type $(Cp_2HfCl)_2L$ have been prepared by reaction of benzil bis(aroyl hydrazones), LH_2, with Cp_2HfCl_2 and n-$BuNH_2$ in THF.[141] The methodology developed using microwave heating required a tenth as much solvent and produced higher yields of the desired product (YIF = 1.1–1.2).

Many reports have focused on the functionalization and derivatization of cyclopentadienyl groups bound to iron in ferrocene complexes. As an example, a series of ferrocenyl esters have been prepared via solvent-free reactions of ferrocenoyl fluoride with substituted phenols after microwave heating for just 1 min.[142] In the case of 4-bromophenol, the YIF was 1.2, but most substrates gave somewhat lower yields than the conventional method. However, the microwave approach did not require the addition of N,N-dimethylaminopyridine that was necessary conventionally. Other reports include a "green" approach to the acylation of ferrocene using the polymeric sulfonic acid Nafion as an acid catalyst[143] and a one-pot approach to the synthesis of 1,5-dioxo-3-substituted[5]ferrocenophanes from 1,1′-diacetylferrocene and aldehydes via a Claisen–Schmidt reaction.[144]

The speed and reliability with which microwave reactions may be performed using scientific apparatus give researchers the opportunity to optimize a particular synthetic method much more rapidly than a conventional approach that involves, for example, an overnight reflux. In a search for the ideal reaction conditions for the scandium-catalyzed acylation of ferrocene, the reaction was performed in a number of organic solvents as well as ionic liquids.[145] The effects of varying reagent concentration, temperature, addition of acid or base, amount of catalyst, type of acylating agent, and whether the reaction was run under an inert atmosphere were all probed. A range of conditions have also been screened for the preparation of 4-aryl-2-ferrocenyl-quinoline derivatives from acetylferrocene.[146] First, five different solvents were assessed, DMF affording the highest product yield. Then the same reaction was carried out at seven different temperatures for different periods of time. The best results were obtained with 10 min of microwave heating at 100 °C.

7.3.3 METALLOCARBORANES

The general structure of a metallocarborane, a half-sandwich compound, is illustrated in Figure 7.7. A wide range of rhenium(I) carboranes, together with some technetium(I) and platinum(II) carboranes, has been prepared using microwave heating. In most cases, isomerization of the carborane cage occurred concomitantly with its coordination. A domestic oven was used to produce two isomers of a $(PMe_2Ph)_2Pt(carborane)$ by heating an ethanol solution of cis-$Pt(PMe_2Ph)_2Cl_2$ and $[7\text{-}Ph\text{-}7,8\text{-}nido\text{-}C_2B_9H_{11}]^-$.[147] A scientific microwave unit has been used to synthesize a series of tricarbonyl rhenacarboranes of the type $[(RR'C_2B_9H_9)Re(CO)_3]^-$ in good yield using water as the solvent.[148-150] A number of the cor-

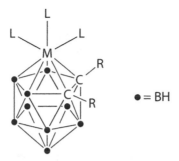

FIGURE 7.7   Structure of a typical metallocarborane.

responding Tc complexes were also prepared in order to explore their potential as radiopharmaceuticals.

7.3.4   METAL ALKYL AND METAL ARYL COMPOUNDS

7.3.4.1   Synthesis Using Domestic Microwave Ovens

One of the earliest reports of microwave-assisted organometallic synthesis involved the preparation of arylmercuric chlorides.[3] Heating an ethanol solution of diphenyldiazene and 2-phenylpyridine with $Hg(OAc)_2$ inside a sealed Teflon vessel led to the formation of the desired arylmercury complexes. Yields were a little lower than conventional approaches, but reaction times were at least 30-times shorter. This chemistry was extended 10 years later when a solvent-free method was developed.[151] Arylmercuric chlorides together with benzoquinone, barbituric acid, or thiobarbituric acid were adsorbed on basic alumina before being exposed to microwave irradiation for 1–3 min to produce derivatives (YIF = 1.3–1.5).

Two cyclometallated iridium(III) complexes of the type Ir(2-arylpyridyl)$_3$ have been prepared by the reaction of $IrCl_3 \cdot H_2O$ with a large excess of the 2-arylpyridine ligand in ethylene glycol.[152] This microwave approach eliminates the need for a dehalogenating reagent or for chromatographic purification of the products. It also gives higher yields than traditional methods using the same starting materials (YIF = 3–7.5).

7.3.4.2   Synthesis Using Scientific Microwave Apparatus

The microwave-promoted hydrozirconation of a wide variety of alkynes with $Cp_2Zr(H)$-(Cl) has been reported, the alkenylchlorobis(cyclopentadienyl) zirconium(IV) prod ucts being very useful reagents for organic synthesis.[6] The insertion reactions were complete within 5–15 min at 100 °C using toluene as the solvent (Scheme 7.20).

Eight platinum(II) complexes bearing either one or two arylalkynyl ligands have been synthesized by heating mixtures of $PtCl_2$, a tertiary phosphine, and an acetylenic compound in THF as a solvent. The resultant $Cl(PR_3)_2Pt-C\equiv C-C_6H_4R$ complexes were obtained in yields ranging from 47% and 76%.[153] The disubstituted arylalkynyl complexes $(PR_3)_2Pt(-C\equiv C-C_6H_4R)_2$ were prepared by treatment of the

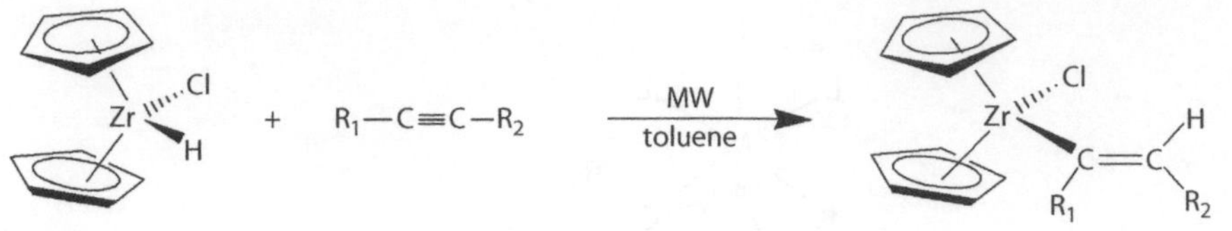

SCHEME 7.20

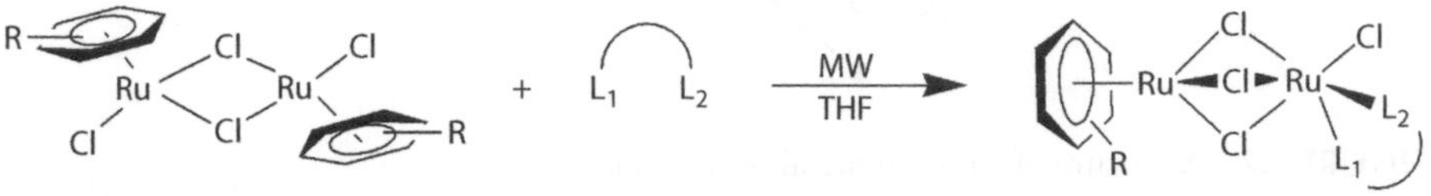

SCHEME 7.21

monosubstituted complexes with additional acetylenic compound in the presence of CuI. When using microwave heating, the isolation of $PtCl_2(PR_3)_2$ intermediates was not necessary, unlike the conventional preparative method. The platinum(II) arylalkynyl complexes $(PPh_3)_2Pt(-C\equiv C-C_6H_4R)_2$ (R = H, Me) have been prepared by heating *cis*-$Pt(N_3)_2(PPh_3)_2$ in the neat alkyne $RC_6H_4C\equiv CH$ as reagent and solvent for 1 h at 100 °C.[40]

The synthesis of a wide range of bis(aryl)zinc(II) complexes has been reported. Heating a THF solution of $(tmp)_2Zn•2MgCl_2•2LiCl$ (tmp = 2,2,6,6-tetramethyl-piperidyl) and the appropriate aromatic compounds at 120 °C for 5 h led to good yields of the desired diarylzinc species.[5] It is noteworthy that functionalities on the aromatic component such as ester, ketone, and cyano groups were left intact, the methodology therefore offering a broad substrate scope.

A series of iridium(III) complexes of the type $Ir(L–L)_3$ have been prepared by reaction of a glycerol solution of $Ir(acac)_3$ and L–L (2-arylpyridine or 2-aryl-1-quinoline) in an open vessel equipped with a reflux condenser.[154] This represented an improvement on the analogous methodology using a domestic microwave oven[152] as less ligand was required, but YIFs were only 0.82–1.2. In the case of 2-phenyl-1-quinoline, a 10% yield of the Ir complex was obtained in 20 min, while only trace amounts were produced after 10 h of conventional reflux. Cyclometallated platinum(II) complexes of the type PtCl(L–L)L′ where L–L = 2-phenylpyridine, 2-(2′-thienyl)pyridine, or benzoquinoline have also been prepared but not with any enhancement in yield (YIF = 0.58–1.0).[155]

7.3.5 MISCELLANEOUS EXAMPLES

7.3.5.1 Displacement of Arene Ligands

Dinuclear ruthenium complexes of the type $(arene)Ru(\mu\text{-}Cl)_3RuCl(L_1–L_2)$ have been prepared from $[(arene)RuCl_2]_2$ starting materials using microwave heating in a scientific unit.[156] Displacement of arene ligands by a diverse set of neutral chelating ligands containing P, S, or N donor atoms was accomplished (Scheme 7.21). The preparation of one complex $(p\text{-}cymene)Ru(\mu\text{-}Cl)_3RuCl[(S,S)\text{-}DIOP]$ was also attempted by conventional means. Refluxing in THF produced a different product, while refluxing in the higher-boiling solvent 2-ethoxyethanol gave the desired product in lower yield than in the case of microwave heating.

7.3.5.2 Synthesis of Carbene Complexes

A number of Fischer carbene complexes of chromium and tungsten have been prepared by conventional and microwave heating (YIF = 0.79–1.1).[157,158] Mono- and dimethylureas were reacted with alkynyl alkoxy Cr or W carbenes using THF as the solvent (Scheme 7.22). The resultant complexes contained a uracil-moiety. A solvent-free approach to the reaction was also attempted, but this gave somewhat lower yields of the desired products.

7.3.5.3 Synthesis of Pincer Complexes

An NCN-pincer palladium(II) complex bearing bulky diphenylhydroxymethylpyrrolidinyl substituents has been prepared (Figure 7.8a).[159] Oxidative addition of the ligand precursor (NCN-Br) to $Pd_2(dba)_3$ was accomplished with both conventional and microwave heating using chloroform as the solvent. The microwave-heating protocol gave a superior yield of the product (YIF = 1.8). Bis(benzimidazolylidene) pincer palladium(II) complexes have been synthesized using DMSO as the solvent (Figure 7.8b).[160] Pyridine-bridged bis(benzimidazolium) salts were first prepared and then reacted with $Pd(OAc)_2$ in an open vessel to yield the cationic CNC-pincer complexes with long-chain alkyl moieties.

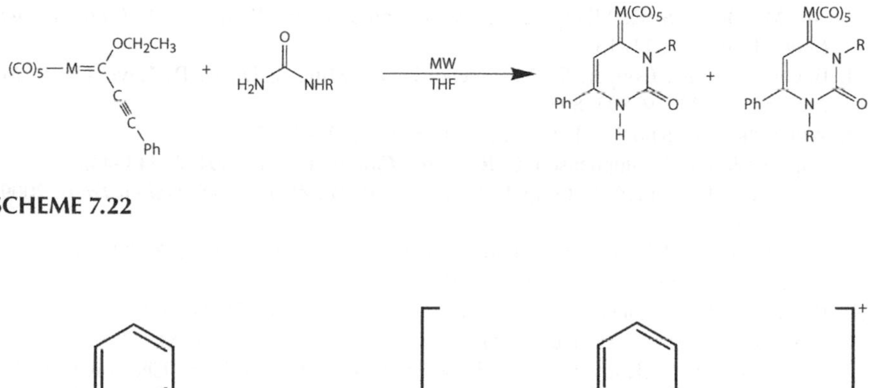

SCHEME 7.22

(a)

(b)

FIGURE 7.8 Structures of (a) an NCN-pincer Pd(II) complex and (b) a CNC-pincer Pd(II) complex.

7.4   CONCLUSION

A variety of inorganic and organometallic complexes have been prepared using microwave heating as a tool, but the field is still relatively young. Researchers have primarily focused their attention on about ten of the more common transition metals; the chemistry of many other metals remains unexplored. As this chapter has shown, microwave heating is a versatile tool that often yields the same product as conventional heating but in higher yields, shorter reaction times, and using less solvent. One particularly noteworthy application is the rapid preparation of transition-metal-containing radiopharmaceuticals for which time and efficiency are of the utmost importance.[128,149,161–164] In a number of the examples discussed in the chapter, different products are observed when using microwave heating as opposed to conventional approaches. Indeed, some transition metal complexes prepared by microwave heating could not be produced via conventional methods despite repeated efforts to do so. This gives a glimpse to the future of microwave-promoted chemistry involving the transition metals, a field that should prove very interesting as it continues to develop.

REFERENCES

1. Gedye, R.; Smith, F.; Westaway, K.; Ali, H.; Baldisera, L.; Laberge, L.; Rousell, J. *Tetrahedron Lett.* 1986, *27*, 279–282.
2. Baghurst, D. R.; Mingos, D. M. P.; Watson, M. J. *J. Organomet. Chem.* 1989, *368*, C43–C45.
3. Ali, M.; Bond, S. P.; Mbogo, S. A.; McWhinnie, W. R.; Watts, P. M. *J. Organomet. Chem.* 1989, *371*, 11–13.
4. Baghurst, D. R.; Cooper, S. R.; Greene, D. L.; Mingos, D. M. P.; Reynolds, S. M. *Polyhedron* 1990, *9*, 893–895.
5. Wunderlich, S.; Knochel, P. *Org. Lett.* 2008, *10*, 4705–4707.
6. Wipf, P.; Janjic, J.; Stephenson, C. R. J. *Org. Biomol. Chem.* 2004, *2*, 443–445.
7. Zhuang, G.-L.; Sun, X.-J.; Long, L.-S.; Huang, R.-B.; Zheng, L.-S. *Dalton Trans.* 2009, 4640–4642.
8. Jung, K. S.; Ro, J. Y.; Lee, J. Y.; Park, S. S. *J. Mater. Sci. Lett.* 2001, *20*, 2203–2205.
9. Perreux, L.; Loupy, A. *Tetrahedron* 2001, *57*, 9199–9223.
10. Barnard, T. M.; Leadbeater, N. E. *Chemical Commun.* 2006, 3615–3616.
11. Leadbeater, N. E.; Shoemaker, K. M. *Organometallics* 2008, *27*, 1254–1258.
12. Berdonosov, S. S.; Kopylova, I. A.; Lebedev, V. Y.; Voronina, N. Y.; Chesnokov, D. E.; Melikhov, I. V. *Neorg. Mater.* 1992, *28*, 1022–1028.
13. Baghurst, D. R.; Mingos, D. M. P. *J. Organomet. Chem.* 1990, *384*, C57–C60.
14. Zhilin, A. Y.; Ilyushin, M. A.; Tselinskii, I. V.; Brykov, A. S. *Russ. J. Appl. Chem.* 2001, *74*, 99–102.
15. Zhilin, A. Y.; Ilyushin, M. A.; Tselinskii, I. V. *Khim. Fiz.* 2002, *21*, 54–56.
16. Greene, D. L.; Mingos, D. M. P. *Transition Met. Chem.* 1991, *16*, 71–72.
17. Baghurst, D. R.; Mingos, D. M. P. *J. Chem. Soc., Dalton Trans.* 1992, 1151–1155.
18. Matsumura-Inoue, T.; Tanabe, M.; Minami, T.; Ohashi, T. *Chem. Lett.* 1994, 2443–2446.
19. Xiao, X.; Sakamoto, J.; Tanabe, M.; Yamazaki, S.; Yamabe, S.; Matsumura-Inoue, T. *J. Electroanal. Chem.* 2002, *527*, 33–40.

20. Liu, Y.; Hammitt, R.; Lutterman, D. A.; Joyce, L. E.; Thummel, R. P.; Turro, C. *Inorg. Chem.* 2009, *48*, 375–385.
21. Bolink, H. J.; Cappelli, L.; Coronado, E.; Graetzel, M.; Nazeeruddin, M. K. *J. Am. Chem. Soc.* 2006, *128*, 46–47.
22. Martineau, D.; Beley, M.; Gros, P. C. *J. Org. Chem.* 2006, *71*, 566–571.
23. Yoshikawa, N.; Masuda, Y.; Matsumura-Inoue, T. *Chem. Lett.* 2000, 1206–1207.
24. Masuda, Y.; Wada, S.; Nakamura, T.; Matsumura-Inoue, T. *J. Alloys Compd.* 2006, *408–412*, 1017–1021.
25. Rau, S.; Schaefer, B.; Gruessing, A.; Schebesta, S.; Lamm, K.; Vieth, J.; Goerls, H.; Walther, D.; Rudolph, M.; Grummt, U. W.; Birkner, E. *Inorg. Chim. Acta* 2004, *357*, 4496–4503.
26. Schaefer, B.; Goerls, H.; Presselt, M.; Schmitt, M.; Popp, J.; Henry, W.; Vos, J. G.; Rau, S. *Dalton Trans.* 2006, 2225–2231.
27. Rutherford, T. J.; Van Gijte, O.; Kirsch-De Mesmaeker, A.; Keene, F. R. *Inorg. Chem.* 1997, *36*, 4465–4474.
28. Anderson, P. A.; Anderson, R. F.; Furue, M.; Junk, P. C.; Keene, F. R.; Patterson, B. T.; Yeomans, B. D. *Inorg. Chem.* 2000, *39*, 2721–2728.
29. Smalley, S. J.; Waterland, M. R.; Telfer, S. G. *Inorg. Chem.* 2009, *48*, 13–15.
30. Pezet, F.; Daran, J.-C.; Sasaki, I.; Aiet-Haddou, H.; Balavoine, G. G. A. *Organometallics* 2000, *19*, 4008–4015.
31. Schwalbe, M.; Schaefer, B.; Goerls, H.; Rau, S.; Tschierlei, S.; Schmitt, M.; Popp, J.; Vaughan, G.; Henry, W.; Vos, J. G. *Eur. J. Inorg. Chem.* 2008, 3310–3319.
32. Martineau, D.; Beley, M.; Gros, P. C.; Cazzanti, S.; Caramori, S.; Bignozzi, C. A. *Inorg. Chem.* 2007, *46*, 2272–2277.
33. Amarante, D.; Cherian, C.; Emmel, C.; Chen, H.-Y.; Dayal, S.; Koshy, M.; Megehee, E. G. *Inorg. Chim. Acta* 2005, *358*, 2231–2238.
34. Matsumura-Inoue, T.; Yamamoto, Y.; Yoshikawa, N.; Terashima, M.; Yoshida, Y.; Fujii, A.; Yoshino, K. *Opt. Mater.* 2004, *27*, 187–191.
35. Yoshikawa, N.; Sakamoto, J.; Matsumura-Inoue, T.; Takashima, H.; Tsukahara, K.; Kanehisa, N.; Kai, Y. *Anal. Sci.* 2004, *20*, 711–716.
36. Mukhopadhyay, S.; Lasri, J.; Charmier, M. A. J.; Guedes da Silva, M. F. C.; Pombeiro, A., J. L. *Dalton Trans* 2007, 5297–5304.
37. Mukhopadhyay, S.; Mukhopadhyay, B. G.; Guedes da Silva, M. F. C.; Lasri, J.; Charmier, M. A. J.; Pombeiro, A. J. L. *Inorg. Chem.* 2008, *47*, 11334–11341.
38. Mukhopadhyay, S.; Lasri, J.; Guedes da Silva, M. F. C.; Charmier, M. A. J.; Pombeiro, A. J. L. *Acta Crystallogr., Sect. E: Struct. Rep. Online* 2007, *E63*, M2656–U281.
39. Smolenski, P.; Mukhopadhyay, S.; Guedes da Silva, M. F. C.; Charmier, M. A. J.; Pombeiro, A. J. L. *Dalton Trans.* 2008, 6546–6555.
40. Mukhopadhyay, S.; Lasri, J.; Guedes da Silva, M. F. C.; Januario Charmier, M. A.; Pombeiro, A. J. L. *Polyhedron* 2008, *27*, 2883–2888.
41. Charmier, A. J. M.; Kukushkin, V. Y.; Pombeiro, A. J. L. *Dalton Trans.* 2003, 2540–2543.
42. Bokach, N. A.; Krokhin, A. A.; Nazarov, A. A.; Kukushkin, V. Y.; Haukka, M.; Frausto da Silva, J. J. R.; Pombeiro, A. J. L. *Eur. J. Inorg. Chem.* 2005, 3042–3048.
43. Lasri, J.; Charmier, M. A. J.; Haukka, M.; Pombeiro, A. J. L. *J. Org. Chem.* 2007, *72*, 750–755.
44. Chowdhry, M. M.; Mingos, D. M. P.; White, A. J. P.; Williams, D. J. *Chem. Commun.* 1996, 899–900.
45. Garg, R.; Saini, M. K.; Fahmi, N.; Singh, R. V. *Transition Met. Chem.* 2006, *31*, 362–367.
46. D'Alessandro, D. M.; Keene, F. R. *Chem. Phys.* 2006, *324*, 8–25.
47. Tardiff, B. J.; Decken, A.; McGrady, G. S. *Inorg. Chem. Commun.* 2008, *11*, 44–46.

48. Mingos, D. M. P.; Baghurst, D. R. *Chem. Soc. Rev.* 1991, *20*, 1–47.
49. Jäger, M.; Kumar, R. J.; Görls, H.; Bergquist, J.; Johansson, O. *Inorg. Chem.* 2009, *48*, 3228–3238.
50. Abrahamsson, M.; Jaeger, M.; Oesterman, T.; Eriksson, L.; Persson, P.; Becker, H.-C.; Johansson, O.; Hammarstroem, L. *J. Am. Chem. Soc.* 2006, *128*, 12616–12617.
51. Abrahamsson, M.; Jaeger, M.; Kumar, R. J.; Oesterman, T.; Persson, P.; Becker, H.-C.; Johansson, O.; Hammarstroem, L. *J. Am. Chem. Soc.* 2008, *130*, 15533–15542.
52. Liu, Y.; Hammitt, R.; Lutterman, D. A.; Thummel, R. P.; Turro, C. *Inorg. Chem.* 2007, *46*, 6011–6021.
53. Abrahamsson, M.; Becker, H.-C.; Hammarstrom, L.; Bonnefous, C.; Chamchoumis, C.; Thummel, R. P. *Inorg. Chem.* 2007, *46*, 10354–10364.
54. Tan, N. Y.; Xiao, X. M.; Li, Z. L.; Matsumura-Inoue, T. *Chin. Chem. Lett.* 2004, *15*, 687–690.
55. Obara, S.; Itabashi, M.; Okuda, F.; Tamaki, S.; Tanabe, Y.; Ishii, Y.; Nozaki, K.; Haga, M. *Inorg. Chem.* 2006, *45*, 8907–8921.
56. Yang, L.; Okuda, F.; Kobayashi, K.; Nozaki, K.; Tanabe, Y.; Ishii, Y.; Haga, M.-A. *Inorg. Chem.* 2008, *47*, 7154–7165.
57. Bhojak, N.; Gudasaria, D. D.; Khiwani, N.; Jain, R. *E-J. Chem.* 2007, *4*, 232–243.
58. Glasson, C. R. K.; Meehan, G. V.; Clegg, J. K.; Lindoy, L. F.; Smith, J. A.; Keene, F. R.; Motti, C. *Chem.—Eur. J.* 2008, *14*, 10535–10538.
59. Juris, A.; Balzani, V.; Barigelletti, F.; Campagna, S.; Belser, P.; von Zelewsky, A. *Coord. Chem. Rev.* 1988, *84*, 85–277.
60. Crosby, G. A.; Watts, R. J. *J. Am. Chem. Soc.* 1971, *93*, 3184–3188.
61. Liu, L.-C.; Lee, C.-C.; Hu, A. T. *J. Porphyrins Phthalocyanines* 2001, *5*, 806–807.
62. Liu, M. O.; Hu, A. T. *J. Organomet. Chem.* 2004, *689*, 2450–2455.
63. Liu, M. O.; Tai, C.-H.; Wang, W.-Y.; Chen, J.-R.; Hu, A. T.; Wei, T.-H. *J. Organomet. Chem.* 2004, *689*, 1078–1084.
64. Safari, N.; Jamaat, P. R.; Pirouzmand, M.; Shaabani, A. *J. Porphyrins Phthalocyanines* 2004, *8*, 1209–1213.
65. Achar, B. N.; Kumar, T. M. M.; Lokesh, K. S. *J. Porphyrins Phthalocyanines* 2005, *9*, 872–879.
66. Bahadoran, F.; Dialameh, S. *J. Porphyrins Phthalocyanines* 2005, *9*, 163–169.
67. Jain, N.; Kumar, A.; Chauhan, S. M. S. *Synth. Commun.* 2005, *35*, 1223–1230.
68. Liu, M. O.; Tai, C.-H.; Hu, A. T. *Mater. Chem. Phys.* 2005, *92*, 322–326.
69. Singh, R.; Geetanjali *Asian J. Chem.* 2005, *17*, 612–614.
70. Jayashankar, L.; Sundar, B. S.; Vijayaraghavan, R.; Betanabhatla, K. S.; Christina, A. J. M.; Athimoolam, J.; Saravanan, K. S. *Pharmacology Online* 2008, 66–68.
71. Dean, M. L.; Schmink, J. R.; Leadbeater, N. E.; Brueckner, C. *Dalton Trans.* 2008, 1341–1345.
72. Shaabani, A. *J. Chem. Res., Synop.* 1998, 672–673.
73. Shaabani, A.; Bahadoran, F.; Bazgir, A.; Safari, N. *Iran. J. Chem. Chem. Eng.* 1999, *18*, 104–107.
74. Shaabani, A.; Bahadoran, F.; Safari, N. *Indian J. Chem., Sect. A: Inorg., Bio-inorg., Phys., Theor. Anal. Chem.* 2001, *40A*, 195–197.
75. Villemin, D.; Hammadi, M.; Hachemi, M.; Bar, N. *Molecules* 2001, *6*, 831–844.
76. Shaabani, A.; Safari, N.; Bazgir, A.; Bahadoran, F.; Sharifi, N.; Jamaat, P. R. *Synth. Commun.* 2003, *33*, 1717–1725.
77. Kahveci, B.; Sasmaz, S.; Ozil, M.; Kantar, C.; Kosar, B.; Buyukgungor, O. *Turk. J. Chem.* 2006, *30*, 681–689.
78. Biyiklioglu, Z.; Kantekin, H. *Transition Met. Chem.* 2007, *32*, 851–856.
79. Biyiklioglu, Z.; Kantekin, H.; Oezil, M. *J. Organomet. Chem.* 2007, *692*, 2436–2440.

80. Kahveci, B.; Oezil, M.; Kantar, C.; Sasmaz, S.; Isik, S.; Koeysal, Y. *J. Organomet. Chem.* 2007, *692*, 4835–4842.
81. Kantekin, H.; Biyiklioglu, Z. *Dyes Pigm.* 2007, *77*, 98–102.
82. Özil, M.; Agar, E.; Sasmaz, S.; Kahveci, B.; Akdemir, N.; Guemruekcueoglu, I. E. *Dyes Pigm.* 2007, *75*, 732–740.
83. Serbest, K.; Degirmencioglu, I.; Uenver, Y.; Er, M.; Kantar, C.; Sancak, K. *J. Organomet. Chem.* 2007, *692*, 5646–5654.
84. Shaabani, A.; Maleki-Moghaddam, R.; Maleki, A.; Rezayan, A. H. *Dyes Pigm.* 2007, *74*, 279–282.
85. Bekircan, O.; Biyiklioglu, Z.; Acar, I.; Bektas, H.; Kantekin, H. *J. Organomet. Chem.* 2008, *693*, 3425–3429.
86. Biyiklioglu, Z.; Acar, I.; Kantekin, H. *Inorg. Chem. Commun.* 2008, *11*, 630–635.
87. Biyiklioglu, Z.; Kantekin, H. *Polyhedron* 2008, *27*, 1650–1654.
88. Biyiklioglu, Z.; Kantekin, H. *J. Organomet. Chem.* 2008, *693*, 505–509.
89. Biyiklioglu, Z.; Kantekin, H.; Acar, I. *Inorg. Chem. Commun.* 2008, *11*, 1448–1451.
90. Celenk, E.; Kantekin, H. *Dyes Pigm.* 2008, *80*, 93–97.
91. Kantar, C.; Akdemir, N.; Agar, E.; Ocak, N.; Sasmaz, S. *Dyes Pigm.* 2008, *76*, 7–12.
92. Kantekin, H.; Biyiklioglu, Z. *Dyes Pigm.* 2008, *77*, 432–436.
93. Kantekin, H.; Biyiklioglu, Z.; Celenk, E. *Inorg. Chem. Commun.* 2008, *11*, 633–635.
94. Kantekin, H.; Celenk, E.; Karadeniz, H. *J. Organomet. Chem.* 2008, *693*, 1353–1358.
95. Lokesh, K. S.; Uma, N.; Achar, B. N. *Polyhedron* 2009, *28*, 1022–1028.
96. Maeda, F.; Uno, K.; Ohta, K.; Sugibayashi, M.; Nakamura, N.; Matsuse, T.; Kimura, M. *J. Porphyrins Phthalocyanines* 2003, *7*, 58–69.
97. Kogan, E. G.; Ivanov, A. V.; Tomilova, L. G.; Zefirov, N. S. *Mendeleev Commun.* 2002, 54–55.
98. Ooi, K.; Maeda, F.; Ohta, K.; Takizawa, T.; Matsuse, T. *J. Porphyrins Phthalocyanines* 2005, *9*, 544–553.
99. Jung, K. S.; Ko, J. P.; Park, S. B.; Park, C. Y.; Min, S. K.; Kwon, J. H.; Oh, I. H.; Park, S. S. *Polymer-Korea* 2001, *25*, 609–616.
100. Jung, K. S.; Kwon, J. H.; Shon, S. M.; Ko, J. P.; Shin, J. S.; Park, S. S. *J. Mater. Sci.* 2004, *39*, 723–726.
101. Jung, K. S.; Kwon, J. H.; Son, S. M.; Shin, J. S.; Lee, G. D.; Park, S. S. *Synth. Met.* 2004, *141*, 259–264.
102. Makhseed, S.; Al-Sawah, M.; Samuel, J.; Manaa, H. *Tetrahedron Lett.* 2009, *50*, 165–168.
103. Maksimov, A. Y.; Ivanov, A. V.; Blikova, Y. N.; Tomilova, L. G.; Zefirov, N. S. *Mendeleev Commun.* 2003, 70–72.
104. Burczyk, A.; Loupy, A.; Bogdal, D.; Petit, A. *Tetrahedron* 2005, *61*, 179–188.
105. Chandrasekharam, M.; Rao, C. S.; Singh, S. P.; Kantam, M. L.; Reddy, M. R.; Reddy, P. Y.; Toru, T. *Tetrahedron Lett.* 2007, *48*, 2627–2630.
106. Luzyanin, K. V.; Kukushkin, V. Y.; Kuznetsov, M. L.; Garnovskii, D. A.; Haukka, M.; Pombeiro, A. J. L. *Inorg. Chem.* 2002, *41*, 2981–2986.
107. Lasri, J.; Charmier, M. A. J.; Guedes da Silva, M. F. C.; Pombeiro, A. J. L. *Dalton Trans.* 2006, 5062–5067.
108. Lasri, J.; Guedes da Silva, M. F. C.; Charmier, M. A. J.; Pombeiro, A. J. L. *Eur. J. Inorg. Chem.* 2008, 3668–3677.
109. Anderson, T. J.; Scott, J. R.; Millett, F.; Durham, B. *Inorg. Chem.* 2006, *45*, 3843–3845.
110. Mueller-Buschbaum, K.; Quitmann, C. C. *Inorg. Chem.* 2006, *45*, 2678–2687.
111. (a) Adams, R. D. In *The Chemistry of Metal Cluster Complexes*; Shriver, D. F.; Kaesz, H. D., Adams, R. D., Eds.; VCH: New York, 1990; pp 121–170. (b) Vargas, M. D.; Nicholls, J. N. *Adv. Inorg. Chem.* 1986, *30*, 123–222.

112. Whittaker, A. G.; Mingos, D. M. P. *J. Chem. Soc., Dalton Trans.* 1995, 2073–2079.
113. Rutherford, T. J.; Keene, F. R. *J. Chem. Soc., Dalton Trans.* 1998, 1155–1162.
114. Hegedus, L. S.; Sundermann, M. J.; Dorhout, P. K. *Inorg. Chem.* 2003, *42*, 4346–4354.
115. Milios, C. J.; Vinslava, A.; Whittaker, A. G.; Parsons, S.; Wernsdorfer, W.; Christou, G.; Perlepes, S. P.; Brechin, E. K. *Inorg. Chem.* 2006, *45*, 5272–5274.
116. Millos, C. J.; Whittaker, A. G.; Brechin, E. K. *Polyhedron* 2007, *26*, 1927–1933.
117. Gass, I. A.; Milios, C. J.; Whittaker, A. G.; Fabiani, F. P. A.; Parsons, S.; Murrie, M.; Perlepes, S. P.; Brechin, E. K. *Inorg. Chem.* 2006, *45*, 5281–5283.
118. Milios, C. J.; Prescimone, A.; Sanchez-Benitez, J.; Parsons, S.; Murrie, M.; Brechin, E. K. *Inorg. Chem.* 2006, *45*, 7053–7055.
119. Johnson, K. D.; Powell, G. L. *J. Organomet. Chem.* 2008, *693*, 1712–1715.
120. Delgado, S.; Sanz Miguel, P. J.; Priego, J. L.; Jimenez-Aparicio, R.; Gomez-Garcia, C. J.; Zamora, F. *Inorg. Chem.* 2008, *47*, 9128–9130.
121. Li, J.-T.; Ma, Y.-S.; Li, S.-G.; Cao, D.-K.; Li, Y.-Z.; Song, Y.; Zheng, L.-M. *Dalton Trans.* 2009, 5029–5034.
122. Jandrasics, E. Z.; Keene, F. R. *J. Chem. Soc., Dalton Trans.* 1997, 153–159.
123. VanAtta, S. L.; Duclos, B. A.; Green, D. B. *Organometallics* 2000, *19*, 2397–2399.
124. Ardon, M.; Hayes, P. D.; Hogarth, G. *J. Chem. Educ.* 2002, *79*, 1249–1251.
125. Ardon, M.; Hogarth, G.; Oscroft, D. T. W. *J. Organomet. Chem.* 2004, *689*, 2429–2435.
126. Coelho, A. C.; Paz, F. A. A.; Klinowski, J.; Pillinger, M.; Goncalves, I. S. *Molecules* 2006, *11*, 940–952.
127. Lee, Y. T.; Choi, S. Y.; Lee, S. T.; Chung, Y. K.; Kang, T. J. *Tetrahedron Lett.* 2006, *47*, 6569–6572.
128. Causey, P. W.; Besanger, T. R.; Schaffer, P.; Valliant, J. F. *Inorg. Chem.* 2008, *47*, 8213–8221.
129. Jung, J. Y.; Newton, B. S.; Tonkin, M. L.; Powell, C. B.; Powell, G. L. *J. Organomet. Chem.* 2009, *694*, 3526–3528.
130. Dyson, P. J.; McIndoe, J. S. *Transition Metal Carbonyl Cluster Chemistry*; Gordon and Breach: Amsterdam, 2000, pp 73–87.
131. Dabirmanesh, Q.; Roberts, R. M. G. *J. Organomet. Chem.* 1993, *460*, C28–C29.
132. Dabirmanesh, Q.; Fernando, S. I. S.; Roberts, R. M. G. *J. Chem. Soc., Perkin Trans. 1* 1995, 743–749.
133. Dabirmanesh, Q.; Roberts, R. M. G. *J. Organomet. Chem.* 1997, *542*, 99–103.
134. Roberts, R. M. G. *J. Organomet. Chem.* 2006, *691*, 2641–2647.
135. Roberts, R. M. G. *J. Organomet. Chem.* 2006, *691*, 4926–4930.
136. Poonia, K.; Maanju, S.; Chaudhary, P.; Singh, R. V. *Transition Met. Chem.* 2007, *32*, 204–208.
137. Whittaker, A. G.; Mingos, D. M. P. *J. Chem. Soc., Dalton Trans.* 2002, 3967–3970.
138. Kuhnert, N.; Danks, T. N. *J. Chem. Res., Synop.* 2002, 66–68.
139. Taylor, C. J.; Motevalli, M.; Richards, C. J. *Organometallics* 2006, *25*, 2899–2902.
140. Harcourt, E. M.; Yonis, S. R.; Lynch, D. E.; Hamilton, D. G. *Organometallics* 2008, *27*, 1653–1656.
141. Sinha, S.; Srivastava, A. K.; Sengupta, S. K.; Pandey, O. P. *Transition Met. Chem.* 2008, *33*, 563–567.
142. Imrie, C.; Elago, E. R. T.; Williams, N.; McCleland, C. W.; Engelbrecht, P. *J. Organomet. Chem.* 2005, *690*, 4959–4966.
143. Birdwhistell, K. R.; Nguyen, A.; Ramos, E. J.; Kobelja, R. *J. Chem. Educ.* 2008, *85*, 261–262.
144. Pedotti, S.; Patti, A. *J. Organomet. Chem.* 2008, *693*, 1375–1381.
145. Berardi, S.; Conte, V.; Fiorani, G.; Floris, B.; Galloni, P. *J. Organomet. Chem.* 2008, *693*, 3015–3020.

146. Tu, S.-J.; Yan, S.; Cao, X.-D.; Wu, S.-S.; Zhang, X.-H.; Hao, W.-J.; Han, Z.-G.; Shi, F. *J. Organomet. Chem.* 2009, *694*, 91–96.
147. Baghurst, D. R.; Copley, R. C. B.; Fleischer, H.; Mingos, D. M. P.; Kyd, G. O.; Yellowlees, L. J.; Welch, A. J.; Spalding, T. R.; O'Connell, D. *J. Organomet. Chem.* 1993, *447*, C14–C17.
148. Green, A. E. C.; Causey, P. W.; Louie, A. S.; Armstrong, A. F.; Harrington, L. E.; Valliant, J. F. *Inorg. Chem.* 2006, *45*, 5727–5729.
149. Armstrong, A. F.; Valliant, J. F. *Inorg. Chem.* 2007, *46*, 2148–2158.
150. Causey, P. W.; Besanger, T. R.; Valliant, J. F. *J. Med. Chem.* 2008, *51*, 2833–2844.
151. Kidwai, M.; Misra, P.; Bhushan, K. R. *Polyhedron* 1999, *18*, 2641–2643.
152. Konno, H.; Sasaki, Y. *Chem. Lett.* 2003, *32*, 252–253.
153. Carlsson, M.; Eliasson, B. *Organometallics* 2006, *25*, 5500–5502.
154. Saito, K.; Matsusue, N.; Kanno, H.; Hamada, Y.; Takahashi, H.; Matsumura, T. *Jpn. J. Appl. Phys., Part 1* 2004, *43*, 2733–2734.
155. Godbert, N.; Pugliese, T.; Aiello, I.; Bellusci, A.; Crispini, A.; Ghedini, M. *Eur. J. Inorg. Chem.* 2007, 5105–5111.
156. Albrecht, C.; Gauthier, S.; Wolf, J.; Scopelliti, R.; Severin, K. *Eur. J. Inorg. Chem.* 2009, 1003–1010.
157. Spinella, A.; Caruso, T.; Pastore, U.; Ricart, S. *J. Organomet. Chem.* 2003, *684*, 266–268.
158. Artillo, A.; Della Sala, G.; De Santis, M.; Llordes, A.; Ricart, S.; Spinella, A. *J. Organomet. Chem.* 2007, *692*, 1277–1284.
159. Gosiewska, S.; Martinez Herreras, S.; Lutz, M.; Spek, A. L.; Havenith, R. W. A.; van Klink, G. P. M.; van Koten, G.; Klein Gebbink, R. J. M. *Organometallics* 2008, *27*, 2549–2559.
160. Tu, T.; Bao, X.; Assenmacher, W.; Peterlik, H.; Daniels, J.; Dotz, K. H. *Chem. Eur. J.* 2009, *15*, 1853–1861.
161. Park, S. H.; Gwon, H. J.; Park, K. B. *Chem. Lett.* 2004, *33*, 1278–1283.
162. Gwon, H. J.; Jang, S. H.; Park, S. H.; Shin, B. C. *J. Korean Chem. Soc.* 2005, *49*, 370–376.
163. Park, S. H.; Gwon, H. J.; Jang, S. H. *Asian J. Chem.* 2007, *19*, 391–395.
164. Green, A. E. C.; Harrington, L. E.; Valliant, J. F. *Can. J. Chem.* 2008, *86*, 1063–1069.

8 Microwave Heating as a Tool for Materials Chemistry

Steven L. Suib and Nicholas E. Leadbeater

CONTENTS

8.1 OVERVIEW AND INTRODUCTION

The aim of this chapter is to show the use of microwave heating for the production of inorganic materials. The materials under dicsussion are separated into specific types such as adsorbents, batteries, ceramics, catalysts, and semiconductors. Attention will focus on work published from 2006 to 2009, although even a complete review of this relatively short period would be relatively expansive. Various unique aspects of microwave heating such as microwave frequency, novel apparatus, power levels, and other parameters will be discussed.

There are several advantages of using microwave heating to produce materials. High product yields and purity are often found, materials with uniform structural properties can be obtained, and novel physical properties can be observed. In addition, conditions that are often required in the synthesis of materials such as high pressures, catalysts, templates, structure directors, and vacuum procedures can often be avoided. As a result, microwave heating can be cleaner, faster, simpler, and economically advantageous as compared to typical thermal heating methods. In many cases, experiments can also be scaled to produce larger amounts of material.

A wide variety of equipment for performing reactions using microwave heating is commercially available, as well as custom-built units in research and development laboratories. Most work is currently carried out in microwave units that operate at a fixed frequency of 2.45 GHz, with both open- and closed-vessel processing being possible. Addition of liquid, gaseous, or even solid reagents can be achieved via entry ports. For scale-up, both batch and continuous-flow reactors can be used. Variable-frequency equipment, while not as prevalent as fixed-frequency analogs, is available. This allows access to a range between 2.4 and 7.0 GHz, with bandwidth sweep rates of 100 to 0.1 s. The capability of this equipment to allow for careful control of reaction parameters is highly advantageous, as well as making chemistry reproducible.

8.2   MICROWAVE HEATING FOR PREPARATION OF ADSORBENTS

A general literature search suggests that microwave syntheses of adsorbents is a somewhat limited area in comparison to other areas such as catalyst or ceramic production. However, microwave heating can have a profound effect on the properties of adsorbents. As an example, when microwave heating was used to activate clays in the presence of hydrochloric acid, both the structural and textural properties of the clay were influenced.[1] Specific properties that were studied include crystallinity of the resultant materials, pore size and distributions, surface areas, and morphology. Thermal stability was studied using differential thermal analyses. BET surface area increased and pores were created, microwave power and reaction time being important parameters. Above 500 W, detrimental effects on porosity and surface area were observed. Pore volumes increased by a factor of 5 when using microwave heating.

Carbon/silica adsorbents or carbosils have been prepared by both conventional and microwave heating.[2] These materials were produced by pyrolysis of CH_2Cl_2 on microporous silica gel surfaces for 30 min to 6 h at 550 °C. The resultant materials were hydrothermally treated with steam or liquid water using either a conventional autoclave or a microwave unit. As with the clay materials,[1] hydrothermal treatment using microwave irradiation leads to a significant increase of surface area and total pore volume of the carbosils, as compared to conventional methods.

A final example of the influence of microwave treatment on adsorption materials involves the synthesis of mesoporous silica materials known as FUD-1.[3] Preparations of FUD-1 used microwave heating both with and without humic acid treatment. The materials were synthesized with a poly(ethylene oxide)-poly-(butylene oxide)-poly(ethylene oxide) triblock copolymer as a template and tetraethylorthosilicate (TEOS) as the source of silica. In addition to being well characterized, the resultant materials were also tested for their ability to adsorb Cd^{2+} ions from solution at a fixed pH of 6. The incorporation of humic acid into FUD-1 led to significantly higher adsorption capacity of Cd^{2+} ions than materials without humic acid. As a result, these materials seem to be promising adsorbents for removal of cadmium and related heavy metal ions from aqueous waste streams.

8.3   MICROWAVE HEATING FOR PREPARATION OF BATTERY MATERIALS

Lithium ion batteries are the focus of much current research in connection with their use in vehicles. Among the cathode materials that have been studied, $LiFePO_4$ is a good candidate for such applications due to its very high theoretical capacity of 170 mAh/g. In addition, $LiFePO_4$ is cheap to make, is an environmentally friendly material, and shows good thermal stability. Carbon is typically added to increase conductivity and a one-step synthesis has been reported for the preparation of carbon-modified nanocrystalline $LiFePO_4$ using microwave heating. The procedure started from a sol-gel precursor of lithium acetate hydrate, ferrous sulfate, phosphoric acid, citric acid, and polyethylene glycol.[4] The precursor was pressed into a small pellet and placed in a porcelain boat before being heated using microwave irradiation. The control of heating time was critical (18 min) as was the carbon content (4.6%) in order to obtain a material that showed optimal electrochemical behavior including a very high specific capacity of 152 mAh/g and good retention of charge. The enhanced electrochemical properties of these materials were attributed to both the excellent electrical conductivity and the small nano-size particles (300 nm) formed in the modification with carbon.

Tin (II) oxide battery materials have been prepared in a few minutes in a sol-gel-assisted protocol using microwave heating.[5] Citric acid was used as a chelating agent and a source of amorphous carbon that stopped aggregation of particles and accelerated the nucleation of SnO_2 particles. The electrochemical performance in battery studies was good and positively influenced by remaining carbon deposits in the SnO_2 particles, similar to the case of the $LiFePO_4$ systems discussed above. The carbon content was found to be inversely proportional to the length of time the samples were irradiated with microwave energy. The in-situ formation of amorphous carbon provides a point for local heating, leading to the formation of nanocrystalline SnO_2 in very short times, optimal properties being observed in materials prepared after heating for 6 min. The specific capacity values of SnO_2 prepared using other methods were lower than those observed using microwave heating, a first cycle discharge capacity of 2156 mAh/g and corresponding charge capacity of 898 mAh/g being measured. More thorough studies suggest that microwave heating and the in-situ incorporation of carbon both lead to different particle sizes and enhanced electrochemical activity. Indeed, the observed capacities were greater than theoretical values. Given the success of this methodology for preparing SnO_2 materials, it may also be useful for synthesis of other metal oxide systems.

Mixed transition metal oxide/lithium containing materials have been prepared in short reaction times using microwave heating.[6] These materials may be useful in secondary rechargeable battery applications. To prepare $Li[Ni_{0.4}Co_{0.2}Mn_{0.4}]O_2$ powders, spherical $[Ni_{0.4}Co_{0.2}Mn_{0.4}](OH)_2$ was first synthesized via coprecipitation. This hydroxide precursor was then heated with LiOH for 10 min using 1200 W microwave irradiation. Highly crystalline materials were formed and could be used to prepare a variety of other lithium-containing battery materials. Intercalation of lithium in such systems is critical and there is a working hypothesis that enhanced porosity and small size crystallinity leads to better performance. Materials were also prepared using conventional heating. While specific discharge capacities in the range of 155–157 mAh/g were observed for materials prepared using both heating methods, after 30 cycles the material made conventionally started to slowly degrade in capacity. No such degradation was observed with the materials prepared using microwave heating. The results are shown in Figure 8.1.

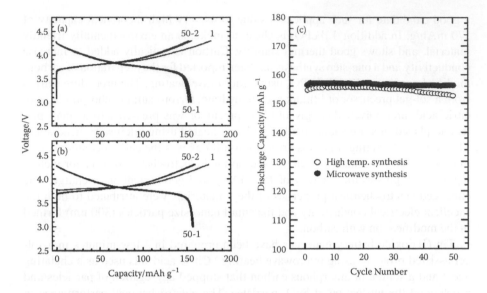

FIGURE 8.1   Continuous charge and discharge curves of Li/Li-[$Ni_{0.4}Co_{0.2}Mn_{0.4}$]O_2 cells at 30 °C; (a) Li[$Ni_{0.4}Co_{0.2}Mn_{0.4}$]O_2 prepared by high-temperature calcination; (b) Li[$Ni_{0.4}Co_{0.2}Mn_{0.4}$]O_2 synthesized using microwave heating (pelletized powder mixture); (c) their corresponding cyclabilities. (Reproduced with permission from Lee, K. S.; Myung, S. T.; Sun, Y. K. *Chem. Mater.*, 2007, *19*, 2727. Copyright American Chemical Society.)

Metal sulfide based systems have also shown potential as battery materials. As a result, there has been a recent interest in the preparation of low-dimensional metal sulfide materials including nano-size wires, helices, ropes, whiskers, tubes, and other morphologies. Tin sulfides in particular can be prepared in many of these forms and how such different morphologies influence physical and chemical properties is under investigation. Microwave heating has been used to prepare these materials. For example, nano-flakes of tin sulfide (SnS and SnS_2) have been prepared rapidly (10–40 min) in an excellent yield (93%).[7] Stannic chloride and stannous chloride were combined with thiourea and the reactions performed in an open vessel using ethylene glycol as solvent and under an inert atmosphere to prevent oxidation. Other routes involved use of elemental sulfur and sodium thiosulfate. The electrochemical performance of these materials as Li-insertion precursors was investigated in a number of electrolytes and was found to be highly sensitive to the solution composition. A stable, reversible capacity higher than 600 mAh/g could be obtained with SnS electrodes. Amorphous SnS_2 electrodes showed a much better reversible behavior compared to the crystalline phase, but worse than that of SnS electrodes.

The importance of nano-size materials for successful battery design is becoming increasingly apparent.[8] One example is in the production of spinel materials having the composition $Li_4Ti_5O_{12}$. An important feature of these materials is that the lattice parameters are almost exactly the same when lithium ions are intercalated and then when they are removed. As a result, insertion and extraction processes involving lithium ions are extremely reversible. This suggests that the framework of $Li_4Ti_5O_{12}$

is very stable and therefore would be well suited for the continual charge and discharge cycles found in lithium batteries. Nano-size crystallites on the order of 40 to 50 nm have been produced by microwave heating using Li_2CO_3 and TiO_2 as starting materials. Both the microwave power and reaction time needed to be optimized in order to obtain materials with the best electrochemical properties. Only when using a relatively high power of 700 W could pure $Li_4Ti_5O_{12}$ spinel structures be formed. A stable voltage plateau, high discharge capacities, and excellent cycling behavior were observed for anodes of $Li_4Ti_5O_{12}$ made this way.

These examples show clearly that microwave heating can be used to produce novel materials for battery precursors. The capacity of these materials and their degradation during cycling need to be maximized and stabilized, respectively. In most cases, stability and enhanced capacity depend on several other parameters in the tested battery composite such as the inclusion of additives such as carbon to increase conductivity, the size and morphologies of the components, as well as the porosity of the materials. Given the rapid reaction rates possible and the ease with which parameters can be varied, microwave heating proves to be a very versatile approach to the research and development of new battery materials.

8.4 MICROWAVE HEATING FOR PREPARATION OF CERAMICS

Ceramic materials are used in a wide variety of applications including biocomposites, inorganic membranes, films, biosensors, glasses, piezoelectrics, fuel cells, batteries, catalyst supports, packaging materials, and others. Multiferric $BiFeO_3$ ceramics have been synthesized by high-temperature magnetic annealing using nanosized precursor powders prepared through microwave combustion.[9] These materials showed excellent magnetic effects as well as electrical polarization when samples were annealed with an external magnetic field strength of 10 T. These strong magnetic annealing procedures are perhaps versatile approaches for the production of new metal oxides, mixed metal oxides, and other materials.

Sol-gel approaches offer a convenient route to ceramic materials. Using conventional heating, these procedures can take up to 48 h to perform. However, when using microwave heating, ceramics can be prepared in a matter of minutes. Such an approach has been used to prepare carbon-based ceramic electrode materials, the carbon originating from graphite powder which was added to the sol–gel mixture.[10] Before microwave heating, ultrasound was used in order to influence the gelation time and dispersion of the graphite particles in the sol–gel mixture. This is critical because, during gelation of the sol, several parameters are changing and the physical properties of the resultant materials can vary markedly. If the heating step was performed in a flat sample holder, burning of the reaction mixture was observed within a few seconds of microwave irradiation. This effect was likely due to escape and subsequent combustion of flammable trapped organics used in the preparation of the sol–gel. If the reaction is performed in a test tube, these undesired effects are attenuated. This example should serve as a note of caution when performing reactions using very highly microwave absorbing components such as graphite or silicon carbide.

A comparison of samples made by conventional thermal methods and those made using microwave heating show no difference in composition. However, the resultant properties when used as an electrode for detection of dopamine are different. Microwave heating approach leads to materials with considerably greater sensitivity than those prepared using other synthetic approaches. Of additional interest is that when the materials were air dried and then used as electrodes for the detection of dopamine, they were found not to have such wide linear ranges of operation and low detection limits compared to their non air-dried counterparts.

Zirconium-doped barium titanate having a stoichiometry of $BaZr_{0.10}Ti_{0.90}O_3$ has been prepared by both conventional and microwave sintering.[11] The cycle time of 4 h using microwave heating is significantly shorter than the 22 h required conventionally. Structural, piezoelectric, dielectric, ferroelectric, and chemical properties of these two sets of materials have been compared. Structural studies show that the microwave heating process allows diffusion of zirconium into the barium titanate to lead to the formation of barium zirconium titanate. These ceramic materials have fine grains as determined by scanning electron microscopy and are dense microstructures. The ceramics prepared using microwave heating also showed better electrical properties, higher resistivity, higher dielectric constants, lower dielectric losses, and reduced dependencies on frequency. Electrical impedance spectroscopy and electric modulus spectroscopy analyses in the frequency range of 40 Hz to 1 MHz and at high-temperatures of 300 °C to 500 °C were performed. In the case of room temperature applications, devices made using the microwave heating route would be more attractive. At higher temperatures samples prepared this way were less stable than those prepared conventionally.

Similar comparisons between conventional and microwave heating have been made in the preparation of nano-sized silicon carbide powders from carbon and silica as starting materials (Figure 8.2).[12] In the conventional method the nanopowders were prepared after 105 min heating at 1500 °C in a coke-bed using an electrical tube furnace. Electron microscopy studies of these powders showed the existence of equiaxed SiC nanopowders with an average particle size of 8.2 nm. Using microwave heating the powders were prepared in 60 min at 1454 °C; the product consisting of a mixture of SiC nanopowders with two average particle sizes of 13.6 and 58.2 nm and particles in the shape of long strands with an average diameter of 330 nm.

Microwave-assisted hydrothermal methods can also be used for preparing ceramics. This involves placing a sealed vessel containing the reagents into the microwave cavity. One example of the use of microwave-assisted hydrothermal synthesis involves the preparation of cubic-shaped $In(OH)_3$ particles from a mixed aqueous solution of $InCl_3$ and urea.[13] These particles were then used to probe the kinetics of the thermal decomposition of $In(OH)_3$ to cubic-In_2O_3.

Microwave heating has been used as a tool for the formation of crystals of lead tungstate ($PbWO_4$).[14] Different morphologies, such as tetragonal flowerlike, amygdaloidal, and multiple-lamellar, were observed. The rapid heating that is possible using microwave irradiation was important for fast nucleation and in the growth of crystals. In addition, the authors suggest that direct coupling of polar molecules in the reaction mixture with the electric field of the microwave irradiation may result in transient, anisotropic microdomains within the reaction mixture, this facilitating the anisotropic growth of $PbWO_4$. The outcome of the reaction was highly dependent on a variety of parameters including pH and the concentration of the surfactant used (cetyltrimethylammonium bromide). The products were found to have unique

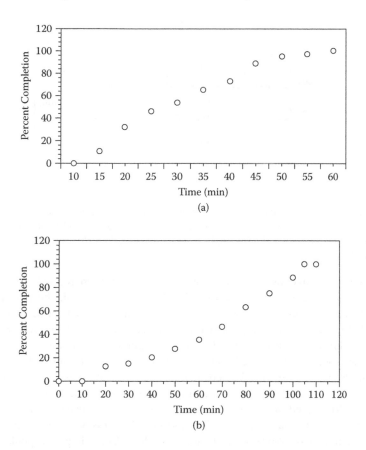

FIGURE 8.2 SiC formation vs time using (a) microwave heating and (b) conventional heating. (Reproduced with permission from Ebadzadeh, T.; Marzban-Rad, E., *Mat. Char.*, 2009, *60*, 69–72. Copyright Elsevier.)

room temperature luminescence properties, showing a high intensity of emission with respect to larger bulk phase materials.

A range of ZnO ceramic materials has been prepared using either hydrazine hydrate and ammonia as reagents in solution in conjunction with microwave heating.[15] Flower-shaped highly crystalline nano-size ZnO particles were produced when hydrazine hydrate was used. In the case of ammonia, spherical agglomerated particles were the primary products. Some of the factors studied in the synthesis were the reaction time, microwave power used, concentration of reagents, and the relative stoichiometry of the starting materials. An added feature of this procedure was that gram quantity pure product could be produced.

Zirconia (ZrO_2) is an important material as an oxygen ion transport agent in fuel cell electrodes, in ceramic processing, and in catalysis. A range of zirconia materials has been prepared using microwave heating using zirconium acetate as a precursor.[16] They were characterized by a variety of methods to study structure and morphology. The average particle size of the nano-size crystallites was found to be 10 nm.

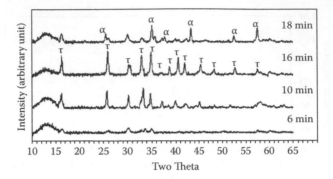

FIGURE 8.3   XRD patterns of precursor powders heated in a microwave furnace for different times (peaks have been identified as (τ) Γ-Al$_2$O$_3$ and (α) α-Al$_2$O$_3$). (Reproduced with permission from Ebadzadeh, T.; Sharifi, L., *J. Am. Cer. Soc.*, 2008, *91*, 3408–3409. Copyright Wiley-Blackwell.)

Ceramic materials based on alumina have also been prepared using microwave heating.[17] There are various phases of alumina and, in general, the nature of the starting materials often controlling the result. Using microwave heating Γ-Al$_2$O$_3$ was prepared (Figure 8.3). The approach was relatively simple and involved heating a mixture of aluminum nitrate and carboxymethyl cellulose. A silicon carbide crucible was used as a heating aid, SiC efficiently absorbing microwave energy due to its high loss factor. Reactions were performed at constant microwave power for periods ranging from 6 to 18 min. At shorter times, and hence lower temperatures (up to 1189 °C) Γ-Al$_2$O$_3$ predominated, scanning electron microscopy showing a plate-like morphology. There was no formation of one of the most stable forms of alumina, the α phase, until temperatures of above 1189 °C were reached. Even then, the peaks for the Γ-Al$_2$O$_3$ phase were still very apparent in the powder diffraction pattern of the product. The fact that the Γ-to-α-Al$_2$O$_3$ transformation occurs at such high temperatures is noteworthy. This has been attributed to the effect of the fast heating possible using microwave irradiation. Since the stable α-Al$_2$O$_3$ phase is formed through nucleation and growth mechanisms, as the heating rate becomes faster the transition point shifts to higher temperatures.

Some procedures involve the use of microwave heating in only one step of a process. For example, platinum nanoparticles have been prepared in a route where one step involves a polyol reduction performed using microwave heating.[18] A versatile sol-gel route had been developed for the synthesis of hierarchical porous silica membranes. These membranes were then used as a support for the highly dispersed platinum nanoparticles. Silica sol and Pt nanoparticles were combined to form a suspension which was then exposed to microwave irradiation for 40 s. Polystyrene latex spheres were finally added as macrotemplates, the entire procedure being performed in one pot. Thin films were then formed by dip coating onto glass slides. The as-prepared films were finally thermally treated conventionally in order to stabilize the inorganic network. The resultant hierarchical porous layers consisted of micropores that were less than 2 nm, ordered mesopores on the order of 4 nm, and macropores of about 70 nm in size. These multifunctional materials could be used in catalysis, as gas separators, or for adsorption. A similar methodology could be envisaged for production of nano-dispersed particles of other metals such as Ni, Pd, and Rh, and could be used to produce membrane supports such as hierarchical single or mixed oxide systems of TiO_2, Al_2O_3, SiO_2–TiO_2, and SiO_2–Al_2O_3.

Ceramic composites of $CaCu_3Ti_4O_{12}$ and $CaTiO_3$ have been produced using both conventional and microwave sintering.[19] Pellets were sintered at 1100 °C for 3 h in a conventional furnace (the heating and cooling rates were 5 °C/min) or at 1050 °C for 30 min in a microwave furnace at a heating rate of 230 °C/min followed by furnace cooling at a rate of approximately 10 °C/min. The focus of this work was on optimizing the non-ohmic and dielectric properties of these materials, and significantly higher nonlinear coefficients were observed in the materials prepared using microwave heating. The report clearly shows that the electrical properties of materials can be influenced by the method of preparation.

Titania (TiO_2) can be used in paints and pigments, photocatalysis, solar cells, and catalysis as well as other areas. TiO_2 has three phases: anatase, rutile, and brookite. Nano-size powdered materials have been prepared by both conventional and microwave hydrothermal synthesis via hydrolysis of $TiOCl_2$.[20] Following characterization, the powders were redispersed in aqueous hydroxypropylcellulose solutions. Further characterization using rheological analysis was undertaken to study the effect of the hydroxypropylcellulose solutions on agglomeration of the TiO_2 particles. Clusters on the order of several hundred nanometers were observed in all cases. When using conventional heating, numerous spherical shaped crystals of a pure rutile phase having an average size of 30 nm were obtained. A small amount of rod-shaped particles of an anatase phase were also observed, being on the order of 80 nm by 20 nm with tight connections to one another. In the case of microwave heating, rutile was also the primary crystalline phase but the crystals were of substantially smaller size (10 nm). A small quantity of anatase with dimensions of 40 nm by 10 nm was also observed. The fact that smaller particle sizes are obtained when using microwave heating suggests that there is an extremely rapid nucleation and crystallization which may be due to localized hot spots that are generated in the reaction mixture.

Nano-composites of porous γ-Al_2O_3-doped porous ZnO have been prepared using microwave heating as well as ultrasound.[21] Changes in the porosity of these materials were observed when the concentration of dopant was varied. The porous nanostructures were formed without the use of any template and consisted of a mixture of micropores (1.3–1.4 nm) and mesopores (3.8–4.5 nm). The surface areas of the materials prepared using microwave heating were higher than those obtained using a sonochemical approach. This is an interesting observation since the particle sizes of the materials prepared using ultrasound were smaller than those obtained when microwave heating was used. This suggests that microwave heating facilitates the creation of mesopores more so that ultrasound. Whether or not this is a general effect has yet to be investigated in detail.

Mixed valent compounds form the basis of a range of advanced materials. Nano-size rods of crystalline Co_3O_4 have been prepared using a microwave-assisted hydrothermal method.[22] An aqueous solution of cobalt (II) chloride and urea was placed into a Teflon-lined autoclave and the contents heated to 110 °C for 6 h using microwave irradiation. After collecting and drying, the precursor was calcined at 300

°C conventionally, this yielding the desired mixed-valence oxide product. Optical spectroscopy studies showed that the absorption bands for the Co_3O_4 nano-rods shift to short wavelengths as compared to bulk materials.

A final example of microwave heating as a tool for preparing ceramics involves preparation of novel foam and plate materials of Fe_2O_3–Li_2O–SiO_2–Al_2O_3 compositions.[23] These mixed metal systems, prepared from a mixture of Fe_3O_4 and petalite $(LiAlSi_4O_{10})$, show excellent heat retention properties as compared to conventional commercial ceramic materials. They also have other potential applications.

As well as being useful for the preparation of ceramics, microwave irradiation can also be used to take advantage of dielectric properties of the products. As a simple example, the novel foam and plate materials, prepared conventionally from a mixture of Fe_3O_4 and petalite $(LiAlSi_4O_{10})$, show interesting heating properties when exposed to microwave energy (Figure 8.4).[24] The Fe_3O_4, being highly microwave absorbing, heats very rapidly and starts to transform to Fe_2O_3. The relatively low thermal conductivity of the composites also leads to their ability to retain heat very effectively, thereby making them potential materials for keeping food hot during delivery. Petalite ceramic materials themselves are microwave-transparent at room temperature. However, as a quantity of Fe_3O_4 is added, the resultant materials become more and more microwave absorbing. By the time a level of 20 wt % Fe_3O_4 is reached, plates made of the composite material can be heated to 100 °C after 1 min of microwave irradiation. These studies show that specific heating rates can be engineered by controlling the composition of materials.

In another example, tubular $ZnO/CoFe_2O_4$ nanocomposites have been prepared via solution growth methods.[25] After dispersion in a phenolic resin, the nanocomposites

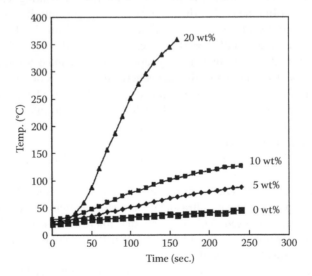

FIGURE 8.4 Temperature profiles of petalite foams containing different amounts of Fe_3O_4 when using 600 W microwave irradiation. (Reproduced with permission from Katsuki, H.; Kamochi, N.; Komarneni, S., *Chem. Mater.*, 2008, *20*, 4803–4807. Copyright American Chemical Society.)

were pasted onto a metallic plate. The materials proved to be very good microwave absorbers and could be potential candidates not only for military stealth technology in reducing the radar cross-section but also for prevention of leakage from devices such as cell phones operating in the microwave region of the electromagnetic spectrum.

8.5 MICROWAVE HEATING FOR PREPARATION OF ZEOLITE MATERIALS

The growth and nucleation of zeolite Y has been controlled by optimization of a process involving spraying a reactant solution into either a receiving solution or air while under microwave irradiation at atmospheric pressure.[25,26] In one method the reactant solution was fed into an ultrasonic nozzle, which generated small droplets which then fell through a microwave-heated zone, this being termed NMW. When mixing of two or more reactant solutions was required, the method could be modified to incorporate an initial in-situ mixing step before passing through the ultrasonic nozzle and then the microwave field, this being termed INM (see Figure 8.5). Colloid formation and nucleation occur in the microwave zone, and the final products can either remain in the microwave zone for further heating or be collected below. The type of nozzle used had significant effects on the outcome. Some nozzles allowed instant mixing while others incorporate vortex mixing or separate shrouds (inner and outer) for introducing the different solutions. The frequency of the nozzle could be varied with the highest frequency used (120 kHz) generating smaller size droplets than lower frequencies. Because the nozzle can get hot during operation, cooling or adjusting the frequency to avoid particle growth could be required. The microwave zone can be adjusted so as to control both the input power as well as the size of the cavity. Varying the cavity size allows for a match to be made between the ionization

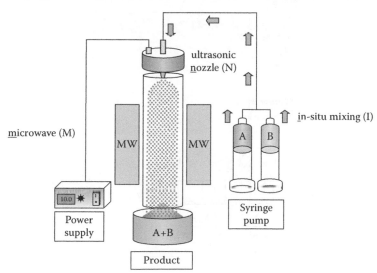

FIGURE 8.5 In situ mixing/ultrasonic nozzle-spray/microwave (INM) method.

potential of the material being heated and the impedance of the cavity. The inner diameter of the microwave cavity used for the INM generation of materials was about 3 feet. Microwave power of less than 500 W was required in order to prevent aggregation of particles. The flow rate of the reagents as well as the geometry of the reaction vessel (straight tube, coil) can also be varied and needed to be optimized. Too long residence times led to aggregation and too short residence times led to incomplete or no reaction.

The INM method has been used successfully to produce nanosize particles of various other materials, examples being mixed metal oxides like $MoVNbTePdO_x$, porous layered manganese oxides, and co-precipitated $Cu/Zn/Al_2O_3$. Each of these materials has particle sizes smaller than those generated using conventional heating. The three materials have been screened for their catalytic activity and found to be active for propane oxidation, selective oxidations of alcohols, and steam reforming of methane reactions, respectively. In the case of $MoVNbTePdO_x$, the catalytic activity of the materials for oxidation of propane to acrolein was found to be three times higher than materials prepared conventionally. This increase in activity has been attributed to the availability of specific crystal faces.

The INM method also allows for the preparation of nanosize synthetic ahktenskite (ε'-MnO_2), which is difficult to prepare in any other way, especially in large amounts (>50 g single batch). Particle sizes are extremely small with diameters of about 30 nm as shown in Figure 8.6 alongside the structure of ahktenskite.

Preliminary studies directed towards the preparation of zeolite A have been undertaken using the INM apparatus.[27] To achieve this, $NaAl_2O_4$ and NaOH were mixed in one clear (sol) solution and tetraethylorthosilicate (TEOS) was introduced as a second reagent stream. The effects of varying the nozzle ultrasound

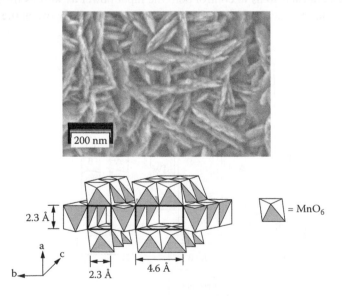

FIGURE 8.6   Bottom: structure of ahktenskite (ε'-MnO_2); Top: SEM of nanoparticles.

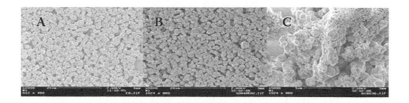

FIGURE 8.7 Preparation of zeolite A. Condition set A: INM technique 0.4 mL/min, N=48 kHz, MW=100 W; Condition set B: INM technique 0.4 mL/min, 48 kHz, MW=50 W; Condition Set C: MW oven, 80 °C, 30 min.

frequency, microwave power input, and flow rate were studied, with the objective of optimizing particle size while at the same time limiting aggregation. As shown in Figure 8.7, smaller, more uniform particle sizes were obtained when using the INM method (Figure 8.7A and B) as opposed to a microwave oven autoclave treatment (Figure 8.7C). On varying the nozzle frequency from 48 to 120 kHz, microwave power from 50 to 200 W, and flow rates from 0.4 to 1 mL/min, particle sizes of the zeolites ranged from 43 to 180 nm. Smaller particles resulted from low microwave power (50 W), low nozzle frequency (48 kHz) and low flow rates (0.4 mL/min). Use of tetramethylammonium-based structure directors resulted in particle sizes as small as 29 nm. At higher nozzle frequencies (120 kHz) the presence of hydroxysodalite impurity phases were observed in X-ray powder diffraction patterns as shown in Figure 8.8.

Multigram quantities of manganese oxide (K-OMS-2) nanomaterials have been prepared using microwave heating and standard reflux glassware.[27,28] The reactions were performed using aqueous/organic solvent mixtures. In the case of a water/DMSO mixture, the crystalline cryptomelane phase was formed within 10 min and the relative

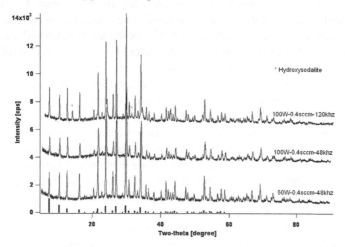

FIGURE 8.8 X-ray powder diffraction data for Zeolite A materials prepared using INM methods. Bottom: N=48 kHz, MW=50 W; middle: N=48 kHz, MW=100 W; top: N=120 kHz, MW=100 W. All at 0.4 mL/min.

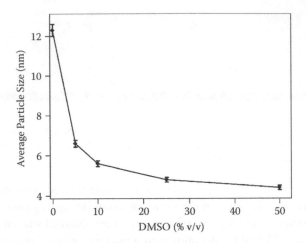

FIGURE 8.9 Average particle size of OMS-2 prepared as a function of DMSO concentration. (Reproduced from Nyutu, E.; Chen, C. H.; Crisostomo, V.; Sithambaram, S.; Suib, S. L., *J. Phys. Chem. C*, 2008, *112*, 6786–6793. Copyright American Chemical Society.)

intensities of diffraction peaks in X-ray power studies remained virtually unchanged with increased reflux time. Of note is that when using conventional heating no phase pure cryptomelane-type OMS-2 materials were formed after 90 min at reflux. The materials, in the size range of 4–20 nm and with very high surface areas up to 227 m²/g, were formed in yields of 70% when using 10% DMSO as cosolvent and >90% when 10% sulfolane was used as a cosolvent. The effects of quantity of cosolvent on particle size were studied. When using DMSO, there is a significant increase in particle size as a function of solvent composition as shown in Figure 8.9. The ability to vary particle size so controllably is important when preparing materials, especially when properties such as catalytic activity need to be correlated to particle size. The effect of varying the cosolvent on the particle size of the resultant OMS-2 materials has also been probed. The order (in increasing diameter of particles) was ethanol > DMSO > N-methyl-2-pyrrolidone > dimethylacetamide > ethyl acetate > water > sulfolane. The OMS-2 materials produced were screened as catalysts for the selective oxidation of p-anisyl alcohol to p-anisyl aldehyde. With the exception of those prepared using 50% DMSO as cosolvent, the OMS-2 materials synthesized using microwave heating showed higher catalytic activities than conventionally prepared analogs.

Microwave heating has also been used to prepare other molecular sieves such as aluminophosphate and silicoaluminophosphate materials. A comparison between conventional and microwave heating in the syntheses of SAPO-11 has been performed, with particular emphasis on study effects on nucleation and crystal growth.[29] Both were enhanced using microwave heating, narrower pore size distributions and more uniform crystal morphologies being observed.

In the synthesis of silicalite, microwave heating was compared with two conventional methods, namely, heating in a bath of ethylene glycol and heating in an electric oven.[30] Using microwave irradiation, the reaction time was significantly reduced due to the rapid heating that is possible. However, the product yield was lower than that obtained conventionally. When the trial was repeated using the same temperature ramp rates for both microwave and conventional heating, the induction time and the crystallization rates were similar. Also, if the temperature ramp rate is carefully controlled, silicalite can be produced using less energy with microwave heating as compared to the conventional process, thereby offering a greener approach.

The effect of the applied microwave power level and delivery mode (continuous versus pulsed) on the synthesis of zeolites SAPO-11, silicalite, and NaY has been studied using a circular waveguide reactor.[31] When compared to continuous microwave delivery, pulsing the microwave power at a rate of 1 s on followed by 2 s off, or 1 s on followed by 3 s off showed no significant effect on the crystallization curve for the preparation of SAPO-11, silicalite, or NaY. However, in the case of SAPO-11, continuous irradiation led to larger particle sizes as compared to pulsed methods. From an energy perspective, significant savings could be made using pulsed power delivery.

In the case of SAPO-11 and silicalite, the effect of applied microwave power on the reaction rate while maintaining the same reaction temperature was also investigated. To perform these experiments, the microwave power used to hold the reaction mixture at a set temperature was varied by applying cooling gas at various rates to the outside of the reaction vessel. In the case of SAPO-11, significant enhancement of the crystallization rate was observed upon increasing the applied microwave power, with little effect on the nucleation time. The relationship between crystallization rate and applied microwave power was found to fit a quadratic curve, indicating a nonlinear process. In the preparation of silicalite, rate of crystallization was found to be independent of the applied microwave power.

The synthesis of silicalite using microwave heating has been monitored in-situ using small and wide angle X-ray scattering (SAXS and WAXS).[32,33] The experiments, performed using the National Synchrotron Light Source at the Brookhaven National Laboratory, involved assembly of a waveguide reactor bearing a sample port through which a thin-walled glass sample vessel could be held in the microwave field (Figure 8.10). The sample vial was equipped with a fiber optic probe for monitoring temperature. Silicalite zeolite precursor solutions were heated at temperatures between 100 and 130 °C for times up to 2 h with SAXS and WAXS patterns being obtained using an X-ray beam which passed through slots in the side of the waveguide. Analysis of the data showed that primary particles of 2 nm diameter formed initially aggregated to 5.8 nm during the course of the heating. In-situ Raman spectroscopy was also used to monitor the reaction at elevated temperatures.

In the synthesis of zeolites β and Y using microwave heating, the effects of varying the vessel size and microwave unit have been assessed.[34] Rates of crystallization were significantly higher when using a multimode microwave unit compared to a smaller monomode system. In addition, the diameter of the reaction vessel had significant effect both on the rate of crystallization and also on the crystal size. There is no single explanation for these results, but the study does highlight the importance of taking into account reactor design when using microwave heating for preparation of materials.

Using scientific microwave equipment it is possible to perform sequential heating steps. A reaction mixture can initially be ramped to one set temperature and then, after an allotted time, ramped either up to a higher temperature or down to a lower point. Such a protocol has been used for the preparation of ordered mesoporous

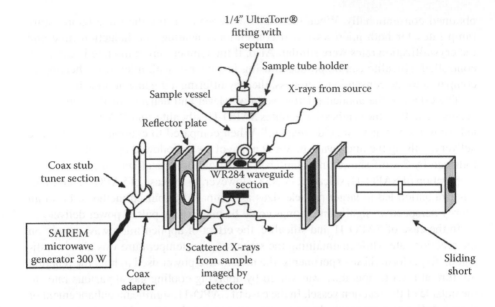

FIGURE 8.10 Schematic of in-situ SAX and WAXS microwave synthesis waveguide apparatus. (Reproduced with permission from Tompsett, G. A.; Panzarella, B.;Conner, W. C.; Yngvesson, K. S.; Lu, F.; Suib, S. L.; Jones, K. W.; Bennett, S., *Rev. Sci. Instr.*, 2006, *77*, 124101/1–124101/10. Copyright American Institute of Physics.)

silicas (OMS), SBA-15 in particular.[35] The synthesis of OMS materials involves an initial step of stirring a gel precursor at between 25 and 50 °C followed by a hydrothermal step at a higher temperature (between 80 and 120 °C). Although the second step has regularly been performed using microwave heating,[36] the initial step has not. In the preparation of SBA-15, the precursors were loaded into a reaction vessel and heated at 40 °C for 2 h in the microwave unit with vigorous stirring. Following this, the temperature was ramped to between 40 and 200 °C for the hydrothermal step, and held for 30 min to 12 h. The effects of temperature and time were assessed. Materials prepared at higher temperatures (160–180 °C) showed better thermal stability than those prepared at 100 °C. In addition, synthesis at higher temperatures also affected the morphology of the SBA-15 materials; samples prepared in the temperature range from 100 to 160 °C possessing rope-like morphology, while those prepared at 180 °C were in the form of discoid particles.

The work was extended to encompass a three-stage heating program. Longer hydrothermal treatment is known to be beneficial for structure consolidation but if this is performed at temperatures above 150 °C, structure deterioration is observed. Using a three-step heating approach, it was possible to take advantage of high-temperature synthesis and simultaneously avoid structure deterioration, Following an initial stage at 40 °C for 2 h, a short hydrothermal treatment at high temperature (1 h at 160 or 180 °C) to expand the pore size and consolidate the structure was followed by a longer stage at lower temperature (6 h at 100 °C). The resultant products have superior thermal stabilities as compared to those prepared using the original two-step approach.

A simple and fast method for the preparation of sulfonic acid-functionalized mesoporous silica (Si-MCM-41) materials has been reported involving microwave heating as a final step.[37] After preparing the materials conventionally, the materials were prepared using a homogeneous precipitation procedure published in which a certain percentage (20, 30, and 40 mol %) of the sodium metasilicate silica source was replaced by 3-mercaptopropyltrimethoxysilane. Microwave heating was used to remove the cetyltrimethylammonium bromide used as a template in the synthesis, while at the same time oxidizing thiol groups to SO_3H moieties. This was achieved by suspending the materials in a mixture of nitric acid and hydrogen peroxide and heating for 5 min at 200 °C. Pellets pressed from the final showed very high proton conductivities that continued to rise with increasing temperature unlike other kinds of proton-conducting membranes such as Nafion where the proton conductivity decreases drastically with increasing temperature.

The trapping of a DB71, a triazo dye, in mesoporous MCM-41 materials has been performed using microwave heating.[38] The materials were prepared using a standard synthesis of MCM-41 but with the inclusion of the dye as a reagent followed by microwave heating at 80 °C for times ranging from 15 min to 1 h. The products prepared using microwave heating had more ordered pore structures than those generated either at room temperature or using conventional heating. A red-shift in the UV-Vis spectrum of the principal absorption band of the dye was observed in the samples prepared using microwave heating, the magnitude of which increased as a function of reaction time.

8.6 EXAMPLES OF OTHER MATERIALS PREPARED USING MICROWAVE HEATING

Metal organic framework (MOF) materials are related to zeolites and molecular sieves and consist of metal ions or clusters coordinated to rigid organic molecules to form one-, two-, or three-dimensional structures. In 2006, three MOFs were prepared, each in less than 1 min, by using microwave heating.[39] The resultant microcrystals had relative uniform size and identical cubic morphology (Figure 8.11). The crystal size could be varied from micrometer down to submicrometer scale by manipulating the concentration of the reactant solution. Yields were in the region of

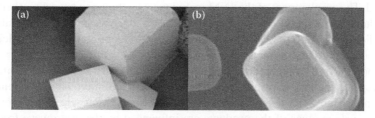

FIGURE 8.11 SEM images of IRMOF-1 synthesized from $Zn(NO_3)_2.6H_2O$ and 1,4-ben-zenedicarboxylate acid ($BDCH_2$) at (a) 0.05 M $BDCH_2$ and (b) 0.0002 M $BDCH_2$. (Reproduced with permission from Ni, Z.; Masel, R. I., *J. Am. Chem. Soc.*, 2006, *128*, 12394–12395. CopyrightAmerican Chemical Society.)

90% as compared to 30% when performing the procedure conventionally. Following this, other MOF materials have been synthesized using microwave heating.[40] It is also possible to scale-up the protocols as well as use robotics for high-throughput synthesis of large numbers of MOF materials.[41,42]

Microwave heating has been used as a tool for preparing a range of gold and silver nanoparticles. A simple one-step process has been developed for preparing materials using poly(N-vinyl-2-pyrrolidone) (PVP) as a both a reducing agent and capping group. Gold nanoparticles could be prepared in as little as 1 min, the particle size being controlled by varying the molecular weight of the PVP and the pH of the reaction medium.[43] The nanoparticles prepared this way have excellent catalytic activity for the reduction of aromatic nitro compounds. Silver-based materials have been prepared in similarly short reaction times using 2,7-dihydroxynaphthalene as a reducing agent and in the presence of a nonionic surfactant, TX-100.[44] The size of the nanoparticles could be varied simply by changing the metal ion-to-surfactant molar ratio, the concentration of dihydroxynaphthalene, and the pH of the reaction medium. The method could also be extended to the synthesis of self-assembled silver nanochains by performing the reaction in a strongly basic medium and at a lower concentration of TX-100.

Along similar lines, microwave heating has been used to prepare stable silver nanofluids in ethanol by reduction of $AgNO_3$ with PVP.[45] Nanofluids are essentially engineered colloidal suspensions of nanoparticles (1–100 nm) in a base fluid. The size of nanoparticles prepared in this methodology was found to be in the range of 30–60 nm, and they were generated after 15 min of microwave heating at 68 °C. Thermal conductivity measurements on the nanofluids showed a significant increase relative to the base fluid. These increases in thermal conductivity found in nanofluids make them attractive heat transfer materials.

The INM method has been used to prepare nanocrystalline spinel nickel ferrite and zinc aluminate particles.[46] The materials were prepared by mixing aqueous solutions of metal nitrate salts in a stoichiometric ratio reflecting that required for the desired products. The resultant solution was introduced into the nozzle together with a solution of NaOH or Na_2CO_3, which led to precipitation of the precursor materials. Passing this through the microwave cavity led to the desired products. At a flow rate of 1 mL/min, when a microwave power setting of 300–400 W was employed together with an ultrasound nozzle frequency of 120 kHz, phase pure products were obtained. The employment of the INM technique was shown to have a marked effect on product formation. When no nozzle was used and the starting materials were simply flowed through the microwave cavity, the product obtained after calcination was impure regardless of microwave power used.

Nanocrystalline tetragonal barium titanate ($BaTiO_3$) particles (30 to 100 nm) have been prepared using microwave irradiation both at fixed microwave frequency and also using variable frequency with 1–5 s sweep times.[47] Particle sizes, morphologies, and surface areas of the products are influenced by the microwave frequency and bandwidth sweep time (Figure 8.12). Materials prepared using the standard frequency of 2.45 GHz yielded particles with a cubic microstructure whereas use of high microwave frequency (5.5 GHz) or variable frequency (3 to 5.5 GHz in 1 s) led to spherical particles with narrow and more uniform particle size distributions. The

2.45 GHz-20 h   5.5 GHz-20 h   4 GHz-20 h   3–5.5 GHz-20 h-1 s   CH-20 h

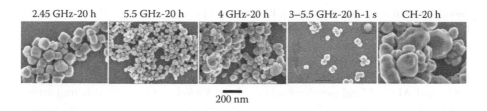

200 nm

FIGURE 8.12 Effects of varying microwave frequency on the morphology of BaTiO₃ particles. (Adapted from Nyutu, E.; Chen, C. H.; Dutta, P. K.; Suib, S. L., *J. Phys. Chem. C*, 2008, *112*, 9659–9667. Copyright American Chemical Society.)

surface areas of the prepared powders decreased with aging time when using microwave frequencies of 4.0 and 5.5 GHz, but increased gradually with extended aging time in variable frequency trials. For comparison purposes, conventional hydrothermal experiments were also performed under similar conditions as in microwave hydrothermal routes. Bimodal particle size distributions were observed.

Nanomaterial growth can be controlled by selective microwave heating as exemplified by the synthesis of CdSe and CdTe in microwave transparent alkane solvents.[48] The high microwave absorptivity of the chalcogenide precursors allowed instantaneous activation and subsequent nucleation and is a clear example of the "specific microwave effect." Regardless of the desired size, narrow dispersity nanocrystals could be isolated in less than 3 min. The reaction did not require a high temperature injection step, a problem encountered with conventional approaches. In addition, the use of a stop-flow reactor allowed for automation of the process for scale-up.

As another example, monodisperse colloidal silica particles inside emulsion droplets can be organized into colloidal crystals by microwave heating.[49] In a water-in-oil emulsion, the water molecules can be heated selectively, leading to evaporation and leaving behind the silica crystals. Starting with emulsion droplets containing silica particles of 240 nm in diameter, spherical colloidal crystals of 18 μm in diameter could be obtained. Compared with conventional approaches, microwave heating reduced considerably the time required for complete consolidation of colloidal particles, the process being complete within 3–20 min, depending on the microwave power used.

At the other extreme, titania nanocrystals have been prepared in ionic liquids as solvents.[50] In this case, the solvent is highly microwave absorbing and heats very rapidly. Using titanium tetraisopropoxide as the TiO₂ precursor and 1-butyl-3-methylimidazolium tetrafluoroborate as the solvent, anatase phase nanocrystals with uniform shape and size were prepared after 3–40 min microwave heating. The materials are low in Tl³⁺ defects, and free from aggregation, these properties being important if the particles are to find application in photocatalysis.

Metal oxide nanowires and nanoparticles have been prepared using direct oxidation of micron-size metal particles in a microwave plasma jet reactor.[51] The reactor could operate at pressures ranging from a few torr to atmospheric pressures and at microwave powers ranging from 300 W to 3 kW. The apparatus produces a plasma of 30–45 cm in length in quartz tubes of 3–5 cm diameter. A metal rod with pointed ends was used to ignite the plasma. Either metal powders were poured directly into the plasma cavity zone and then allowed to flow down by gravity with gas flow or alternatively the metal source could be injected into the plasma jet by means of pressurized gas or a mechanical dispenser. The resulting powders were collected at the bottom of the plasma tube. Materials prepared using the reactor included SnO_2, ZnO, TiO_2, and Al_2O_3. Metal powder feeds of about 5 g/min were used, this translating to a production capacity of 5 kg of metal oxide per day when operated continuously (Figure 8.13.)

Microwave plasma technology has also been used extensively for the preparation of diamond materials via chemical vapor deposition (CVD).[52] Plasma systems offer uniform films over larger substrate areas than hot filament CVD growth, with the possibilty of industrial scale up by using more powerful reactors. Since microwave plasmas are detached from any reactor surface, no impurities from reactor construction materials enter the film bulk during deposition. In contrast, when using hot filament assisted CVD, incorporation of filament materials into the film is a problem during deposition. When using microwave plasma CVD, unpolished natural diamond seeds are often used to initiate growth. In one example, the process was studied in the temperature range from 850 to 1200 °C.[53] A gas mixture of methane, hydrogen, and oxygen was used for the deposition of diamond. The morphology of the products

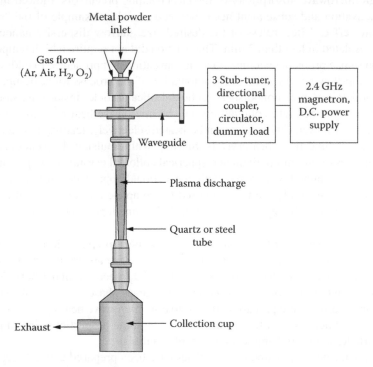

FIGURE 8.13 Schematic of the microwave plasma jet reactor used for preparation of metal oxide nanowires and nanoparticles. (Reproduced with permission from Kumar, V.; Kim, J. H.; Pendyala, C.; Chernomordik, B.; Sunkara, M. K., *J. Phys. Chem. C*, 2008, *112*, 17750–17754. Copyright American Chemical Society.)

was found to be temperature sensitive, there being a transition from polycrystalline to single crystal as the temperature is increased. Single crystal diamond growth was observed to proceed via a step-growth mechanism. Using larger apparatus, homogeneous films up to 75 mm in diameter have been grown, with microwave plasmas at 2.45 GHz.[54] Interestingly, plasma excited at 915 MHz can deposit diamond on a surface area up to 4.5 times larger than this when using conditions conducive to fast growth and high film quality. As a result, low-frequency plasma processes are very interesting from an economic and industrial point of view.

8.7 CONCLUSIONS

The examples presented here clearly show that there is a revolution occuring in the use of microwave heating for the preparation of inorganic materials. The application of microwave heating to the preparation of sols, gels, solutions, solids, dry powders, areosols, emulsions, and others is ongoing.[55] There are numerous commercially available microwave units for performing reactions, and accurate control of temperature, use of high pressure, and monitoring individual steps are possible. Simple procedures, rapid heating, decreased energy costs, unique transformations, ease of scalability, and other factors are major advantages that are becoming apparent when using microwave heating, and although the field is only in its infancy, the future looks bright.

REFERENCES

1. Korichi, S.; Elias, A.; Mefti, A. *Appl. Clay Sci.*, 2009, *42*, 432–438.
2. Skubiszewska-Zieba, J. *Adsorption*, 2008, *14*, 695–709.
3. Cides da Silva, L. C.; Abate, G.; Oliveira, N. A.; Fantini, M. C. A.; Masini, J. C.; Mercuri, L. P.; Olkhovyk, O.; Jaroniec, M.; Matos, J. R. *Studies in Surface Science and Catalysis*, 2005, *156*, 941–950.
4. Zhang, Y.; Feng, H.; Wu, X.; Wang, L.; Zhang, A.; Xia, T.; Dong, H.; Liu, M. *Electrochim. Acta*, 2009, *54*, 3206–3210.
5. Subramanian, V.; Burke, W. W.; Zhu, H.; Wei, B. *J. Phys. Chem. C*, 2008, *112*, 4550–4556.
6. Lee, K. S.; Myung, S. T.; Sun, Y. K. *Chem. Mater.*, 2007, *19*, 2727–2729.
7. Patra, C. R.; Odani, A.; Pol, V. G.; Aurbach, D.; Gedanken, A. *J. Solid State Electrochem.*, 2006, *11*, 186–194.
8. Li, J.; Jin, Y. L.; Zhang, X. G.; Yang, H., *Sol. State Ion.*, 2007, *178*, 1590–1594.
9. Luo, W.; Wang, D.; Wang, F.; Liu, T.; Cai, J.; Zhang, L.; Liu, Y., *Appl. Phys. Lett.*, 2009, *94*, 202507/1–202507/3.
10. Abbaspour, A.; Ghaffarinejad, A. *Anal. Chem.*, 2009, *81*, 3660–3664.
11. Mahajan, S.; Thakur, O. P.; Bhattacharya, D. K.; Sreenivas, K., *J. Am. Cer. Soc.*, 2009, *92*, 416–423.
12. Ebadzadeh, T.; Marzban-Rad, E., *Mat. Char.*, 2009, *60*, 69–72.
13. Koga, N.; Kimizu, T., *J. Am. Cer. Soc.*, 2008, *91*, 4052–4058.
14. Wang, G.; Hao, C., *Mat. Res. Bull.*, 2009, *44*, 418–421.
15. Krishnakumar, T.; Jayaprakash, R.; Pinna, Nicola; Singh, V. N.; Mehta, B. R.; Phani, A. R., *Mat. Lett.*, 2009, *63*, 242–245.
16. Hembram, K. P. S. S.; Rao, G. M., *J. Nanosci. Nanotechnol.*, 2008, *8*, 4159–4162.

17. Ebadzadeh, T.; Sharifi, L., *J. Am. Cer. Soc.*, 2008, *91*, 3408–3409.
18. Yacou, C.; Fontaine, M. L.; Ayral, A.; Lacroix-Desmazes, P.; Albouy, P. A.; Julbe, A., *J. Mat. Chem.*, 2008, 18, 4274–4279.
19. Ramirez, M. A.; Bueno, P. R.; Longo, E.; Varela, J. A., *J. Phys. D: Appl. Phys.*, 2008, 41, 152004/1–152004/5.
20. Baldassari, S.; Corradi, A. Bonamartini; Bondioli, F.; Ferrari, A. M.; Romagnoli, M.; Villa, C., *J. Eur. Cer. Soc.*, 2008, *28*, 2665–2671.
21. Bhattacharyya, S.; Gedanken, A., *J. Phys. Chem. C*, 2008, *112*, 13156–13162.
22. Li, W. H., *Mat. Lett.*, 2008, *62*, 4149–4151.
23. Katsuki, H.; Kamochi, N.; Komarneni, S., *Chem. Mater.*, 2008, *20*, 4803–4807.
24. Cao, J.; Fu, W.; Yang, H.; Yu, Q.; Zhang, Y.; Liu, S.; Sun, P.; Zhou, X.; Leng, Y.; Wang, S., et al., *J. Phys. Chem. B*, 2009, 113, 4642–4647.
25. (a) Gaffney, A. M.; Espinal, L.; Le, D.; Suib, S. L. Method for preparing catalysts and the catalysts produced thereby, European Patent 06250531.8, April 20, 2006. (b) Suib, S. L.; Espinal, L.; Nyutu, E. K., Process and apparatus to synthesize materials, U.S. Patent application 20060291827.
26. Espinal, L.; Malinger, K.; Espinal, A. E.; Gaffney, A. M.; Suib, S. L., *Adv. Funct. Mat.*, 2007, *17*, 2572–2579.
27. Corbin, D. R.; Sacco, A. J.; Suib, S. L.; Zhang, Q., Process for the production of nano-sized zeolite A. U.S. Patent 7,014,837 B2, March 21, 2006.
28. Nyutu, E.; Chen, C. H.; Crisostomo, V.; Sithambaram, S.; Suib, S. L., *J. Phys. Chem. C*, 2008, *112*, 6786–6793.
29. Gharibeh, M.; Tompsett, G. A.; Conner, W. C., *Top. Catal.*, 2008, *49*, 157–166.
30. Choi, K. Y.; Tompsett, G.; Conner, W. C., *Green Chem.*, 2008, *10*, 1313–1317.
31. Gharibeh, M.; Tompsett, G.. A.; Yngvesson, K. S.; Conner, W. C., *J. Phys. Chem. B*, 2009, *113*, 8930–8940
32. Tompsett, G. A.; Panzarella, B.;Conner, W. C.; Jones, K. W., *Mat. Res. Soc. Sympos. Proc.*, 2006, 900E.
33. Tompsett, G. A.; Panzarella, B.; Conner, W. C.; Yngvesson, K. S.; Lu, F.; Suib, S. L.; Jones, K. W.; Bennett, S., *Rev. Sci. Instr.*, 2006, *77*, 124101/1–124101/10.
34. Panzarella, B.; Tompsett, G. A.; Yngvesson, K. S.; Conner, W. C., *J. Phys. Chem. B*, 2007, *111*, 12657–12667.
35. Celer, E. B.; Jaroniec, M., *J. Am. Chem. Soc.*, 2006, *128*, 14408–14414.
36. Tompsett, G. A.; Conner, W. C.; Yngvesson, K. S. *Chem. Phys. Chem.*, 2006, *7*, 296–319.
37. Marschall, R.; Rathousky, J.; Wark, M., *Chem. Mater.*, 2007, *19*, 6401–6407.
38. Ren, T. Z.; Yuan, Z. Y.; Su, B. L., *Coll. Surf. A: Physicochem. Eng. Asp.*, 2007, 300, 88–93.
39. Ni, Z.; Masel, R. I., *J. Am. Chem. Soc.*, 2006, *128*, 12394–12395.
40. (a) Campbell, N. L.; Clowes, R.; Ritchie, L. K.; Cooper, A. I. *Chem. Mater.*, 2009, *21*, 204–206. (b) Yoo, Y.; Lai, Z.; Jeong, H. K. *Microporous Mesoporous Mater.*, 2009, *123*, 100–106. (c) Seo, Y. K.; Hundal, G.; Jang, I. T.; Hwang, Y. K.; Jun, C. H.; Chang, J. S. *Microporous Mesoporous Mater.*, 2009, *119*, 331–337. (d) Lin Z., Wragg, D. S.; Morris, R. E. *Chem Commun.*, 2006, 2021–2023. (e) Jhung, S. H.; Lee, J.-H. Yoon; J. W.; Serre, C.; Férey, G.; Chang, J.-S. *Adv. Mater.*, 2006, *19*, 121–124. (f) Choi, J. Y.; Kim, J.; Jhung, S. H.; Kim, H.-Y. Chang, J.-S.; Chae, H. K. *Bull. Korean Chem. Soc.*, 2006, *27*, 1523–1524.
41. Choi, J.-S.; Son, W.-J.; Kim, J.; Ahn, W.-S. *Microporous Mesoporous Mater.*, 2008, *116*, 727–731.
42. Campbell, N. L.; Grace, J.; Dewson, R. W.; Rebilly, J.-N.; Bradshaw, D.; Carter, B.; Cooper, A. I.; Rosseinsky, M. J. *mms.technologynetworks.net/posters/0457*
43. Kundu, S.; Wang, K.; Liang, H., *J. Phys. Chem. C*, 2009, *113*, 5157–5163.
44. Kundu, S.; Wang, K.; Liang, H., *J. Phys. Chem. C*, 2009, *113*, 134–141.

45. Singh, A. K.; Raykar, V. S., *Coll. Polym. Sci.*, 2008, *286*, 1667–1673.
46. Nyutu, E.; Conner, W.; Auerbach, S.; Chen, C.; Suib, S., *J. Phys. Chem.* C, 2008, *112*, 1407–1414.
47. Nyutu, E.; Chen, C. H.; Dutta, P. K.; Suib, S. L., *J. Phys. Chem. C*, 2008, *112*, 9659–9667.
48. Washington, A. L., II; Strouse, G. F., *J. Am. Chem. Soc.*, 2008, *130*, 8916–8922.
49. Kim, S. H.; Lee, S. Y.; Yi, G. R.; Pine, D. J.; Yang, S. M., Microwave Assisted Self-Organization of Colloidal Particles in Confining Aqueous Droplets, *J. Am. Chem. Soc.*, 2006, 128, 10897–10904.
50. Ding, K.; Miao, Z.; Liu, Z.; Zhang, Z.; Han, B.; An, G.; Miao, S.; Xie, Y., *J. Am. Chem. Soc.*, 2007, *129*, 6362–6363.
51. Kumar, V.; Kim, J. H.; Pendyala, C.; Chernomordik, B.; Sunkara, M. K., *J. Phys. Chem. C*, 2008, *112*, 17750–17754.
52. For recent reviews see: (a) Butler, J. E.; Mankelevich, Y. A.; Cheesman, A.; Ma, J.; Ashfold, M. N. R., *J. Phys.: Cond. Matter*, 2009, *21*, 364201. (b) Mokuno, Y.; Chayahara, A.; Yamada, H.; Tsubouchi, N., *Mat. Sci. For.*, 2009, *615–617*, 991–994.
53. Tyagi, P. K.; Misra, A.; Unni, K. N. N.; Rai, P.; Singh, M. K.; Palnitkar, U.; Misra, D. S.; Le Normand, F.; Roy, M.; Kulshreshtha, S. K., *Diamond Relat. Mater.*, 2006, *15*, 304–308.
54. Silva, F.; Hassouni, K.; Bonnin, X.; Gicquel, A. *J. Phys.: Condens. Matter*, 2009, *21*, 364202.
55. SLS acknowledges support of NSF GOALI grant CBET Award # 0827800 for financial support.

9 Microwave Heating as a Tool for the Biosciences

Grace S. Vanier

CONTENTS

9.1 ABBREVIATIONS

Boc	*tert*-Butoxycarbonyl
BOP	(Benzotriazol-1-yloxy)tris(dimethylamino)phosphonium hexafluorophosphate
DBF	Dibenzofulvene
DBU	1,8-Diazabicyclo[5.4.0]undec-7-ene
DCM	Dichloromethane
DIC	*N,N'*-Diisopropylcarbodiimide
DIEA	*N,N*-Diisopropylethylamine
DMF	*N,N*-Dimethylformamide
DNA	Deoxyribonucleic acid
Fmoc	9-Fluorenylmethoxycarbonyl
HATU	*O*-(7-Azabenzotriazol-1-yl)-*N,N,N',N'*-tetramethyluronium hexafluorophosphate
HBTU	*O*-(Benzotriazol-1-yl)-*N,N,N',N'*-tetramethyluronium hexafluorophosphate
HCTU	*O*-(6-Chlorobenzotriazol-1-yl)-*N,N,N',N'*-tetramethyluronium hexafluorophosphate
HF	Hydrofluoric acid
HOBt	1-Hydroxybenzotriazole
HPLC	High-performance liquid chromatography
NMM	*N*-Methylmorpholine
NMP	*N*-Methylpyrrolidinone
NMR	Nuclear magnetic resonance
PAL	5-[3,5-Dimethoxy-4-(aminomethyl)phenoxy]pentanoic acid
PEG	Poly(ethylene glycol)
PNA	Peptide nucleic acid
PyBOP	(Benzotriazol-1-yloxy)tripyrrolidinophosphonium hexafluorophosphate
SPPS	Solid-phase peptide synthesis
TBTU	*O*-(Benzotriazol-1-yl)-*N,N,N',N'*-tetramethyluronium tetrafluoroborate
TFA	Trifluroracetic acid

9.2 INTRODUCTION

Since the first reports in 1986,[1,2] microwave heating has found wide use in organic chemistry as evidenced by the number of books[3–8] and review articles[9–17] in this field. The interest is due to the advantages microwave heating can bring over conventional approaches. Reactions can be complete in minutes instead of hours and often product yields are higher. In addition, microwave heating is very well suited to help address the tenets of sustainability, it being used widely to facilitate organic synthesis in environmentally benign solvents and for the preparation of biodegradable target molecules.[18–22] However, as this book shows, the use of microwave heating is not limited solely to organic chemistry. One of the newest and perhaps fastest-growing applications of microwave-assisted chemistry is in the area of biosciences. This chapter will highlight the use of microwave irradiation as a tool to facilitate a variety of bio-related reactions including the synthesis of peptides[23,24] and the preparation of samples for proteomic analysis,[25-27] as well as present some of the newest developments in the area of microwave-assisted biosciences. In each example presented, the use of microwave energy provides some advantage over conventional methods.

There are two mechanisms by which microwave energy leads to heating; dipole rotation and ionic conduction. If a molecule possesses a dipole moment then, when it is exposed to microwave irradiation, the dipole attempts to align with the applied electric field. Since the electric field is oscillating, the dipoles constantly try to realign to follow this. Molecules have time to align with the electric field but not to follow the oscillating field exactly. This continual motion of the molecules is translated into heat. If a molecule is ionic then the electric field component of the microwave irradiation moves the ions back and forth through the sample. This movement again generates heat. While the exact mechanisms by which microwave energy interacts with biomolecules are not fully understood, their macromolecular structures suggest some intriguing possibilities. The amide bond that is present in all peptides and proteins has two resonance forms that lead to a high dipole moment that is double that of water and can interact strongly with microwave energy. One of the major problems encountered during peptide synthesis conventionally is the aggregation of the growing peptide chain due to hydrogen bonding between peptide moieties in the backbone. Microwave energy could potentially interact with the peptide backbone in such a way that would disrupt the hydrogen bonding and, in so doing, limit aggregation (Figure 9.1).

Proteins also can adopt secondary structures that are stabilized by hydrogen bonds, examples being alpha helices and beta sheets. To form an alpha helix, every amide N–H group hydrogen bonds to the amide carbonyl moiety of the amino acid four residues away. This creates a stacking of peptide bond dipoles in the same direction, generating a macrodipole moment through the helix. Indeed, the presence of several alpha helices in proteins can lead to an overall dipole within a protein that is 100 times greater than that of water. As a consequence, when a protein is placed in a microwave field, alpha helices could potentially begin to absorb the energy in such a way that the tertiary structure of the protein is perturbed, thus exposing parts of the protein that are not accessible in the rest state or when using conventional heating (Figure 9.2).

The structure of DNA may also make it suited for potential interaction with microwave energy. The backbone features nucleotides with repeating sugar-bound phosphate groups connected through ester linkages. The phosphate groups are charged and thus highly polar. Attached to each of the sugar moieties is a nucleotide base that forms hydrogen bonding to a complementary base generating the two-stranded DNA double helix. The absorption of microwave energy by the DNA backbone will induce molecular motion that may potentially have enough energy to overcome the nucleobase hydrogen bonding causing the DNA to unwind or melt. The microwave-assisted melting of DNA exposes the backbone for a variety of chemistries including hybridization and polymerase chain reaction processes.

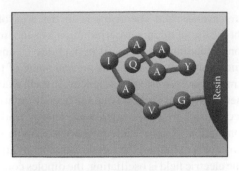

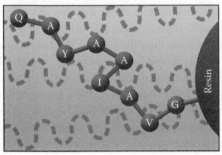

FIGURE 9.1 Schematic of microwave energy disrupting the aggregation in a peptide. (Reproduced with permission from Palasek, S. A.; Cox, Z. J.; Collins, J. M. *J. Pept. Sci.* 2007, *13*, 143–148. Copyright John Wiley and Sons, Inc.)

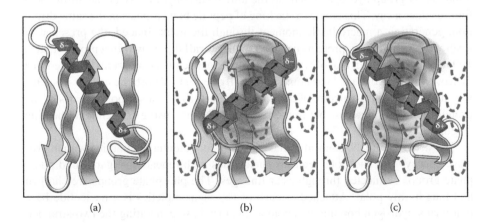

(a) (b) (c)

FIGURE 9.2 A generic protein before (a) and potentially during irradiation with microwave energy (b) and (c). Forms (b) and (c) interconvert resulting in the helix oscillating in the microwave field.

9.3 SYNTHESIS OF PEPTIDES

Peptides are defined as short amino acid sequences. They serve a variety of bio-chemical functions and have found application as active pharmaceutical ingredients (APIs) for oncology, cardiac, and immune-mediated and metabolic diseases, as well as in the development of diagnostics and vaccines. As of 2007, about half of the approved peptide APIs were synthesized using solid-phase synthesis methods, and nearly all of the peptide-derived products still in clinical development were also prepared with solid-phase techniques.[28] Microwave-promoted peptide synthesis was first reported in 1992.[29] Using a domestic microwave oven, three test peptides were prepared with a two- to threefold rate enhancement over conventional synthesis and with no observable racemization. This effect was most pronounced during the cou-pling of sterically hindered β-branched amino acids. While this work was a major breakthrough for peptide synthesis, accurate temperature measurement is not pos-sible using an unmodified domestic microwave oven and reproducibility with this type of system can be very difficult. It was not until nearly a decade later, when scientific microwave apparatus became readily available, that the work realized its full potential. Scientific microwave units feature integrated temperature and micro-wave power control, together with a homogeneous microwave field that allows for much more even heating than in a domestic microwave oven. The initial report of the use of such a unit for peptide synthesis came about from a piece of work focused around the incorporation of the sterically bulky α-aminoisobutyric acid (Aib) moi-ety into a sequence using solution-phase conditions.[30] Six Aib-containing dipeptides were prepared and a number of parameters studied. These included the use of two different coupling reagents, PyBOP/HOBt and HBTU/HOBt, and performing the reaction either at room temperature, at 55 °C in an oil bath, or at 55–60 °C using microwave heating. The results were quite dramatic. When using PyBOP/HOBt, the time required for the coupling using microwave heating was 12 times less than the conventional approaches and gave 15% higher yield. In the case of HBTU/HOBt the coupling time was 32 times shorter using microwave heating and gave 66% and 46% increases in yield over the room temperature and conventionally heating approaches, respectively. The use of scientific microwave apparatus as a tool for solid-phase synthesis of peptides has also been reported, again with a focus on the coupling of sterically hindered amino acids.[31] Four different coupling reagents, PyBOP, Mukaiyama's reagent (2-fluoro-1-methyl-pyridinium tosylate), TBTU, and HATU, were selected for the synthesis of the dipeptides Fmoc-Ala-Ile-NH$_2$ and Fmoc-Thr-Ile-NH$_2$ and the tripeptide Fmoc-Thr-Val-Ile-NH$_2$ on Rink amide resin. The coupling reactions were performed using microwave heating, and the deprotection steps were performed at room temperature under standard conditions. The best results were obtained with PyBOP where the coupling was performed 20 min at 110 °C in DMF to give 60% yield of the dipeptides and 48% yield of the tripeptide. Both HPLC and ¹H NMR were used to confirm that racemization was not an issue, despite the high temperature used during the coupling steps. However, a drawback of the elevated temperatures was that the couplings had to be performed in sealed tubes, these not being readily compatible with solid-phase synthesis due to the lack of a frit for easy filtration. As a result, each reaction mixture had to be transferred into and out of the tube, thus decreasing the coupling efficiency due to resin loss.

Both the coupling and deprotection steps in solid-phase peptide synthesis can be performed using microwave heating.[32] The acyl carrier protein (ACP) fragment 65–74 was synthesized with 1 min microwave deprotection and 2 min microwave coupling protocols to give the peptide in high purity. The corresponding conventional synthesis resulted in multiple deletions and a lower crude purity.

Microwave heating has been used for coupling and deprotection steps in the synthesis of a nonapeptide.[33] The initial microwave experiment was performed with cycles of fixed power irradiation of 30 W for 30 s followed by rigorous water cooling of the reaction vessel and its contents to room temperature for 2 min. The deprotection required three cycles (7.5 min each) and the coupling required four cycles (10 min each), giving a total synthesis time of 2.5 h, some five times faster than the analogous conventional approach. However, the final yield and purity of the peptide prepared using microwave heating was lower, this being attributed to the high temperature reached during the 30 s microwave step (nearly 100 °C). A modified synthesis performed by precooling the reaction mixture to 0 °C before the microwave steps resulted in higher yield and purity of the final peptide, the maximum temperature reached being only 64 °C. To facilitate the process, MicroKan capsules were utilized. These are small porous containers in which solid-phase synthesis can be performed. The advantage of their use is that it is possible to avoid resin-transfer manipulations in multistep processes since the support can be washed and reactions performed without the need to remove the resin from the MicroKan. Under optimized conditions, the peptide was obtained in 95% yield and 83% purity. The corresponding conventional approach gave only 80% yield and 78% purity.

Taking the concept one stage further, resin loading and reagent modification steps can also benefit from microwave heating. Three different linkers and three different supports have been used to prepare three peptides, one short and two medium length, including a phosphopeptide.[34] All of the steps were performed at 60 °C using microwave heating. Loading of the support was reduced from an overnight process to 20 min, reductive amination was reduced from 48 h to 2 × 10 min stages, acylation of secondary amines was reduced from 2 × 90 min stages to 2 × 10 min, couplings were reduced from 45 min to 2–10 min, deprotection was reduced from 11 min to 4 min, and the final cleavage was reduced from 2 h to 5 min. Microwave heating was particularly beneficial for the slowest steps, namely, reductive amination and peptide couplings.

One of the major concerns when using microwave heating for peptide synthesis is an increase in unwanted side reactions. Racemization of activated amino acids during coupling steps has been observed conventionally and is a particular problem with cysteine and histidine residues. To probe this in more detail, a model peptide that contained each of the 20 naturally occurring amino acids has been prepared using both conventional and microwave approaches.[35] The degree of racemization was measured by GC–MS analysis of the resulting amino acids from hydrolysis in 6N DCl/D$_2$O. Conventional synthesis of the peptide (VYWTSPFMKLIHEQCNRADG-NH$_2$) at room temperature resulted in a product of 68% crude purity with less than 1.5% racemization for all amino acids. An initial microwave approach was performed with deprotection steps of 30 s at 50 W followed by 180 s at 50 W (T_{max} = 80 °C) and coupling steps of 5 min at 40 W (T_{max} = 80 °C) resulting in an increase of the crude purity to 76%, but also a significant increase in the racemization of Cys and His. The use of HOBt as an additive in the coupling, PyBOP as an alternative activator, and more hindered bases during coupling (TMP and NMM)[36] were all explored but attempts to reduce the racemization under microwave conditions failed. However, lowering the maximum coupling temperature of the Cys and His residues from 80 °C to 50 °C by using a two-stage heating method (120 s at 0 W followed by 240 s at 40 W) did successfully reduce the level of racemization. It can be reduced even further by performing the coupling of Cys and His at room temperature and the others using microwave heating. Once incorporated into the peptide, racemization of Cys and His residues is not observed in subsequent couplings performed at 80 °C. This suggests that the issues associated with racemization are limited to the use of the activated ester forms of Cys and His.

The 20-mer peptide prepared also contained an Asp–Gly sequence, this potentially leading to aspartamide formation, another well documented potential side reaction when performing peptide synthesis. This sequence-dependent side reaction occurs during the Fmoc deprotection step of Asp-containing peptides where the adjacent residue is Gly, Asn, Ser, or Ala. Each subsequent deprotection after the inclusion of Asp will result in an iterative decrease in the peptide purity. In the initial synthesis of the test peptide, deprotection with 20% piperidine at 80 °C led to considerable amounts of aspartimide formation. Addition of 0.1 M HOBt to the deprotection mixture together with piperidine reduced the aspartimide formation and resulted in a 10% increase in crude purity. The use of piperazine in place of piperidine resulted in an even further decrease in the aspartimide formation, and no deletion products were observed despite the decreased pKa of the deprotection reagent.

In 2003 the first automated microwave peptide synthesizer was introduced. The system is designed to perform all of the steps required for peptide synthesis including deprotection, coupling, peptide cleavage, and all necessary washing stages, each using microwave heating. The unit has been used to prepare a variety of different peptides with biological activity. Examples include human relaxin peptides[37,38] Alzheimer's disease amyloid-β peptides[39,40] human melanocortin receptors,[41] human urotension-II,[42] insulin,[43–45] nicotinic acetylcholine receptors,[46] proline-rich antibacterial peptides,[47,48] human apolipoprotein C-I,[49] and glucagon-like peptide-1,[50] as well as in the study of nucleophosmin mutations in acute myeloid leukemia,[51] probing GPCR-peptide interactions[52] and controlling the interfacial properties of a solid.[53] The apparatus has also been used for the synthesis of peptide nucleic acid hybrids, a new class of biomolecules that selectively target DNA/RNA sequences with high binding specificity and affinity.[54,55]

In addition to use in the coupling of Fmoc-protected amino acids on solid-phase, microwave heating can also be used to accelerate the coupling reactions using Boc-protected analogs. The use of Boc-protected peptide synthesis is preferential in a number of cases, in particular in the preparation of certain difficult peptide sequences and nonnatural peptide analogs that are base-sensitive. It is also useful in conjunction with native chemical ligation for the synthesis of large peptides and proteins. As an example, kalata B1, which is a member of the cysteine rich cyclotide family of peptides, has been prepared using a Boc-protected strategy both by automated microwave synthesis and a conventional approach.[56] Previous syntheses were carried out using manual approaches, which typically take around 10–14 days for chain assembly, cleavage, and purification, depending on the resin substitution, quality, and the efficiency of the coupling steps. In the protocol using microwave heating, coupling steps were performed with a standard in-situ neutralization protocol using HBTU/DIEA, the reaction mixture being irradiated at 35 W for 5 min ($T_{max} = 87$ °C). Deprotection steps were performed in 2 x 1 min stages at room temperature. Kalata B1 was prepared in 20% yield. This compares favorably with the conventional approach, which yields only 7% of the desired product when using an analogous coupling and deprotection strategy, but all the reactions are performed at room temperature. Other cysteine rich peptides were also prepared to show the generality of the approach. In addition, the methodology could be scaled up from 0.1 mM to 0.5 mM scale. This was demonstrated in the case of kalata B12, a new cyclotide, which was obtained as a single product with a purity of >95%.

The degree of loading on resins can be a significant factor in the efficiency of solid-phase peptide synthesis, both conventionally and using microwave heating. For example, in the synthesis of the nonapeptide Ac-Gly-Cys-Asp-Pro-Asp-Arg-His-Cys-Ala-NH_2, the use of low loading resins (<0.5 mmol/g) yielded the product in good purity.[57] However, when higher-loading resins (0.94 mmol/g) were used, the product was isolated in only 42% purity. These results were obtained using microwave heating but without stirring. Repeating the synthesis with stirring greatly enhanced the process, and the product was obtained in 76% purity. Thus, when using high loading resins, stirring or agitating the reaction mixture is imperative in order to maintain a homogeneous concentration of the activated coupling species around the resin beads.

Acid chlorides offer an alternative to the standard activated acids or mixed anhydrides used in peptide synthesis to couple hindered residues. Microwave heating for just 30 s allowed for the synthesis of a range of dipeptides using Fmoc-Phe-Cl as a starting material and preactivated zinc dust as an additive.[58] The role of the zinc is not fully understood but, as it is reported to neutralize hydrochloride amino acid salts, it is conceivable that it acts purely as a neutral acid scavenger in these reactions. The methodology could be extended to the preparation of longer peptides and to the coupling of both acyclic and cyclic α,α-dialkylamino acids, a process that conventionally suffers from long reaction times and poor yields due to increased steric bulk. A similar approach has also been used to incorporate the extremely hindered N-benzylaminoisobutyric acid moiety (NBn-Aib) into dipeptides.[59] For example, the coupling of NBn-Aib-OEt with Fmoc-Phe-Cl in the presence of five equivalents of zinc dust in dichloromethane at 90 °C for 2 h using microwave heating yielded dipeptide Fmoc-Phe-NBn-Aib-OEt in 81% conversion and 61% isolated yield (Scheme 9.1).

Microwave heating can be used to facilitate peptide synthesis using other nontraditional coupling methodologies. For example, N-Fmoc-protected(α-aminoacyl) benzotriazoles have been used in the preparation of tri-, tetra-, penta-, hexa-, and heptapeptides on solid-support with an average crude yield of 71% (Scheme 9.2).[60] Acylbenzotriazoles are easily prepared, chirally stable analogs of acid halides that are relatively water stable. Coupling reactions of N-protected(α-aminoacyl)

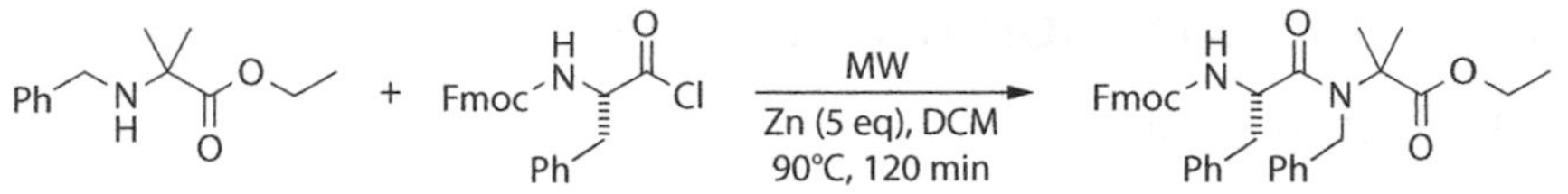

SCHEME 9.1 Microwave-assisted zinc-mediated peptide coupling of *N*-benzyl-α,α-disubstituted amino acids.

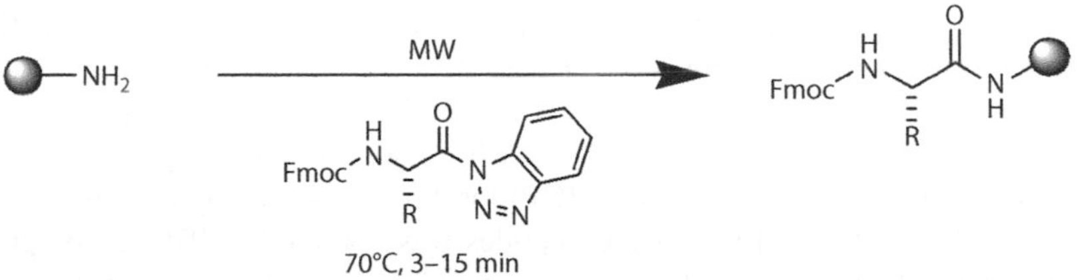

70°C, 3–15 min

SCHEME 9.2 Solid-phase peptide synthesis utilizing *N*-Fmoc-protected(α-aminoacylbenzotriazoles).

benzotriazoles with unprotected amino acids proceed with minimal epimerization under mild conditions and do not require the addition of an activator or additive such as HOBt. The methodology can be extended by preparing and using *N*-Fmoc-protected(α-dipeptidyl)benzotriazoles as reagents to yield tri-, tetra-, penta-, hexa-, and heptapeptides.[61] The approach has also been applied to preparation of eight difficult peptide sequences including Leu-enkephalin and the amyloid-β fragment (34–42) in crude yields of 34–68%.[62]

Microwave heating has been used to facilitate the coupling of α,α-dialkylated amino acids with acid fluorides. This methodology proved particularly valuable when attempting to prepare peptabiols.[63] This class of peptides, which possess membrane-disrupting abilities leading to antibiotic activity, contain a high proportion of sterically hindered α,α-dialkylated amino acids. Since acid fluorides are not readily available, the strategy taken for the synthesis of the peptabiols was to use standard coupling reagents and conditions for all the standard amino acids and acid fluoride-based couplings for the α,α-dialkylated amino acids, each performed for only 5 min using the automated microwave peptide synthesizer. This proved successful for the preparation of shorter peptabiols but for longer chains the yields of the couplings with the α,α-dialkylated amino acids were found to be low, and thus it was not possible to obtain the desired product in acceptable crude purity. To overcome this, a double coupling protocol was employed whereby whenever an α,α-dialkylated amino acid was included into the chain, the coupling was performed twice in succession in an attempt to increase the level of substitution. This proved successful as demonstrated by the synthesis of the 15-mer ampullosporin I in 22% isolated yield despite the fact that it contains three consecutive Aib residues in the sequence. The methodology can also be performed conventionally at room temperature, but not surprisingly, the synthesis time has to be extended to 1 h per step instead of 5 min using the microwave approach.

9.4 SYNTHESIS OF MODIFIED PEPTIDES

9.4.1 LABELED PEPTIDES

Labeling peptides with fluorescent tags allows for assessment of their structure and biological function using techniques such as fluorescence resonance energy transfer (FRET). As a result, significant effort has been directed towards the efficient synthesis of peptides that contain a fluorophore at one end and a fluorescence quencher at the other. The synthesis of such peptides can be challenging due to the low reactivity of some of the common fluorophores and quenchers. Microwave heating has been used to overcome this. In one example, two peptides (a 6-mer and a 10-mer) were prepared using a microwave-heating approach on solid-phase.[64] Optimized conditions involved preactivation with DIC/HOBt for 2 min followed by microwave heating at 60 °C for 10 min. Each of the peptides was obtained in 97% crude purity. While still attached to the solid support, the peptides were then each split into two aliquots and each labeled with a different FRET couple, [5(6)-carboxyfluorescein/dabcyl] and 2-aminobenzoic acid/3-nitrotyrosine, using the same DIC/HOBt protocol for 10 min. After cleavage from the support, the desired products were obtained in an average isolated yield of 75% and crude purities of >90%.

Coumarins also have extensive and diverse applications as fluorescent probes. A series of coumarin-labeled peptides have been prepared using microwave heating.[65] Di-, tri-, tetra-, and hexapeptides have been coupled with coumarin-derivatized lysine at 70 °C for 10 min to give labeled peptides in 18–45% yields.

An alternative labeling strategy is to incorporate organometallic moieties into the peptide framework. Metal carbonyl complexes exhibit characteristic bands in the infrared spectrum due to C–O stretches, this area of the spectrum being relatively free of signals from functionalities of bio(macro)molecules. They are then able to be used as monitors for concentration and spatial distribution of the peptide in a biological system. In addition, the functionalization of peptides with organometallic groups can facilitate membrane transfer and delivery of a metal complex to a specific cellular target, thereby opening avenues for interesting pharmacological studies. Microwave heating has been used to attach a derivative of cyclopentadienyl tricarbonylmanganese (CpMn(CO)$_3$) to a solid-supported peptide, itself prepared using microwave heating (Scheme 9.3).[66] Cleavage of the derivatized peptide from resin followed by HPLC analysis showed only a single peak indicating the high purity of the labeled peptide as well as the stability of the metal complex under microwave promoted solid-phase peptide synthesis and cleavage conditions. The cytotoxicity of the derivatized peptide was probed by using MCF-7 human breast cancer cells in the concentration range of 10–100 μM. The cell viability was not changed relative to an untreated control. The labeling methodology has subsequently been extended to the preparation of four other manganese-labeled bioconjugates, each of which were then internalized in HT-29 human colon cancer cells and found to be nontoxic at concentrations up to 100 μM.[67]

Lipidated proteins are important for signal transduction processes but the preparation and study of peptide analogs has been hampered by synthetic challenges as well

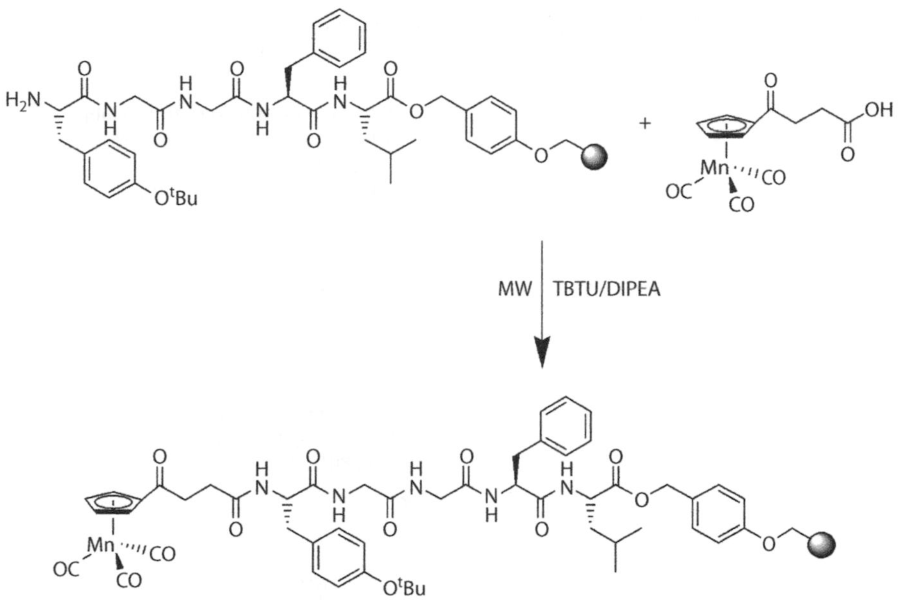

SCHEME 9.3 Preparation of a cyclopentadienyl tricarbonylmanganese-derivatized peptide.

as peptide instability. A carbon isostere of palmitoylated cysteine has been used to prepare a highly lipidated peptide that can be considered as an acid stable analog of the C-terminus of the H-Ras heptapeptides 180–186.[68] The conventional preparation of the peptide resulted in low yields so microwave heating was used as an alternative for both the deprotection and coupling steps (Scheme 9.4). The solid-supported synthesis proceeded smoothly and, following cleavage, the desired peptide was obtained in 30% overall yield.

9.4.2 SYNTHESIS OF CYCLIC PEPTIDES

Cyclic peptides are of biological interest because they are typically resistant to digestion. This is a trait that makes them particularly suitable as peptide-based drugs. Synthesis of lactam-bridged cyclic peptides can be performed either in solution-phase or using solid-supported techniques. Solution phase cyclization must be performed under dilute conditions to avoid dimerization, and requires the use of a minimal excess of reagents to avoid extensive purification. As a result, these reactions can require a lot of time and often result in poor yields. Using solid-phase synthesis, the cyclization reactions can be performed using an excess of the reagents, improving the yield. SynPhase™ lanterns have been used in conjunction with microwave heating to prepare a series of head-to-tail cyclic peptides.[69] Seven peptides varying in length from 4 to 10 residues were synthesized using conventional peptide synthesis. They all featured alternating cationic (Lys) and nonpolar (Leu) groups, as well as a Phe residue in the middle of the sequence and a Glu on the C-terminus. The cyclization

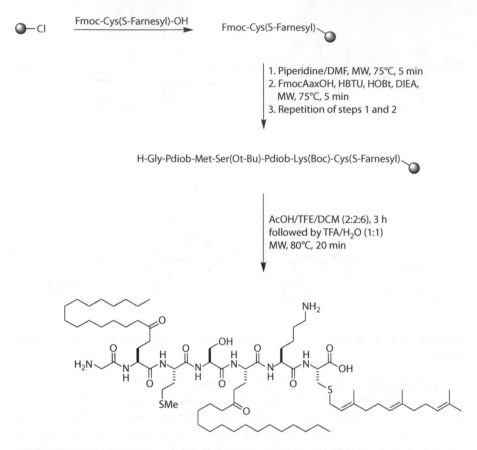

SCHEME 9.4 Solid-phase synthesis of the H-Ras heptapeptide isostere.

step was performed with PyBOP/HOBt/DIEA in one of three ways; at room temperature for 24 h, using conventional heating at 50 °C for 1 h, or using microwave heating (50 °C for 2 × 15 min or 4 × 15 min). At room temperature the cyclodecapeptide was prepared in 97% crude purity. For the shorter peptide sequences, the cyclization of those containing an even number of amino acids was moderately successful (26–38% crude purity) while for odd congeners poor results were obtained (0–19% crude purity). Performing the cyclization using conventional heating at 50 °C for 1 h resulted in higher product conversions but the reaction still was not complete. There was also significant decomposition of the PyBOP coupling agent. When using microwave heating (2 × 15 min) the amount of cyclized product increased relative to the previous trials but the reactions were still incomplete at the end of the protocol. Increasing the reaction time to 4 × 15 min led to complete conversion to the desired cyclic peptides in high purity in all but one case.

The formation of intramolecular disulfide bonds serves as another means to cyclize peptides. The ability to perform this reaction also opens avenues for study of the important role of disulfide bonds in the folding and stability of peptides. Microwave heating has proven valuable for the efficient synthesis of a range of disulfide bridged conopeptides, a class of neurotoxins that selectively bind to subtypes of nicotinic acetylcholine receptors.[70] Of particular interest was α-conotoxin MII (α-CtxMII), a 16-residue bicyclic peptide containing two disulfide bonds. Four linear precursor derivatives of α-CtxMII were prepared on a solid support using a conventional room temperature approach and also with microwave heating. The latter furnished the peptides in higher yield and purity. Three cyclization strategies were then tested. In the first approach both disulfide bonds were formed using solution-phase chemistry at room temperature. The desired bicyclic peptide was formed in good yield, the reaction taking 5–72 h to reach completion. In the second approach both cyclizations were performed with the peptide immobilized on a solid support. Using microwave heating, the first cyclization could be performed efficiently, but in the case of the second, only a low yield of the desired product was obtained. In the third approach, the first cyclization was undertaken on the support and then the second performed in solution, both steps using microwave heating. Again the first cyclization was successful but now the second step could also be effectively undertaken and the desired peptide prepared in good yield in approximately 1 h.

While disulfide bridges offer a means to control the secondary structure of a peptide, they can be metabolically and chemically unstable. Replacement of the disulfide bond with a dicarba isostere proves an effective way to maintain the structural effects seen with disulfide bonds without the issues of chemical sensitivity. One route to cyclized dicarba analogs is through the use of ring closing metathesis (RCM) of the appropriate unsaturated starting peptide. In an effort to prepare probes for the delineation of functional (redox active) and structural roles of native disulfide bonds in conotoxins, such a method has been used for the synthesis of dicarba hybrids of α-conotoxin IMI (CTx-IMI).[71] To explore the metathesis chemistry two CTx-IMI derivatives were prepared: Fmoc-Gly-Hag-Cys(Trt)-Ser-Asp-Pro-Arg-Hag-Ala-Trp-Arg-Cys(Trt)-NH-resin and Fmoc-Gly-Cys(Trt)-Hag-Ser-Asp-Pro-Arg-Cys(Trt)-Ala-Trp-Arg-Hag-NH-resin where Hag = l-allylglycine. Treatment of the first peptide with 10 mol% Grubbs second-generation metathesis catalyst in DCM for 1 h at 100 °C using microwave heating resulted in complete conversion to the unsaturated [2,8]-dicarba analog of CTx-IMI. Complete conversion to product was also observed in the case of the second peptide using identical reaction conditions, this giving the unsaturated [3,12]-dicarba analog of CTx-IMI. Hydrogenation with Wilkinson's catalyst followed by cleavage from the resin and oxidation provided the saturated [2,8]-dicarba- and [3,12]-dicarba-CTx-IMI. Performing the metathesis reactions conventionally using the first-generation Grubbs catalyst (50 mol%) resulted in the formation of the unsaturated [2,8]-dicarba-CTx-IMI in less than 10% conversion after 72 h at 50 °C. Use of the second-generation catalyst (10 mol%) increased the conversion to about 70%, but full conversion could not be obtained even with a high catalyst loading (50 mol%). Metathesis of the isomeric [3,12]-dicarba-CTx-IMI using conventional heating under a variety of conditions failed to produce the desired cyclized product.

An RCM approach in conjunction with microwave heating has found other applications in cyclic peptide synthesis. Novel A-chain dicarba analogs of human relaxin-3 (H3 relaxin, INSL7) have been prepared and found to have applications in the treatment of stress and obesity.[38] Artificial α-helices that feature a dicarba hydrogen-bond surrogate have been synthesized using RCM between appropriately placed olefins on the peptide chain.[72] Five di-olefin derivatized peptides were first prepared and then used in a comparative study between conventional and microwave heating. Two of the peptides contained a *tert*-butyl protecting group, and one of the substrates contained a trityl protected His residue. Both of these sterically demanding protecting groups are known to be disruptive to RCM reactions. Two of the peptides were functionalized with olefin moieties such that the RCM reaction could result in the formation of different size macrocycles and therefore different conformations. A wide variety of reaction conditions were probed, variables including the catalyst and solvent used as well as the time and temperature at which the reaction was run. Dichlorobenzene was found to be the solvent of choice for each of the peptides studied, and each of the RCM reactions was complete within 5 min using microwave heating. Second-generation Hoveyda-Grubbs and Grubbs catalysts were found to be the most active of those screened, the optimal reaction temperatures being 200 and 120 °C, respectively. Of interest is that the second-generation Grubbs catalyst was not catalytically active when the reactions were performed using conventional heating. Each of the macrocycles was obtained in good conversion (76–84%) with one exception, a 10-membered macrocycle, that could not be prepared under either microwave or conventional heating. The methodology has been extended to the synthesis of hydrogen-bond surrogate α-helices that inhibit HIV-1 fusion[73] and a similar approach has also been used to prepare macrocyclic β-strand templates.[74]

Linking the C and N termini to form cyclic peptides is particularly challenging given the inherent difficulties in bringing together two ends of a linear chain. The incorporation of turn residues such as D-amino acids, Aib, proline, pseudoproline dipeptides, and N-alkylated amides is one strategy used to confine the open chain and facilitate the cyclization. An alternative strategy has been to use a metathesis-labile tether.[75] In this approach, RCM is used to form the tether, the cyclization at the N and C termini is performed, and then ring-opening metathesis (ROM) is used to release the tether. This approach has been used successfully in the preparation of mahafacyclin B (Scheme 9.5). The key cyclization step required to form this polypeptide occurs in only 30% yield under conventional high dilution conditions. To prepare the tethered analog of mahafacyclin B for use in the cyclization step, native Phe at positions 2 and 6 were replaced with l-allylglycine. A variety of reaction conditions were screened for the RCM step using conventional heating. The best product conversion (~50%) was obtained when using second-generation Grubbs catalyst and performing the reaction at 80 °C for 15 h in dichlorobenzene with the addition of 10% LiCl-DMF. However, when using microwave heating the reaction was complete in 100% conversion (>97% yield) after 1 h at 100 °C, the catalyst and additive being the same as in the conventionally heated case. The conformationally constrained peptide then underwent head-to-tail cyclization in 55% yield and subsequent ROM in the presence of butene cleaved the tether and yielded the target mahafacyclin backbone in 42% yield.

In addition to ring-closing metathesis, nucleophilic aromatic substitution[76] and Heck coupling reactions[77] can also be used to efficiently form cyclic peptides using microwave irradiation.

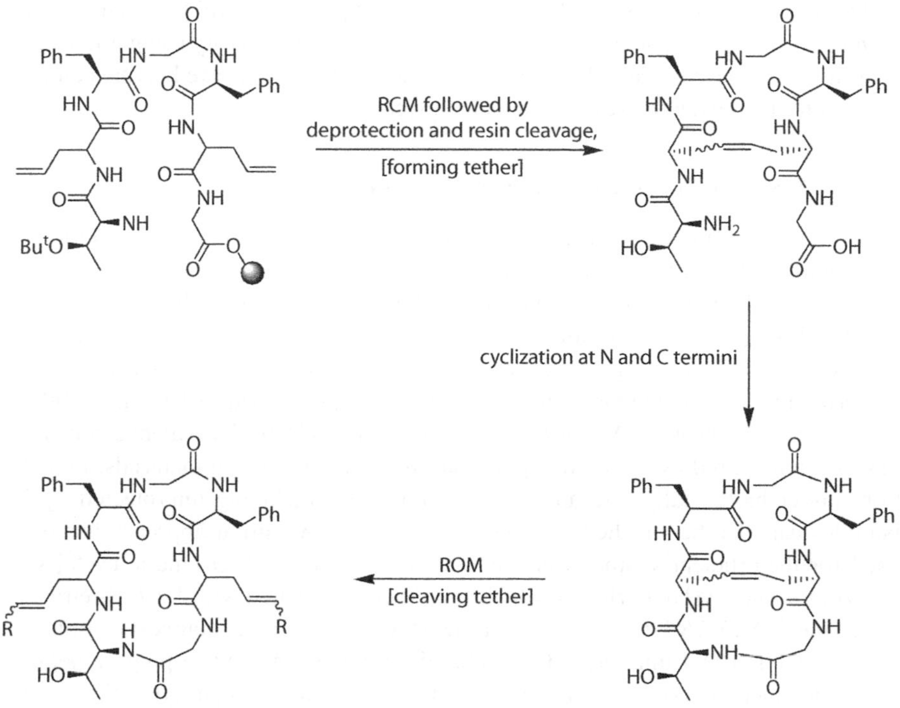

SCHEME 9.5 Use of ring-closing and ring-opening metathesis as an aid to cyclization of a peptide at the C and N termini.

9.4.3 SYNTHESIS OF GLYCOPEPTIDES

Over half of the proteins in the human body are post-translationally modified; these modifications including glycosylation, phosphorylation, methylation, and lipidation. Synthetic models of these post-translationally modified proteins can be used to probe the structure and function of proteins, and ultimately can be used as diagnostics or as therapeutics. The solid-phase peptide synthesis of glycopeptides typically requires a large excess of the sugar-bound amino acid derivatives and long reaction times. By using microwave heating, the reaction times can be dramatically decreased, and couplings can be performed using only 1.5 equivalents of a single glycosylated amino acid building block.[78] This has been exemplified in the synthesis of glycopeptide MUC1, composed of 20 amino acids, 5 of which bear O-glycans. Using a TentaGel support, the standard amino acids were coupled using a microwave heating protocol of 10 min at 50 °C, and the glycosylated amino acids coupled in 20 min at 50 °C. The synthesis was complete in 7 h and gave MUC1 in 44% yield. Changing the support used from TentaGel to a polyethylene glycol polyacrylamide copolymer allowed for the glycopeptide to be prepared in an even higher yield.[79] Using conventional heating (50 °C) the peptide was formed in less than 15% yield and even though the synthesis could be performed at room temperature in 41% yield, the total time required was 99 h.

In another example CSF114(Glc), a glucosylated peptide useful for diagnosis and monitoring of multiple sclerosis, was prepared in 74% crude purity using microwave heating as compared to less than 20% conventionally.[80] Microwave heating can also be used for preparing fluorescently labeled glycopeptides.[81]

9.4.4 SYNTHESIS OF DENDRIMERS AND POLYMERS

The use of dendrimeric materials offers a means for simultaneous presentation of biologically relevant ligands and can be used to enhance the interaction of weakly binding ligands. Peptide-based ligands have been attached to dendrimeric supports using a 1,3-dipolar cycloaddition (click) reaction between peptide azides and the dendrimer acetylene using microwave heating.[82] Conventionally this chemistry is performed at room temperature for 16 h but can be performed in 10 min at 100 °C using microwave heating. A series of di-, tetra-, octa-, and hexadecavalent dendrimers have been prepared using small peptide-based azides as starting materials. In addition, larger biologically relevant biomacromolecules including unprotected cyclic peptides can also be attached to dendrimer supports. A similar approach has been used for the efficient synthesis of monomeric, dimeric, and tetrameric cyclo[Arg-Gly-Asp-d-Phe-Lys] dendrimers, which were then conjugated with 1,4,7,10-tetraaza-dodecane-N,N',N'',N'''-tetraacetic acid (DOTA) for use as a diagnostic tools for tumor targeting and imaging.[83] Click chemistry has been used to prepare peptide based polymer materials. Using azido-phenylalanyl-alanyl-propargyl amide as the monomer, copper-catalyzed cycloaddition polymerization has been performed both at room temperature and using microwave heating.[84] After 3 days at room temperature, a polymer with M_n of 27,600 Da was obtained. When using microwave heating, 30 min at 100 °C yielded a polymer with M_n of 38,000 Da. Performing the reaction for 30 min at 100 °C conventionally resulted in a very low molecular weight polymer ($M_n = 7050$) showing the advantage of the microwave heating protocol. This methodology has been expanded for use in the synthesis of biodegradable polymers that contain peptide sequences susceptible to enzymatic digestion.[85] The molecular weight of the polymers can be tailored between 4,500 and 14,000 Da by varying the concentration of the reaction mixture.

9.4.5 SYNTHESIS OF PEPTIDOMIMETICS

9.4.5.1 β-Peptides

β-Peptides are oligomers of β-amino acids that have excellent proteolytic and metabolic stability, which makes them attractive as peptide-based drugs. They also can adopt discrete and predictable secondary structures, the most widely studied of which is the 14-helix. The Fmoc solid-phase peptide synthesis of β-peptides often suffers from poor efficiency of both the deprotection and coupling steps especially after the fifth residue. Microwave heating has been used to facilitate the synthesis of *trans*-2-aminocyclohexane carboxylic acid (ACHC)-containing β-peptides.[86] A hexamer was selected as the initial target since, at two, this contains the minimum number of ACHC residues to induce 14-helix formation (Figure 9.3). When incorporating

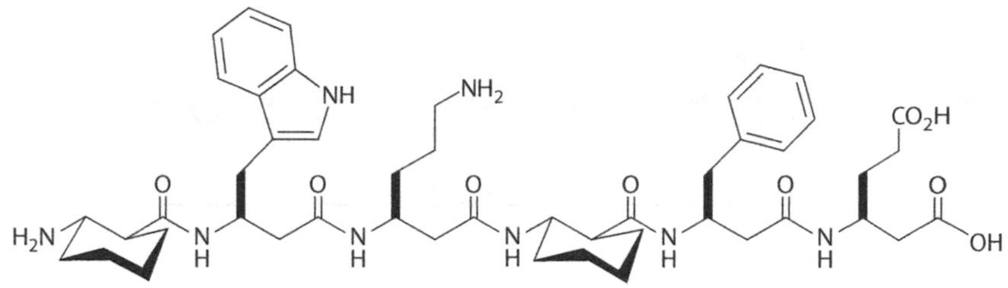

FIGURE 9.3   Helical β-peptide. (Reproduced with permission from Murray, J. K.; Gellman, S. H. *Org. Lett.* 2005, *7*, 1517–1520. Copyright American Chemical Society.)

ACHC into the peptide, a double-coupling/double-deprotection strategy was taken. Using conventional approaches, the hexamer was prepared in only 55% purity, the major impurity arising from ACHC deletion. The use of microwave heating (4 min deprotection at 60 °C and 6 min coupling at 50 °C) gave the desired β-peptide in 80% purity, and the ACHC1 deletion byproduct was reduced to only 5%. The protocol could be further improved by switching the solvent in which the coupling of the ACHC components were performed from DMF to NMP and using 0.8 M LiCl as an additive. Under these conditions the hexamer was synthesized in 94% purity and 81% yield. The protocol has been extended to the synthesis of a deca-β-peptide that incorporates three ACHC residues. Conventional synthesis of this peptide provides only 21% purity, but using microwave heating, it could be obtained in 88% purity and 65% yield. The advantage of microwave heating over conventional methods has been further shown by the synthesis of Z28, a 28-membered β-peptide that was the longest synthesized at that time.[87]

The preparation of β-peptide libraries has also been the subject of significant research effort. Using polystyrene macrobeads attempts have been made to prepare hexamers using microwave heating.[88] However, when using the conditions optimized for single reactions, the desired products were formed only in very low yield, this being attributed to poor diffusion of the reagents into the polymer matrix. Allowing the reagents to diffuse into the macrobeads for 1 h, followed by a 4 min heating protocol for deprotection and then standing for 2 h followed by a 6 min coupling step, increased the efficacy of the procedure. It could be improved even more by repetition of a cycle of 2 min of heating followed by 10 min of cooling (3 cycles per deprotection and 6 cycles per coupling). Using this procedure, a 100-member library of octa-β-peptides has been prepared in an average purity of 65% using a split-pool approach. The protocol works equally well in a 96-well format, the library being generated with less than 10% variation in crude purity, indicating even heating across the plate.[89]

9.4.5.2   Peptoids

Peptoids are oligomers of *N*-substituted glycines that, similar to β-peptides, have enhanced stability toward proteolysis and can be used in the design of new biological probes and in drug discovery. Peptoids are typically synthesized via the submonomer approach that features iterative acylation and amination reactions as shown in Scheme 9.6. This can take up to 3 h per monomer unit, and longer peptoid sequences exacerbate this problem. Monomer incorporation can be reduced to

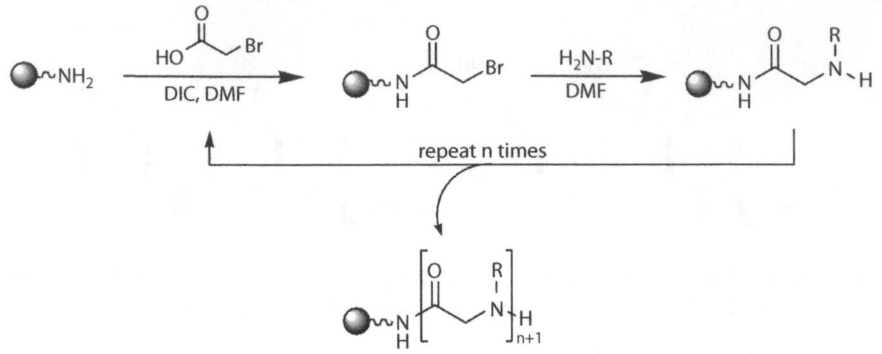

SCHEME 9.6   Schematic of solid-phase submonomer peptoid synthesis method.

under 1 min using microwave heating.[90] Using a domestic microwave system, both the acylation and amination reactions could be performed using 2 × 15 s irradiations at 10% of 1000 W and allowing the reaction mixture to cool between the two. The maximum temperature reached during each 15 s irradiation was 35 °C. A series of six nonamers have been prepared in 60–90% crude purity, which is comparable to those obtained when performing the chemistry conventionally at 37 °C (45 min for acylation and 1 h for amination). A protocol for the incorporation of less reactive amines has been developed using a scientific microwave apparatus, the acylation step being performed at 35 °C for 25 s and the amination at 95 °C for 90 s.[91] A series of homopentamers was prepared in 56–93% crude purity. While no real advantage was found when using unhindered primary amines, peptoid synthesis involving electronically deactivated benzylamines could be improved significantly through the use of microwave heating.

As an alternative to the submonomer approach to synthesize peptoids, amino acid-based monomers bearing a secondary amine functionality at the N-terminus can be coupled directly to each other using standard peptide coupling protocols.[64] A heptapeptoid has been prepared this way in a single day and in 98% purity using microwave heating (60 °C for 20 min) for the coupling step. This approach can also be used for the synthesis of peptoid dendrimers,[92] fluorescein-labeled cationic peptoid conjugates,[93] and α-peptide-β-peptoid chimeras.[94]

9.5   MICROWAVE-ASSISTED PROTEOMICS

Proteomics can be broadly defined as the identification, localization, and functional analysis of the protein make-up of cells. One of the major challenges faced, however, is lengthy sample preparation. Several methods have been developed to address this issue including the use of immobilized enzymes[95,96] and acid-labile surfactants.[97] Microwave heating can also play a valuable role.

9.5.1   GEL STAINING

The fixing, staining, and destaining of sodium dodecyl sulfate-polyacrylamide (SDS-PAGE) gels and polyvinylidene-difluoride (PVDF) membranes was one of the first applications of microwave heating in the biosciences and saw widespread use in the 1990s. More recently the staining and destaining of both one- and two-dimensional SDS gels has been explored using a number of stains.[98] Gels of 0.75 mm and 1.5 mm thickness were used and, in a domestic microwave, heated for two 30 s periods at 700 W of microwave energy. In the case of the 0.75 mm gels, the time required for staining with Coomassie Blue was reduced by 10-fold over conventional approaches, whereas with SYPRO® Ruby (a fluorescent ruthenium-based stain) a 60-fold reduction in time was possible without compromising the intensity of the bands. Similar results were observed for the 1.5 mm gels. Extending the protocol for use with PVDF membranes was less successful since loss of band intensity was observed, as well as some heat-related damage to the membranes themselves. In addition to the effects of microwave heating on the staining procedure, the impact on the subsequent mass spectrometric analysis was also assessed. An increased number of peptides could be extracted after proteolytic digestion of bands from the gels visualized using the microwave-promoted protein staining methodology. This is attributed to the heat-induced denaturation of proteins causing greater accessibility to the key cleavage sites by the protease, as well as an increase in the permeability of the gel allowing for greater recovery of the resultant peptides. A similar approach has been used to analyze glycoproteins, a process that can often be difficult due to unfavorable interactions between the glycan and the stain.[99]

9.5.2 THE AKABORI REACTION

Identification of the amino acid at the C-terminus of proteins is possible using the Akabori reaction, devised over half a century ago. Conventionally it involves heating the protein with anhydrous hydrazine under reflux at 125 °C for several hours. The amino acid at the C-terminus of the peptide is liberated and can be distinguished from the remaining amino acid residues that have been converted to hydrazides. Using a domestic microwave it is possible to reduce the reaction time to 5–10 min, the dipeptide Trp-Phe, tripeptide Tyr-Gly-Gly, tetrapeptide Pro-Phe-Gly-Lys, heptapeptide Ala-Pro-Arg-Leu-Arg-Phe-Tyr, and a N-terminal modified tripeptide *N*-acetyl-Met-Leu-Phe being used as test proteins.[100] The methodology has also been applied to the sequencing of cyclic peptides, resulting in selective opening and subsequent cleavage of the substrates in only a few minutes.[101]

9.5.3 AMINO ACID SEQUENCE DETERMINATION BY PROTEIN HYDROLYSIS

Hydrolysis and subsequent amino acid analysis can be used to determine the composition of a protein as well as quantify concentration. Essential to the methodology is the initial hydrolysis step during which the peptide bonds are broken under acidic conditions at high temperatures to generate free amino acids. Conventional protein hydrolysis is typically performed with 6 N HCl at 110 °C for 24 h. The first published study on the use of microwave heating for acid hydrolysis of proteins appeared in 1987, ribonuclease A and insulin B being used as substrates.[102] Solutions were heated in a domestic microwave oven for periods of 1 to 7 min. Amino acid recoveries after heating for 3, 4, and 5 min corresponded well to both the known composition of the substrates and the values obtained from conventional hydrolysis. However, the method lacked a means for measuring the temperature of the vessel contents. The technique has continued to attract attention.[103] The vapor-phase hydrolysis of methionyl human growth hormone (m-HGH, Protropin®) has also been performed using microwave heating.[104] The process was carried out using 6 N HCl at 178–180 °C and 140 psi. A range of sample sizes from 20 to 80 µmol were heated for times between 2 and 12 min. A comparative study between conventional and microwave hydrolysis of beta-casein has been undertaken.[25–27] Figure 9.4 shows the relative recoveries of amino acids and, while the two methods are comparable, using microwave heating was significantly faster.

Microwave-promoted acid hydrolysis has been used for analysis of food samples.[105] Starting initially with bovine serum albumin, different reaction conditions were screened. Heating in a sealed vessel for 5 min (pressure of 80 psi) led to good recovery of most amino acids but complete hydrolysis of amide bonds to valine,

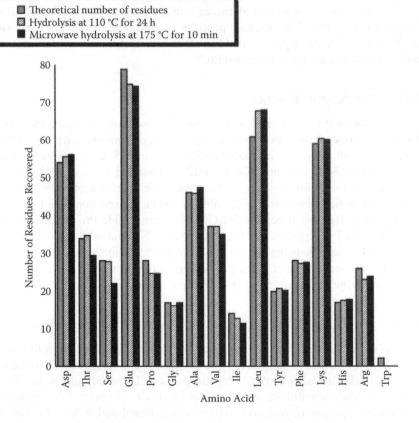

FIGURE 9.4 A comparison of amino acid recoveries after hydrolysis using microwave-mediated hydrolysis and conventional hydrolysis methods for the protein beta-casein. (Reproduced with permission from Sandoval, W. N.; Pham, V. C.; Lill, J. R. *Drug Discov. Today* 2008, *13*, 1075–1081. Copyright Elsevier.)

leucine, and isoleucine required extension of the reaction time to 10 min. Hydrolysis of durum wheat semolina and Pecorino Romano cheese was successful, giving comparable results to conventional hydrolysis but in a fraction of the time. Determining the collagen content in meat-based products[106] and the analysis of wall painting samples are other applications of the method.[107]

9.5.4 MICROWAVE-ASSISTED PROTEOLYSIS

To obtain detailed structural information, proteins can be selectively cleaved into smaller polypeptide fragments by controlled enzymatic digestion. The resulting mixtures can then be analyzed by various mass spectroscopic techniques and, from this, the structure of the original protein determined. Digestion times depend on the nature of the proteins and can vary from hours to days. It is necessary to ensure that sufficient quantities of the peptides are generated such that the detection limit of the analytical techniques is surpassed. Trypsin is the most commonly used enzyme because it specifically hydrolyzes peptide bonds at the carboxyl side of lysine and arginine residues, except when either is followed by proline. Microwave heating is an effective tool for performing proteolytic digestions, and several proteins—including cytochrome c, ubiquitin, myoglobin, lysozyme, and recombinant human interferon α-2b (rh-IFN α-2b)—have been studied as substrates.[108] In the case of cytochrome c, excellent results were obtained within 10 min at 60 °C, no nonspecific cleavage being observed. In the absence of a protease, and under the same conditions no degradation of the protein is observed, this indicating that the results are due solely to proteolysis. The digestion can also be performed with endoproteinase lysine-C and with in-gel protein samples, good coverage of the proteins being observed. A kinetic study of the trypsin digestion of IFN α-2b showed that in the first few minutes microwave heating had a beneficial effect but this diminished rapidly after about 30 min. The drop in activity is attributed to denaturation of the protein at elevated temperatures reached. In contrast, the conventional approach, performed at 37 °C did not show reaction products until after 20 min of incubation and the rate of the digestion continued to increase after 30 min. In an attempt to mimic conventionally the temperature profile seen when using microwave heating, the same digestion was performed using metal block preheated to 60 °C. The rate of enzymic cleavage was found to be essentially identical, suggesting that the rapid increase in temperature is responsible for the rate acceleration seen when using microwave heating.

The application of microwave heating to the in-gel trypsin digestion of proteins has been demonstrated.[109] Lysozyme, chicken egg albumin, bovine albumin, conalbumin, and ribonuclease were separated by 1D SDS-PAGE, stained with Coomassie Blue, excised from the gel, and digested by trypsin using both microwave and conventional heating. The conventional digest was performed at 37 °C for 16 h and the digest using microwave heating was performed for 5 min in a domestic oven at power settings of 195 W or 325 W. The peptide mass values were searched against the Swiss-Prot database using the MASCOT peptide mass fingerprint search program. The microwave-promoted digest at 195 W gave more matched fragments than the conventional method for all proteins except conalbumin where the number of matched fragments was 10% lower. The protocol using microwave heating also gave the highest probability-based MOWSE score returned from the MASCOT MS/MS ion search program.

Microwave heating has found application in quantitative analysis of the fragments resulting from enzymatic digestion of glycated hemoglobin (HbA1c).[110] Both trypsin and endoproteinase Glu-C (Glu-C) were trialed as proteases and a number of reaction parameters including the temperature, buffer concentration, and digestion time were screened. In the case of trypsin, 50 °C for 20 min was found to be optimal. Above 55 °C the efficiency of the digestion substantially decreased, probably due to heat-induced deactivation of the protease. Use of a 100 mM buffer was optimal, but results did not deviate significantly at lower concentrations. Using the optimized conditions, the digestion efficiency using microwave heating was about 20% higher than conventional overnight digestion at 37 °C. The optimum operating temperature for Glu-C was 40 °C, but even at its zenith the efficiency was considerably lower than the conventional method. The trypsin used in this study was modified by reductive methylation to increase its activity and stability; however, the same modifications cannot be made to Glu-C. The analysis of the Glu-C digest showed no evidence of peptides that could have originated from the protease itself; therefore, it is more likely that deactivation of the enzyme resulted from heat-induced denaturation rather than autolysis.

The digestion and analysis of complex protein mixtures is more challenging than that of purified proteins. Microwave heating can prove useful in achieving this. Using total yeast lysate and a sample of human urinary proteins as substrates, digestions can be performed in solution in a total of 6 min with efficiencies comparable to conventional methods.[111] Performing in-gel digests takes a longer time (25 min) and higher peptide recovery can be obtained.

Enzymatic digestions are typically performed in an aqueous buffer solution, but the addition of a small amount of organic solvent can be used to denature the protein partly, which improves digestion efficiency. Trypsin digestion of myoglobin, cytochrome c, lysozyme, and ubiquitin has been performed using microwave heating with addition of organic solvents.[112] Initially, comparisons between digestion at 60 °C for 10 min and conventional digestion at 37 °C for 6 h were made, running the reactions in buffer. For each protein examined, microwave-promoted digestion gave comparable or higher efficiencies than the conventional method. Both efficiency and sequence coverage increased when acetonitrile was added to the reaction mixture. If methanol is used as an additive, as the quantity increases so the protease activity decreases indicating that it is being deactivated. This is not surprising given the fact that methanol inhibits many enzyme-catalyzed processes. The digest was also performed in a methanol/chloroform/water (49%/49%/2%; pH 8) solution, but with such a high percentage of organic solvent that the protease is greatly deactivated. A very low abundance of tryptic fragments was observed when using microwave heating, and none were present in the digests performed conventionally.

The microwave-promoted digestion of cytochrome c has been performed in the presence of magnetite beads surface-functionalized with net negatively charged species.[113] When using trypsin as the protease, an 86% sequence coverage could be obtained after only 30 s of microwave heating. The digestion of myoglobin in the presence of the beads required 1 min and the addition of 30% acetonitrile to give a 90% sequence coverage. The beads could serve two roles. They readily absorb micro-wave energy, thus rapidly heating the reaction mixture. In addition, their negatively charged surface can adsorb proteins with opposite charges onto the surfaces through electrostatic interactions. The adsorbed proteins could feasibly unfold on the hot beads, thereby making them more susceptible to enzymatic digestion. To exploit this phenomenon, a nonfat milk sample with trace amounts of cytochrome c was screened as a substrate for tryptic digestion in the presence of the beads. Milk proteins carry net negative charges while cytochrome c carries a net positive charge in buffer solu-tion at pH 8. The digestion products of cytochrome c were obtained cleanly and with minimal peptide fragments from the milk proteins. This suggests that the functional-ized magnetite beads can act as concentrating affinity probes to enrich trace amounts of specific proteins in complex samples. Trypsin was also immobilized onto the mag-netite beads noncovalently and used to digest cytochrome c. The proteolysis was com-plete within 1 min, a sequence coverage of 87% being obtained.

Trypsin has also been immobilized on magnetic microspheres and these used for proteolytic digests in conjunction with microwave heating.[114] Cytochrome c was digested in 15 s with 92% sequence coverage. By comparison, conventional heating gave only 78% coverage after 12 h at 37 °C. The immobilized trypsin can be recycled seven times without loss of enzymatic activity. The protocol can be extended to other substrates including myoglobin and bovine serum albumin.[115,116] A slightly longer reaction time was required (45 s) for the proteolysis of human pituitary extract in order to identify 9709 unique peptides and 485 proteins.[117] In the case of lens pro-teins, a digestion time of 1 min resulted in the formation of 26 new proteins while the conventional method gave only 11.[118]

Other supports have found applications in microwave-promoted digestion. Zirconia-coated magnetic nanoparticles have been used for digestion and enrich-ment of phosphoproteins,[119] and titania-coated magnetic iron oxide nanoparticles used in the characterization of pathogenic bacteria.[120]

A limitation often encountered in proteomics is the fact that it is hard to ana-lyze proteins and protein biomarkers reliably at concentrations below 100 ng/mL. Protein quantification at or below this level using liquid chromatography/tandem mass spectrometry (LC/MS/MS) has been developed but involves an immunoaffin-ity enrichment step such as immunoprecipitation (IP). In order to maximize mass spectrometry (MS) sensitivity, the protein enrichment step is followed by a prote-olytic cleavage step used to generate a surrogate peptide with better mass spectro-metric properties. This cleavage step can be performed with microwave heating.[121] As an example, two rat cardiotoxicity biomarkers, Myl3 and NTproBNP, have been screened as substrates. The IP step was undertaken conventionally in a 96-well plate and then the microwave-promoted proteolysis with trypsin was performed in indi-vidual Eppendorf tubes. For Myl3, microwave heating for 50 min at 55 °C with a pro-tease to protein ratio (weight of trypsin:weight protein) of 1000:1 gave about the same recovery (80%) as the conventional method (15 h at 37 °C with a protease to protein ratio of 100:1). Incubation for 4 h at 37 °C was also tried, but the absolute recovery was only 36%, indicating the time for the conventional methodology cannot be cur-tailed. Validation statistics for the microwave-promoted method (accuracy within

11% and precision within 12%) were similar to those for the conventional approach (accuracy within 12.9% and precision within 13.2%). The digestion of NTproBNP was much more difficult than that of Myl3 when using either microwave or conventional heating. The highest practical protease to protein ratio of 4000:1 was used to give 65% recovery after 50 min at 55 °C in the microwave heating approach, similar to the 63% recovery obtained with the conventional method. The validation statistics for the microwave method were slightly better than those obtained with the conventional method. Overall, the protocol incorporating microwave heating allowed for the limits of quantification to reach as low as 0.95 ng/mL for Myl3 and 100 pg/mL for NTproBNP and also significantly reduced the validation time to a single day.

One bottleneck in the previously described procedure was that the microwave step required three separate runs to process 60 samples. The use of a multiwell plate for the microwave digestion would eliminate the need to transfer the samples from the plate to individual tubes. The use of 48-well silicon carbide plate for the digestion of insulin chain B using microwave heating has been reported.[122] Initially, 12 identical digests were performed at 50 °C for 30 min using a protease to protein ratio of 1:5 in order to test the uniformity of the well plate. The digestion efficiency ranged from 84% to 91% and was independent of location, indicating even heating across the plate. The effect on the digestion efficiency of varying the protease to protein ratio was also probed. As the ratio was decreased from 1:5 to 1:25 the efficiency of the digestion also decreased (95% to 69%).

9.5.5 CHEMICAL PROTEOLYSIS

Chemical proteolysis is often used as an alternative to enzymatic digestion for the identification of proteins. It is particularly useful when attempting to analyze insoluble materials such as membrane proteins. These can be solubilized when using additives such as urea, detergents, or other salts, but these can inhibit protease activity and also interfere with mass spectrometric analysis. Another application of chemical proteolysis is in the analysis of proteins typically resistant to enzymatic digestion. Microwave-assisted acid hydrolysis (MAAH) has been used to characterize proteins and also map a subset of a proteome.[123] Initial studies focused around the model transmembrane protein, bacteriorhodopisin. The methodology involved suspending the protein powder in aqueous TFA and heating for 10 min. Analysis of the hydrolysate resulted in almost complete sequence coverage by the peptides detected, including the identification of two posttranslational modification sites. These results were compared to those obtained in hydrolysis at 110 °C for 4 h with 70% formic acid or 25% TFA using conventional heating (Figure 9.5). Hydrolysis with formic acid resulted in significant N-terminal formylation of the resultant peptides, and hydrolysis with 25% TFA generated only one peak from the N-terminus due to aggregation of the protein. Acid concentration and microwave heating time could be easily manipulated to control the extent of hydrolysis. At a low acid concentration (0.3 M TFA) and short heating time (2 min), predominantly C- and N-terminus fragments were generated. Hydrolysis at higher concentration (3 M TFA) and longer heating time (10 min) resulted in considerably more internal fragments being obtained. Hydrocholoric acid was examined as an alternative to TFA but found to be inferior both in terms of

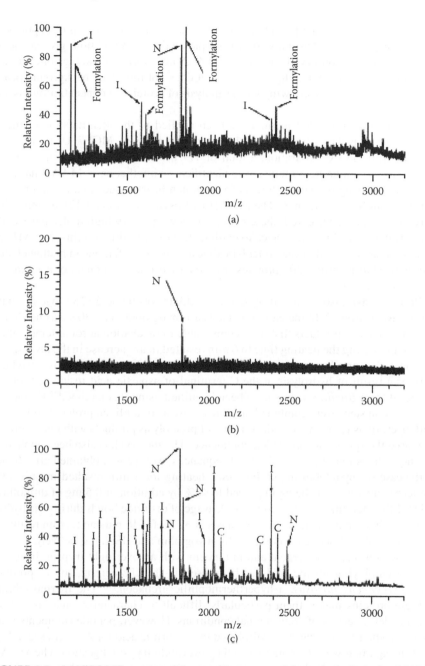

FIGURE 9.5 MALDI MS spectra of the peptides generated from the hydrolysis of 200 ng of bacteriorhodopsin under different conditions: (a) 70% formic acid heated at 110 °C for 4 h, (b) 25% TFA heated at 110 °C for 4 h, and (c) 25% TFA digested by microwave irradiation for 10 min. "I" indicates peptides resulting from internal fragmentation. "C" and "N" indicate C- and N-terminal peptides, respectively. (Reproduced with permission from Zhong, H.; Marcus, S. L.; Li, L. *J. Am. Soc. Mass Spectrom.* 2005, *16*, 471–481. Copyright Elsevier.)

the generation of internal peptide fragments and also the production of peptides of optimal length for tandem mass spectroscopic analysis. In addition, glycine cleavage specificity was observed when using TFA. A noteworthy advantage of chemical proteolysis over enzymatic digestion is that it does not introduce background peaks from protease autolysis, allowing for the analysis of proteins at concentrations as low as 1 ng/μL.

The optimized MAAH method has been applied to the identification of membrane proteins isolated from human breast cancer cell line MCF7. Using MALDI MS/MS with 1D-LC separation, a total of 119 proteins, including 41 membrane-associated or membrane proteins, were identified. In another case, MAAH has been used for analyzing proteins extracted from human heart tissues that are not readily soluble in an SDS solution.[124] The MAAH hydrosylate contained 121 proteins that were not otherwise observed, these being very low and very high molecular weight proteins that bear direct relevance to cardiac disease. Building on this, MAAH has been used together with three proteolytic digests (one in buffer, one in methanol, and one in SDS) to identify 1204 proteins from the cytosolic component of a zebra fish liver tissue sample.[125]

Proteins have also been hydrolyzed using dilute formic acid (2%) together with microwave heating.[126] In the case of horse heart myoglobin, only 30 s was needed to generate cleavage products that were comparable to incubation at room temperature for 8 h. Increasing the heating time to 6 min resulted in an increase in the lower-mass fragment ions and fewer missed cleavages. The majority of the cleavage products were as a result of hydrolysis at the C-terminal of aspartic acid residues. Proteins extracted from *Bacillus* spores have been identified using this method.[127] Extraction and digestion steps were combined into one to give a streamlined protocol. While the acid digestion was incomplete, the mixture of proteolysis products with intact protein increased the species identification confidence. The method has also been applied to the in-gel digestion of proteins.[126] Total sequence coverage was obtained after 6 min in the case of myoglobin in-gel. For BSA, heating for 8 min resulted in only 23% sequence coverage, this being improved to 46% by addition of 0.5 mg of dithiothreitol (DTT) to facilitate in situ reductive cleavage of disulfide bonds during the hydrolysis step. In comparison, in-gel proteolysis at 37 °C for 16 h using trypsin results in only 78% and 37% sequence coverage for myoglobin and BSA, respectively.

In the presence of detergent and in tandem with an overnight trypsin digestion, MAAH employing formic acid has been used in the characterization of proteins from *Pyrococcus furiosus*, a hyperthermophilic archaeon.[128] The thermostability of these proteins makes them particularly difficult to characterize and proteolysis failed under a variety of conventional conditions. However, peptide fragments were obtained when *P. furiosus* was hydrolyzed in 6% formic acid for 2.75 min seconds at 40 °C using microwave heating followed by overnight trypsin digestion. The MAAH step served to provide the protease with greater access to potential cleavage sites. The combination approach allowed for the identification of over 900 proteins, this representing 44% of the proteome.

In an effort to accelerate the process of protein identification even further, microwave digestion with formic acid has been coupled to ESI-MS and LC-ESI-MS analytical tools.[129] This took advantage of the speed, simplicity, and specificity of acid

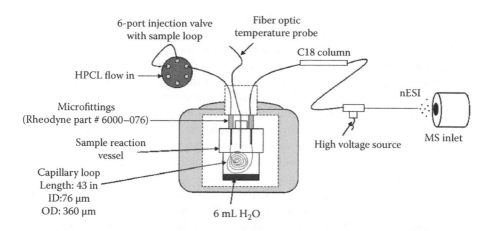

6-port injection valve
with sample loop

Fiber optic
temperature probe

C18 column

HPCL flow in

nESI

Microfittings
(Rheodyne part # 6000–076)

Sample reaction
vessel

High voltage source

MS inlet

Capillary loop
Length: 43 in
ID:76 µm
OD: 360 µm

6 mL H$_2$O

FIGURE 9.6 Schematic of the microwave digestion flow cell for online hydrolysis. The reaction loop was 5 µL in total internal volume and was operated at a flow rate of 1 µL/min giving a total residence time of 5 min. (Reproduced with permission from Hauser, N. J.; Basile, F. *J. Proteome Res.* 2008, 7, 1012–1026. Copyright American Chemical Society.)

hydrolysis at aspartic acid residues as well as facilitating the analysis of proteins in small volumes without transfer losses.[129] The specially designed microwave flow-cell contained a 5 µL sample loop that was preheated to 130 °C (Figure 9.6), and the protein in 12.5% formic acid was passed through at a rate of 1 µL/min, giving a total digestion time of 5 min. Myoglobin was used as a test substrate and exposed to the online cleavage conditions, both with and without microwave heating (i.e., at room temperature). In the absence of microwave heating, only the intact protein was observed. When using microwave heating, a number of signals due to digestion products were observed and none of the intact protein detected. The 5 min digestion furnished an 88% sequence coverage for myoglobin. One of the disadvantages of this method is that the hydrolysis protocol generates peptide fragments that are inherently longer than those obtained from proteolysis, this complicating their identification by collision induced dissociation (CID) and making manual identification necessary. Because the longer peptide fragments are typically highly charged (above +3), studies showed that electron transfer dissociation (ETD) proved a more effective method for their characterization and eliminated the need for manual identification.[130]

The digestion of α-lactalbumin, a protein that contains 13 aspartic acid residues and 4 intra-chain disulfide bonds has also been attempted using the flow-cell, but without success.[129] No detectable hydrolyzed peptides were obtained. The presence of the disulfide bonds was thought to be either hindering the MAAH or else leading to the production of large peptide chains cross-linked by disulfide bonds. By adding DTT directly to the protein and formic acid mixture prior to injection into the microwave flow cell it was possible to cleave the disulfide bonds at the same time as the hydrolysis took place. Having shown that this worked successfully in the case of insulin (which contains two disulfide bonds), the digestion of α-lactalbumin was repeated to give 90% sequence coverage.

Acetic acid has been used in an MAAH approach to the rapid identification of viruses.[131] Digestion with acetic acid is selective for aspartic acid. The digestion of bacteriophage MS2 using 50% acetic acid and 1% Triton X-100 (a nonionic surfactant) has been explored using microwave heating at different power inputs (100, 130, 160, 190, and 220 W). Samples were irradiated at the selected power until the temperature reached 108 °C (typically less than 15 s) and then held at that temperature for an additional 15 s before cooling and analyzing the product mixture using mass spectrometry. As the applied microwave power increased, the intact protein ion in the product mixture decreased while signals of ions from digested fragments increased. A microwave power of 190 W proved optimal. When the power was increased to 220 W signal intensities of the digestion products were significantly reduced. Digestion of bacteriophage MS2 coat protein provided 100% sequence coverage with high confidence and also allowed for the identification of a number of the strains of the RNA bacteriophage family. With a slight modification to the reaction conditions (microwave irradiation with 12.5% acetic acid at 140 °C for 20 min), the methodology has been applied to the identification of yeast ribosomal proteins, about 73% of the proteome being identified.[132] Microwave acid digestion of ribonuclease A and B could not be realized without reduction of the disulfide bonds.[133] Addition of DTT directly to the protein solution with acetic acid followed by microwave heating for 5 min successfully reduced the disulfide bonds and cleaved the peptide backbone allowing for excellent sequence coverage. Under these conditions the N-linked glycan of ribonuclease B was intact with minimal hydrolysis within the carbohydrate chain, but the O-linked glycan of alpha crystallin A was cleaved during the protocol. Using ovalbumin as a test substrate the fate of post translational modifications during the MAAH procedure has also been studied.[134] While acetylation sites were preserved, hydrolysis of phosphate groups occurred at detectable rates.

9.5.6 Hydrolysis for N- and C-Terminal Sequencing

A widely used method for sequencing amino acids in proteins is Edman degradation, also known as N-terminal sequencing.[135] While useful, the procedure does have drawbacks. As the peptide length increases, the efficiency of the reaction decreases; peptides of approximately 50–60 amino acids being the upper limit. It does not work if the N-terminal amino acid has been chemically modified or if it is concealed on the interior of the protein and the positions of disulfide bridges cannot be determined. As a result, there have been a number of attempts to develop alternative N-terminal sequencing techniques. One reported example involved the use of MAAH.[136] Heating proteins in the presence of 6 N HCl for 2 min predominantly formed two sets of polypeptides. One set contained the N-terminal amino acid and the other contained the C-terminal amino acid. No internal polypeptides were formed (i.e., all polypeptides contained either the N- or the C-terminus). Mass spectrometry allowed for ready identification of each amino acid in the protein by measuring mass differences between adjacent peaks within the same series. The method could be used for sequencing the hemoglobin analog HbG Coushatta,[137] and an adaptation was used to sequence an unusually stable cyclic peptide kalata B2.[138] This was resistant to enzymatic degradation, but microwave heating for 10 min with 6 N HCl was effective in generating cleavage fragments. In conjunction with capillary electrophoresis, the characterization and identification of cytochrome *c* was also possible.[139]

9.5.7 Analysis and Characterization of Post-translational Modifications

9.5.7.1 N-Terminal Pyroglutamate

Many proteins containing an N-terminal glutamine are blocked with a pyroglutamyl group that is a result of cyclization by the enzyme glutamine cyclase. Since Edman degradation requires the presence of a free amine, the pyroglutamyl moiety precludes this and the group needs to be removed to allow N-terminal sequencing. Pyroglutamate amino peptidase (PGAP) is a thermostabile aminopeptidase responsible for removing the N-terminal pyroglutamate. Microwave heating can be used to facilitate this as evidenced by the removal of pyroglutamate from the heavy chain of anti-OX40 ligand.[26] The PGAP digestion was performed at 90 °C both using microwave heating and in a thermocycler for 1–60 min followed by N-terminal sequencing. The initial yield of the deblocked protein was obtained from the sequencing data of the first five residues. Not surprisingly, microwave heating led to higher initial yields and faster rates of reaction.

9.5.7.2 Phosphorylation

Phosphorylation is an important post-translational modification, but the identification and positional mapping of the phosphate groups is complicated by their lability during typical proteomic analysis. A selective method for mapping phosphorylation sites has been developed and involves a microwave-promoted β-elimination/Michael addition protocol.[26] It selectively converts phosphoserine and phosphothreonine to more stable *S*-ethylcysteine and β-methyl-*S*-ethylcysteine, respectively. Under basic conditions the phosphate moiety is lost via β-elimination, and the resulting Michael acceptor is then immediately trapped with a nucleophile, in this case an alkyl thiol. Conventionally, this process is performed at 60 °C for 1–3 h, but using microwave heating the process is performed at 100 °C for 2 min, giving comparable results. The process works for a range of substrates including phosphoproteins embedded in transfer membranes.

9.5.7.3 Deglycosylation

Like phosphorylation, glycosylation is an important post-translational modification of proteins that facilitates the regulation of many cellular processes and intracellular signaling events. Glycosylation occurs either via the amide nitrogen of asparagine (*N*-linked) or via the hydroxyl of serine, threonine, and to a lesser extent hydroxyproline and hydroxylysine (*O*-linked). Deglycosylation is an important process for proteomic analysis because it reduces the heterogeneity of the protein sample and decreases smearing during protein separation by SDS-PAGE. Typical deglycosylation techniques require at least 2 h and often suffer from incomplete conversion. Tryptic *N*-glycosylated peptides of horseradish peroxidase have been treated with TFA for 1 min using microwave heating to affect partial cleavage of the oligosaccharides.[140] The most labile group was α-l-fucose, and most cleaved peptides still contained one *N*-acetylglucosamine (GlcNAc). Despite achieving only partial cleavage, enough monosaccharide composition information could be obtained in the short reaction time. The method has also been applied to the partial acid hydrolysis of glycoproteins ribonuclease B, avidin, α1-acid glycoprotein, and fetuin.[141] Partial and complete cleavage products of the oligosaccharide from the intact glycoprotein were observed within 2 min of microwave heating in the presence of TFA.

Complete deglycosylation of N-linked glycans from proteins can be performed with the enzyme PNGase-F without any acid-induced cleavage of the protein. Conventional digestions with PNGase-F typically require incubation at 37 °C for 24 h to reach completion. Using microwave heating the process can be performed in 10–60 min.[142] In the case of Avastin, 10 min at 37 °C resulted in complete deglycosylation (Figure 9.7). After 10 min using conventional heating in a water bath, glycosylated protein was still present and, even after 1 h deglycosylation was not complete. In the case of Herceptin, DOTA conjugated antibodies, and RNase B complete removal of the N-linked glycans was possible in 1 h or less. A modification of this method has been used to perform the deglycosylation of a variety of glycoproteins in the presence of functionalized diamond nanoparticles in order to facilitate the MALDI-TOF-MS analysis of the neutral underivatized glycans.[143]

9.6   OTHER BIOSCIENCE APPLICATIONS OF MICROWAVE HEATING

As well as proving valuable in peptide synthesis and proteomics, microwave heating has also found applications in allied fields. This includes microwave-promoted enzymatic extraction of protein-bound selenium,[144] dissociation of Protein A from the IgG matrix,[145] ink staining for protein quantification on nitrocellulose membranes,[146] and tissue processing for pathology applications[147–149] as well as the study of the effects of microwave irradiation on the folding and denaturing of proteins[150–152] Microwave heating has also been used extensively for enzyme catalysis in organic synthesis.[153]

Microwave heating has also been used as an alternative to conventional approaches for DNA-related applications. One example is in the polymerase chain reaction (PCR) for DNA amplification.[154,155] Using controlled microwave heating it is possible to reduce cycle times by approximately half. This is because the target temperatures can be reached very rapidly and then held there easily. As a result, incubation times can be shortened over the conventional counterparts since, in the case of the latter, time for equalization of the temperature is required. In the case of a 220-base pair segment of the chromosomal 23S rRNA gene of the thermophilic campylobacters, amplification was performed by adopting a protocol involving 15 s heating, 30 s microwave off, 15 s heating, and then transferring to 54 °C water bath for 1 min.[154] Optimum power settings were 100 W for the first heating stage and 130 W for the second. The process was repeated 25–35 times, and the overall amplification was found to be 70% as efficient as that performed conventionally. It is also possible to perform PCR on the mL scale using microwave heating. Conventionally, the slow distribution of heat together with the importance of short process times and reproducibility limits

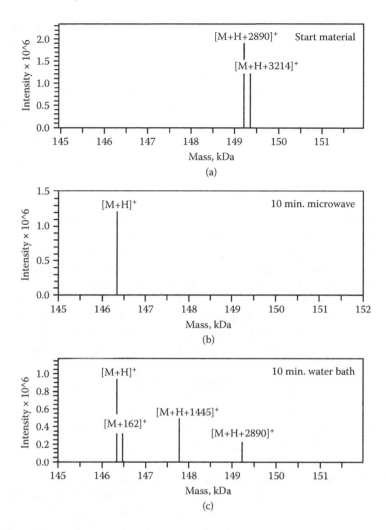

FIGURE 9.7 (a) Deconvoluted mass spectrum of non-reduced, glycosylated Avastin; (b) nonreduced Avastin after 10 minute incubation in the microwave at 37 °C with PNGase F at an enzyme:substrate ratio of 1:50; (c) nonreduced Avastin after 10 minute incubation in a water bath at 37 °C with PNGase F at an enzyme:substrate ratio of 1:50. (Reproduced with permission from Sandoval, W. N.; Arellano, F.; Arnott, D.; Raab, H.; Vandlen, R.; Lill, J. R. *Int. J. Mass Spectrom.* 2007, *259*, 117–123. Copyright American Chemical Society.)

the volume for most reactions to 0.2 mL. A 53 bp fragment from human chromosome 13 has been amplified using the power/time method, repeating the process 33 times.[155] Working on the 2.5 mL scale, the microwave instrument was programmed to heat for 45 s at 90 °C and 175 W initial power, cooled for 50 s, and then heated for 85 s at 60 °C with 15 W maximum power during the first thermocycle. In the 32 subsequent thermocycles, the first step was modified to heat for 35 s at 88 °C. Overall, an amplification efficiency of 92–96% was obtained and the total processing time was

1 h 34 min. The method could be scaled up to process 15 mL using similar conditions but extending reaction times and cooling times somewhat to take account for the larger volume.

The preparation of oligonucleotides has become increasingly important due to the use of the products as primers, probes, sequencing aids, and diagnostics. While the synthesis of oligonucleotides, and in particular that of DNA, has seen numerous improvements, the procedures still suffer from the large excess of phosphoramidite building blocks used during coupling steps. Microwave heating has been used for DNA-chain extension with the aim of overcoming this problem.[156] Phosphoramidite coupling with a standard thymidine building block was performed at 30 °C for 6 min under microwave irradiation and the number of equivalents of the phosphoramidite could be reduced 2.5 times. In the case of the less reactive aminomethylthymidine building block, direct comparison of microwave and conventional heating at 30 °C showed an increase in conversion in the case of the formed.

Deprotection of synthetic oligonucleotides can also be performed using microwave heating.[157] The conventional method for removal of standard N-protecting groups (dA^{Bz}, dC^{Bz}, dG^{ibu}) involves heating at 55 °C for 17 h in the presence of 29% aqueous ammonia. Comparable results can be obtained after 5 min at 170 °C using microwave heating. Deprotection of dC^{Bz} is the slowest and dictates the reaction time. The method has been applied to the deprotection of oligonucleotides up to 50-mer in length and no thermodegradation due to the elevated temperature was observed.

Gene therapy is currently performed by one of three routes, namely, using viral vectors, non-viral biochemical vectors, or physical methods. Each of these has advantages and disadvantages and there is increasing focus on the design and development of new methods for gene delivery. Microwave heating has been used for the delivery of oligonucleotides.[158] Starting with a luciferase expression plasmid, repetitive cycles of 5 s heating followed by 45 s resting were performed using either 240 and 420 W as the input microwave power, the reactions being run in six-position well-plates. Maximum efficiency of luciferase expression was found after 12 cycles at 420 W, giving eight-times higher luminescence than the control samples in an activity assay. However, considerable difference in activity was observed between samples from different wells, presumably due to uneven heating. Cell viability was also measured and found to reduced to 85% of the control levels indicating delivery of transgenes with microwave heating induces some cell damage. However, immunostaining for caspase 3 showed expression in very few cells demonstrating the microwave heating does not induce apoptosis. Interestingly, using a microwave power of 240 W for 12 or 15 cycles or 420 W for 6 cycles showed increased metabolic activity.

In an effort to examine the effects of microwave irradiation on DNA, a DNA hybridization experiment featuring an end-point assay that utilized fluorescence quenching to detect DNA melting and annealing has been performed.[159] It used two sets of DNA, the first containing the fluorophore and fluorescence quencher and the second being unlabeled (Figure 9.8). Upon melting of the two sets and then cooling, the strands would be expected to anneal in a statistical fashion to either

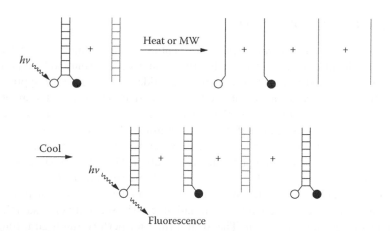

FIGURE 9.8 Schematic of the fluorescence-quenching assay used to determine if microwave irradiation (MW) is capable of melting dsDNA. Open circle = fluorophore (fluorescein); closed circle = fluorescence quencher). (Reproduced with permission from Edwards, W. F.; Young, D. D.; Deiters, A. *Org. Biomol. Chem.* 2009, 7, 2506–2508. Copyright Royal Society of Chemistry.)

their previous partner or the other complementary strand. Coupling of the fluorescein labeled DNA with unlabeled DNA would result in a fluorescence signal. A similar result would be expected if the DNA failed to rehybridize. Conventional heating was used to establish both the melting point of the DNA and a baseline for fluorescence. Experiments using the microwave apparatus were performed in a jacketed reaction vessel to keep the temperature well below the melting point in an attempt to differentiate effects arising from thermal heating and direct interaction with microwave irradiation. Prehybridized oligomers were precooled to −20 °C and then irradiated with 300 W microwave energy until the temperature reached 20 °C followed by cooling back to −20 °C. The same experiments were performed using conventional heating. No increase of fluorescence was observed in the samples heated conventionally but those exposed to microwave energy showed a fluorescence signal identical to that of the control. The experiment was repeated with different DNA/DNA duplexes with varying microwave power inputs. While the oligo pairs had different melting temperatures, they melted at virtually the same microwave power (108 W ± 8 W). A number of possible explanations for this unexpected result have been presented. Direct energy transfer into the oligonucleotide could possibly increase as a function of the number of nucleotides, thus maintaining the melting temperature at the same microwave power despite the increased degree of hybridization in longer DNA sequences. The alignment of the anions and cations with microwave field may cause a repulsion of the negative charges on the DNA, destabilizing the duplex and facilitating DNA melting. Lastly, as the DNA could begin to unravel, the resulting single strands may absorb a greater amount of microwave energy due to their higher dielectric permittivity, accelerating the melting process.

9.7 FUTURE PERSPECTIVES

The applications of microwave heating in peptide synthesis have grown tremendously since its introduction in 1992. It will undoubtedly continue to be a powerful method for clean and efficient production of peptides for a variety of applications. One of the major disadvantages of solid-phase peptide synthesis is the amount of waste generated, particularly from the washing steps. Washing is critical to ensure the purity of the final peptide and cannot be eliminated. Therefore, new chemistry is necessary for peptide synthesis that features the use of nontoxic solvents such as water and safer reagents for coupling and deprotection. The unique characteristics of microwave heating can aid in the development of these new chemistries.

Microwave technology still has enormous potential in the field of proteomics sample preparation. Current hardware limitations restrict the number of samples that can be processed in a single run. The ideal format for performing high-throughput proteomics is a 96-well plate, but heating a multiwell plate evenly using microwave heating can be a challenge. Temperature gradients across a plate lead to irreproducible and inconsistent results. Recent efforts have been made to address the uniform heating issue through the use of silicon carbide plates. However, an ideal platform would support any plate design including a variety of well sizes and dimensions with even microwave heating across the plate. The development of such a system will open a wide variety of applications including high-throughput proteomics sample preparation, assay development, and protein folding experiments. In addition to protein-based applications, a multiwell plate system would also expand the scope to oligonucleotide chemistry.

In order to take advantage fully of microwave heating for protein chemistry it is necessary to understand the behavior of proteins in the microwave field. While theories abound, there is no concrete evidence to support them. One of the best methods for observing the secondary structure of a protein is circular dichroism (CD). Incorporating microwave irradiation either in tandem or in-situ with CD measurements should provide some interesting insight into the effect microwave irradiation has on protein structure.

REFERENCES

1. Gedye, R.; Smith, F.; Westaway, K.; Ali, H.; Baldisera, L.; Laberge, L.; Rousell, J. *Tetrahedron Lett.* 1986, *27*, 279–282.
2. Giguere, R. J.; Bray, T. L.; Duncan, S. M.; Majetich, G. *Tetrahedron Lett.* 1986, *27*, 4945–4948.
3. Hayes, B. L. *Microwave Synthesis: Chemistry at the Speed of Light*; CEM Publishing: Matthews, 2002.
4. Loupy, A., Ed. *Microwaves in Organic Synthesis*; Wiley-VCH: Weinheim, Germany, 2002.
5. Kappe, C. O.; Stadler, A. *Microwaves in Organic and Medicinal Chemistry*; Wiley-VCH: Weinheim, Germany, 2005.
6. Lidström, P.; Tierney, J. P., Eds. *Microwave-Assisted Organic Synthesis*; Blackwell: Oxford, 2005.
7. Loupy, A., Ed. *Microwaves in Organic Synthesis*; Wiley-VCH: Weinheim, Germany, 2006.

8. Kappe, C. O.; Dallinger, D.; Murphee, S. S. *Practical Microwave Synthesis for Organic Chemists: Strategies, Instruments, and Protocols*; Wiley-VCH: Weinheim, Germany, 2009.

9. Larhed, M.; Moberg, C.; Hallberg, A. *Acc. Chem. Res.* 2002, *35*, 717–727.

10. Lew, A.; Krutzik, P. O.; Hart, M. E.; Chamberlin, A. R. *J. Comb. Chem.* 2002, *4*, 95–105.

11. Wathey, B.; Tierney, J.; Lidstrom, P.; Westman, J. *Drug Discov. Today* 2002, *7*, 373–380.

12. Kappe, C. O. *Angew. Chem. Int. Ed. Engl.* 2004, *43*, 6250–6284.

13. Man, A. K.; Shahidan, R. *J. Macromol. Sci. A* 2007, *44*, 651–657.

14. Kappe, C. O. *Chem. Soc. Rev.* 2008, *37*, 1127–1139.

15. Caddick, S.; Fitzmaurice, R. *Tetrahedron* 2009, *65*, 3325–3355.

16. Kappe, C. O.; Dallinger, D. *Mol. Diversity* 2009, *13*, 71–193.

17. Santagada, V.; Frecentese, F.; Perissutti, E.; Fiorino, F.; Severino, B.; Caliendo, G. *Mini-Rev. Med. Chem.* 2009, *9*, 340–358.

18. Strauss, C. R.; Varma, R. S. In *Microwave Methods in Organic Synthesis* 2006; Vol. 266; pp 199–231.

19. Dallinger, D.; Kappe, C. O. *Chem. Rev.* 2007, *107*, 2563–2591.

20. Martinez-Palou, R. *J. Mex. Chem. Soc.* 2007, *51*, 252–264.

21. Polshettiwar, V.; Varma, R. S. *Chem. Soc. Rev.* 2008, *37*, 1546–1557.

22. Polshettiwar, V.; Varma, R. S. *Acc. Chem. Res.* 2008, *41*, 629–639.

23. Collins, J. M.; Leadbeater, N. E. *Org. Biomol. Chem.* 2007, *5*, 1141–1150.

24. Sabatino, G.; Papini, A. M. *Curr. Opin. Drug Discov. Devel.* 2008, *11*, 762–770.

25. Lill, J. R.; Ingle, E. S.; Liu, P. S.; Pham, V.; Sandoval, W. N. *Mass Spectrom. Rev.* 2007, *26*, 657–671.

26. Sandoval, W. N.; Pham, V.; Ingle, E. S.; Liu, P. S.; Lill, J. R. *Comb. Chem. High Throughput Screening* 2007, *10*, 751–765.

27. Sandoval, W. N.; Pham, V. C.; Lill, J. R. *Drug Discov. Today* 2008, *13*, 1075–1081.

28. Verlander, M. *Int. J. Pept. Res. Ther.* 2007, *13*, 75–82.

29. Yu, H. M.; Chen, S. T.; Wang, K. T. *J. Org. Chem.* 1992, *57*, 4781–4784.

30. Santagada, V.; Fiorino, F.; Perissutti, E.; Severino, B.; De Filippis, V.; Vivenzio, B.; Caliendo, G. *Tetrahedron Lett.* 2001, *42*, 5171–5173.

31. Erdelyi, M.; Gogoll, A. *Synthesis* 2002, 1592–1596.

32. Collins, J. M.; Collins, M. J. In *Microwaves in Organic Synthesis*; Loupy, A., Ed.; Wiley-VCH: Weinheim, Germany, 2006; pp 898–930.

33. Bacsa, B.; Desai, B.; Dibo, G.; Kappe, C. O. *J. Pept. Sci.* 2006, *12*, 633–638.

34. Brandt, M.; Gammeltoft, S.; Jensen, K. J. *Int. J. Pept. Res. Ther.* 2006, *12*, 349–357.

35. Palasek, S. A.; Cox, Z. J.; Collins, J. M. *J. Pept. Sci.* 2007, *13*, 143–148.

36. Unpublished results.

37. Hossain, M. A.; Rosengren, K. J.; Haugaard-Jonsson, L. M.; Zhang, S.; Layfield, S.; Ferraro, T.; Daly, N. L.; Tregear, G. W.; Wade, J. D.; Bathgate, R. A. D. *J. Biol. Chem.* 2008, *283*, 17287–17297.

38. Hossain, M. A.; Rosengren, K. J.; Zhang, S.; Bathgate, R. A.; Tregear, G. W.; van Lierop, B. J.; Robinson, A. J.; Wade, J. D. *Org. Biomol. Chem.* 2009, *7*, 1547–1553.

39. Drew, S. C.; Masters, C. L.; Barnham, K. J. *J. Am. Chem. Soc.* 2009, *131*, 8760–8761.

40. Drew, S. C.; Noble, C. J.; Masters, C. L.; Hanson, G. R.; Barnham, K. J. *J. Am. Chem. Soc.* 2009, *131*, 1195–1207.

41. Grieco, P.; Cai, M.; Liu, L.; Mayorov, A.; Chandler, K.; Trivedi, D.; Lin, G.; Campiglia, P.; Novellino, E.; Hruby, V. J. *J. Med. Chem.* 2008, *51*, 2701–2707.

42. Bin Zhang, H.; Chi, Y. S.; Huang, W. L.; Ni, S. J. *Chinese Chem. Lett.* 2007, *18*, 902–904.

43. Tofteng, A. P.; Jensen, K. J.; Schaffer, L.; Hoeg-Jensen, T. *Chembiochem* 2008, *9*, 2989–2996.
44. Schäffer, L.; Brand, C. L.; Hansen, B. F.; Ribel, U.; Shaw, A. C.; Slaaby, R.; Sturis, J. *Biochem. Biophys. Res. Commun.* 2008, *376*, 380–383.
45. Shabanpoor, F.; Hughes, R. A.; Bathgate, R. A.; Zhang, S.; Scanlon, D. B.; Lin, F.; Hossain, M. A.; Separovic, F.; Wade, J. D. *Bioconjug. Chem.* 2008, *19*, 1456–1463.
46. Armishaw, C.; Jensen, A. A.; Balle, T.; Clark, R. J.; Harpsoe, K.; Skonberg, C.; Liljefors, T.; Stromgaard, K. *J. Biol. Chem.* 2009, *284*, 9498–9512.
47. Cassone, M.; Vogiatzi, P.; La Montagna, R.; De Olivier Inacio, V.; Cudic, P.; Wade, J. D.; Otvos, L., Jr. *Peptides* 2008, *29*, 1878–1886.
48. Noto, P. B.; Abbadessa, G.; Cassone, M.; Mateo, G. D.; Agelan, A.; Wade, J. D.; Szabo, D.; Kocsis, B.; Nagy, K.; Rozgonyi, F.; Otvos, L., Jr. *Protein Sci.* 2008, *17*, 1249–1255.
49. James, P. F.; Dogovski, C.; Dobson, R. C. J.; Bailey, M. F.; Goldie, K. N.; Karas, J. A.; Scanlon, D. B.; O'Hair, R. A. J.; Perugini, M. A. *J. Lipid Res.* 2009, *50*, 1384–1394.
50. Chi, Y.; Zhang, H.; Huang, W.; Zhou, J.; Zhou, Y.; Qian, H.; Ni, S. *Bioorg. Med. Chem.* 2008, *16*, 7607–7614.
51. Grummitt, C. G.; Townsley, F. M.; Johnson, C. M.; Warren, A. J.; Bycroft, M. *J. Biol. Chem.* 2008, *283*, 23326–23332.
52. Harterich, S.; Koschatzky, S.; Einsiedel, J.; Gmeiner, P. *Bioorg. Med. Chem.* 2008, *16*, 9359–9368.
53. Mosse, W. K.; Koppens, M. L.; Gengenbach, T. R.; Scanlon, D. B.; Gras, S. L.; Ducker, W. A. *Langmuir* 2009, *25*, 1488–1494.
54. Svensen, N.; Diaz-Mochon, J. J.; Bradley, M. *Tetrahedron Lett.* 2008, *49*, 6498–6500.
55. Ivanova, G. D.; Arzumanov, A.; Abes, R.; Yin, H.; Wood, M. J.; Lebleu, B.; Gait, M. J. *Nucleic Acids Res.* 2008, *36*, 6418–6428.
56. Cemazar, M.; Craik, D. J. *J. Pept. Sci.* 2008, *14*, 683–689.
57. Coantic, S.; Subra, G.; Martinez, J. *Int. J. Pept. Res. Ther.* 2008, *14*, 143–147.
58. Tantry, S. J.; Rao, R. V. R.; Babu, V. V. S. *Arkivoc* 2006, *i*, 21–30.
59. Cianci, J.; Baell, J. B.; Harvey, A. J. *Tetrahedron Lett.* 2007, *48*, 5973–5975.
60. Katritzky, A. R.; Khashab, N. M.; Yoshioka, M.; Haase, D. N.; Wilson, K. R.; Johnson, J. V.; Chung, A.; Haskell-Luevano, C. *Chem. Biol. Drug Des.* 2007, *70*, 465–468.
61. Katritzky, A. R.; Yoshioka, M.; Narindoshvili, T.; Chung, A.; Khashab, N. M. *Chem. Biol. Drug Des.* 2008, *72*, 182–188.
62. Katritzky, A. R.; Haase, D. N.; Johnson, J. V.; Chung, A. *J. Org. Chem.* 2009, 2028–2032.
63. Hjorringgaard, C. U.; Pedersen, J. M.; Vosegaard, T.; Nielsen, N. C.; Skrydstrup, T. *J. Org. Chem.* 2009, *74*, 1329–1332.
64. Fara, M. A.; Diaz-Mochon, J. J.; Bradley, M. *Tetrahedron Lett.* 2006, *47*, 1011–1014.
65. Katritzky, A. R.; Yoshioka, M.; Narindoshvili, T.; Chung, A.; Johnson, J. V. *Org. Biomol. Chem.* 2008, *6*, 4582–4586.
66. Peindy N'Dongo, H. W.; Neundorf, I.; Merz, K.; Schatzschneider, U. *J. Inorg. Biochem.* 2008, *102*, 2114–2119.
67. N'Dongo, H. W. P.; Ott, I.; Gust, R.; Schatzschneider, U. In *4th International Symposium on Bioorganometallic Chemistry ISBOMC 08,* Missoula, MT, 2008; pp 823–827.
68. Cini, E.; Lampariello, L. R.; Rodriduez, M.; Taddei, M. *Tetrahedron* 2009, *65*, 844–848.
69. Monroc, S.; Feliu, L.; Planas, M.; Bardaji, E. *Synlett* 2006, 1311–1314.
70. Galanis, A. S.; Albericio, F.; Grotli, M. *Peptide Sci.* 2009, *92*, 23–34.
71. Robinson, A. J.; Elaridi, J.; Van Lierop, B. J.; Mujcinovic, S.; Jackson, W. R. *J. Pept. Sci.* 2007, *13*, 280–285.
72. Chapman, R. N.; Arora, P. S. *Org. Lett.* 2006, *8*, 5825–5828.
73. Wang, D.; Lu, M.; Arora, P. S. *Angew. Chem. Int. Ed. Engl.* 2008, *47*, 1879–1882.

74. Abell, A. D.; Alexander, N. A.; Aitken, S. G.; Chen, H.; Coxon, J. M.; Jones, M. A.; McNabb, S. B.; Muscroft-Taylor, A. *J. Org. Chem.* 2009, *74*, 4354–4356.
75. Illesinghe, J.; Guo, C. X.; Garland, R.; Ahmed, A.; van Lierop, B.; Elaridi, J.; Jackson, W. R.; Robinson, A. *J. Chem. Commun.* 2009, 295–297.
76. Grieco, P.; Campiglia, P.; Gomez-Monterrey, I.; Lama, T.; Novellino, E. *Synlett* 2003, 2216–2218.
77. Byk, G.; Cohen-Ohana, M.; Raichman, D. *Biopolymers* 2006, *84*, 274–282.
78. Matsushita, T.; Hinou, H.; Kurogochi, M.; Shimizu, H.; Nishimura, S. I. *Org. Lett.* 2005, *7*, 877–880.
79. Matsushita, T.; Hinou, H.; Fumoto, M.; Kurogochi, M.; Fujitani, N.; Shimizu, H.; Nishimura, S. I. *J. Org. Chem.* 2006, *71*, 3051–3063.
80. Rizzolo, F.; Sabatino, G.; Chelli, M.; Rovero, P.; Papini, A. M. *Int. J. Pept. Res. Ther.* 2007, *13*, 203–208.
81. Kowalczyk, R.; Harris, P. W. R.; Dunbar, R. P.; Brimble, M. A. *Synthesis* 2009, 2210–2222.
82. Rijkers, D. T. S.; van Esse, G. W.; Merkx, R.; Brouwer, A. J.; Jacobs, H. J. F.; Pieters, R. J.; Liskamp, R. M. J. *Chem. Commun.* 2005, 4581–4583.
83. Dijkgraaf, I.; Rijnders, A. Y.; Soede, A.; Dechesne, A. C.; van Esse, G. W.; Brouwer, A. J.; Corstens, F. H. M.; Boerman, O. C.; Rijkers, D. T. S.; Liskamp, R. M. J. *Org. Biomol. Chem.* 2007, *5*, 935–944.
84. van Dijk, M.; Mustafa, K.; Dechesne, A. C.; van Nostrum, C. F.; Hennink, W. E.; Rijkers, D. T. S.; Liskamp, R. M. J. *Biomacromolecules* 2007, *8*, 327–330.
85. van Dijk, M.; Nollet, M. L.; Weijers, P.; Dechesne, A. C.; van Nostrum, C. F.; Hennink, W. E.; Rijkers, D. T.; Liskamp, R. M. *Biomacromolecules* 2008, *9*, 2834–2843.
86. Murray, J. K.; Gellman, S. H. *Org. Lett.* 2005, *7*, 1517–1520.
87. Petersson, E. J.; Schepartz, A. *J. Am. Chem. Soc.* 2008, *130*, 821–823.
88. Murray, J. K.; Farooqi, B.; Sadowsky, J. D.; Scalf, M.; Freund, W. A.; Smith, L. M.; Chen, J.; Gellman, S. H. *J. Am. Chem. Soc.* 2005, *127*, 13271–13280.
89. Murray, J. K.; Gellman, S. H. *J. Comb. Chem.* 2006, *8*, 58–65.
90. Olivos, H. J.; Alluri, P. G.; Reddy, M. M.; Salony, D.; Kodadek, T. *Org. Lett.* 2002, *4*, 4057–4059.
91. Gorske, B. C.; Jewell, S. A.; Guerard, E. J.; Blackwell, H. E. *Org. Lett.* 2005, *7*, 1521–1524.
92. Diaz-Mochon, J. J.; Fara, M. A.; Sanchez-Martin, R. M.; Bradley, M. *Tetrahedron Lett.* 2008, *49*, 923–926.
93. Unciti-Broceta, A.; Diezmann, F.; Ou-Yang, C. Y.; Fara, M. A.; Bradley, M. *Bioorg. Med. Chem.* 2009, *17*, 959–966.
94. Bonke, G.; Vedel, L.; Witt, M.; Jaroszewski, J. W.; Olsen, C. A.; Franzyk, H. *Synthesis* 2008, 2381–2390.
95. Massolini, G.; Calleri, E. *J. Sep. Sci.* 2005, *28*, 7–21.
96. Duan, J.; Liang, Z.; Yang, C.; Zhang, J.; Zhang, L.; Zhang, W.; Zhang, Y. *Proteomics* 2006, *6*, 412–419.
97. Umar, A.; Dalebout, J. C.; Timmermans, A. M.; Foekens, J. A.; Luider, T. M. *Proteomics* 2005, *5*, 2680–2688.
98. Nooatyy, V. J.; Dacanay, A.; Kelly, J. F.; Ross, N. W. *Rapid Commun. Mass Spectrom.* 2002, *16*, 272–280.
99. Marchetti-Deschmann, M.; Kemptner, J.; Reichel, C.; Allmaier, G. *J. Proteomics* 2009, *72*, 628–639.
100. Bose, A. K.; Ing, Y. H.; Lavlinskaia, N.; Sareen, C.; Pramanik, B. N.; Bartner, P. L.; Liu, Y. H.; Heimark, L. *J. Am. Soc. Mass Spectrom.* 2002, *13*, 839–850.
101. Pramanik, B. N.; Ing, Y. H.; Bose, A. K.; Zhang, L. K.; Liu, Y. H.; Ganguly, S. N.; Bartner, P. *Tetrahedron Lett.* 2003, *44*, 2565–2568.

102. Chen, S. T.; Chiou, S. H.; Chu, Y. H.; Wang, K. T. *Int. J. Pept. Protein Res.* 1987, *30*, 572–576.
103. Chiou, S. H.; Wang, K. T. In *Current Research in Protein Chemistry: Techniques, Structure, and Function*; Villafranca, J. J., Ed.; Academic Press: San Diego, 1990, pp 3–10.
104. Gilman, L. B.; Woodward, C. In *Current Research in Protein Chemistry: Techniques, Structure, and Function*; Villafranca, J. J., Ed.; Academic Press: San Diego, 1990, pp 23–36.
105. Marconi, E.; Panfili, G.; Bruschi, L.; Vivanti, V.; Pizzoferrato, L. *Amino Acids* 1995, *8*, 201–208.
106. Messia, M. C.; Di Falco, T.; Panfili, G.; Marconi, E. *Meat Sci.* 2008, *80*, 401–409.
107. Colombini, M. P.; Fuoco, R.; Giacomelli, A.; Muscatello, B. *Stud. Conserv.* 1998, *43*, 33–41.
108. Pramanik, B. N.; Mirza, U. A.; Ing, Y. H.; Liu, Y. H.; Bartner, P. L.; Weber, P. C.; Bose, A. K. *Protein Sci.* 2002, *11*, 2676–2687.
109. Juan, H. F.; Chang, S. C.; Huang, H. C.; Chen, S. T. *Proteomics* 2005, *5*, 840–842.
110. Vesper, H. W.; Mi, L.; Enada, A.; Myers, G. L. *Rapid Commun. Mass Spectrom.* 2005, *19*, 2865–2870.
111. Sun, W.; Gao, S.; Wang, L.; Chen, Y.; Wu, S.; Wang, X.; Zheng, D.; Gao, Y. *Mol. Cell. Proteomics* 2006, *5*, 769–776.
112. Lin, S. S.; Wu, C. H.; Sun, M. C.; Sun, C. M.; Ho, Y. P. *J. Am. Soc. Mass Spectrom.* 2005, *16*, 581–588.
113. Chen, W. Y.; Chen, Y. C. *Anal. Chem.* 2007, *79*, 2394–2401.
114. Lin, S.; Lin, Z.; Yao, G.; Deng, C.; Yang, P.; Zhang, X. *Rapid Commun. Mass Spectrom.* 2007, *21*, 3910–3918.
115. Lin, S.; Yun, D.; Qi, D.; Deng, C.; Li, Y.; Zhang, X. *J. Proteome Res.* 2008, *7*, 1297–1307.
116. Lin, S.; Yao, G.; Qi, D.; Li, Y.; Deng, C.; Yang, P.; Zhang, X. *Anal. Chem.* 2008, *80*, 3655–3665.
117. Yao, G.; Qi, D.; Deng, C.; Zhang, X. *J. Chromatogr. A* 2008, *1215*, 82–91.
118. Miao, A.; Dai, Y.; Ji, Y.; Jiang, Y.; Lu, Y. *Biochem. Biophys. Res. Commun.* 2009, *380*, 603–608.
119. Lo, C. Y.; Chen, W. Y.; Chen, C. T.; Chen, Y. C. *J. Proteome Res.* 2007, *6*, 887–893.
120. Chen, W. J.; Tsai, P. J.; Chen, Y. C. *Anal. Chem.* 2008.
121. Berna, M.; Ackermann, B. *Anal. Chem.* 2009, *81*, 3950–3956.
122. Stencel, L. M.; Kormos, C. M.; Avery, K. B.; Leadbeater, N. E. *Org. Biomol. Chem.* 2009, *7*, 2452–2457.
123. Zhong, H.; Marcus, S. L.; Li, L. *J. Am. Soc. Mass Spectrom.* 2005, *16*, 471–481.
124. Gebremedhin, M.; Zhong, H.; Wang, S.; Weinfeld, M.; Li, L. *Rapid Commun. Mass Spectrom.* 2007, *21*, 2779–2783.
125. Wang, N.; Mackenzie, L.; De Souza, A. G.; Zhong, H.; Goss, G.; Li, L. *J. Proteome Res.* 2007, *6*, 263–272.
126. Hua, L.; Low, T. Y.; Sze, S. K. *Proteomics* 2006, *6*, 586–591.
127. Swatkoski, S.; Russell, S. C.; Edwards, N.; Fenselau, C. *Anal. Chem.* 2006, *78*, 181–188.
128. Lee, A. M.; Sevinsky, J. R.; Bundy, J. L.; Grunden, A. M.; Stephenson, J. L. *J. Proteome Res.* 2009.
129. Hauser, N. J.; Basile, F. *J. Proteome Res.* 2008, *7*, 1012–1026.
130. Hauser, N. J.; Han, H.; McLuckey, S. A.; Basile, F. *J. Proteome Res.* 2008, *7*, 1867–1872.
131. Swatkoski, S.; Russell, S.; Edwards, N.; Fenselau, C. *Anal. Chem.* 2007, *79*, 654–658.
132. Swatkoski, S.; Gutierrez, P.; Ginter, J.; Petrov, A.; Dinman, J. D.; Edwards, N.; Fenselau, C. *J. Proteome Res.* 2007, *6*, 4525–4527.

133. Li, J. X.; Shefcheck, K.; Callahan, J.; Fenselau, C. *Int. J. Mass Spectrom.* 2008, *278*, 109–113.
134. Swatkoski, S.; Gutierrez, P.; Wynne, C.; Petrov, A.; Dinman, J. D.; Edwards, N.; Fenselau, C. *J. Proteome Res.* 2008, *7*, 579–586.
135. Edman, P.; Begg, G. *Eur. J. Biochem.* 1967, *1*, 80–91.
136. Zhong, H.; Zhang, Y.; Wen, Z.; Li, L. *Nat. Biotechnol.* 2004, *22*, 1291–1296.
137. Guo, N.; Higgins, T. N. *Clin. Biochem.* 2009, *42*, 99–107.
138. Nair, S. S.; Romanuka, J.; Billeter, M.; Skjeldal, L.; Emmett, M. R.; Nilsson, C. L.; Marshall, A. G. *Biochim. Biophys. Acta* 2006, *1764*, 1568–1576.
139. Yassine, M. M.; Guo, N.; Zhong, H.; Li, L.; Lucy, C. A. *Anal. Chim. Acta* 2007, *597*, 41–49.
140. Lee, B. S.; Krishnanchettiar, S.; Lateef, S. S.; Gupta, S. *Rapid Commun. Mass Spectrom.* 2005, *19*, 1545–1550.
141. Lee, B. S.; Krishnanchettiar, S.; Lateef, S. S.; Lateef, N. S.; Gupta, S. *Rapid Commun. Mass Spectrom.* 2005, *19*, 2629–2635.
142. Sandoval, W. N.; Arellano, F.; Arnott, D.; Raab, H.; Vandlen, R.; Lill, J. R. *Int. J. Mass Spectrom.* 2007, *259*, 117–123.
143. Tzeng, Y. K.; Chang, C. C.; Huang, C. N.; Wu, C. C.; Han, C. C.; Chang, H. C. *Anal. Chem.* 2008, *80*, 6809–6814.
144. Peachey, E.; McCarthy, N.; Goenaga-Infante, H. *J. Anal. Atom. Spectrom.* 2008, *23*, 487–492.
145. Zhu-Shimoni, J.; Gunawan, F.; Thomas, A.; Vanderlaan, M.; Stults, J. *J. Immunol. Methods* 2009, *341*, 59–67.
146. Wu, X. P.; Cheng, Y. S.; Liu, J. Y. *J. Proteome Res.* 2007, *6*, 387–391.
147. Munoz, T. E.; Giberson, R. T.; Demaree, R.; Day, J. R. *J. Neurosci. Methods* 2004, *137*, 133–139.
148. Long, D. J., 2nd; Buggs, C. *J. Mol. Histol.* 2008, *39*, 1–4.
149. Temel, S. G.; Minbay, F. Z.; Kahveci, Z.; Jennes, L. *J. Neurosci. Methods* 2006, *156*, 154–160.
150. Bohr, H.; Bohr, J. *Phys. Rev. E* 2000, *61*, 4310–4314.
151. Bohr, H.; Bohr, J. *Bioelectromagnetics* 2000, *21*, 68–72.
152. George, D. F.; Bilek, M. M.; McKenzie, D. R. *Bioelectromagnetics* 2008, *29*, 324–330.
153. Roy, I.; Gupta, M. N. *Current Sci.* 2003, *85*, 1685–1693.
154. Fermer, C.; Nilsson, P.; Larhed, M. *Eur. J. Pharm. Sci.* 2003, *18*, 129–132.
155. Orrling, K.; Nilsson, P.; Gullberg, M.; Larhed, M. *Chem. Commun.* 2004, 790–791.
156. Grunefeld, P.; Richert, C. *Nucleos. Nucleot. Nucl.* 2006, *25*, 815–821.
157. Culf, A. S.; Cuperlovic-Culf, M.; Laflamme, M.; Tardiff, B. J.; Ouellette, R. J. *Oligonucleotides* 2008, *18*, 81–92.
158. Doran, T. J.; Lu, P. J.; Vanier, G. S.; Collins, M. J.; Wu, B.; Lu, Q. L. *Gene Ther.* 2009, *16*, 119–126.
159. Edwards, W. F.; Young, D. D.; Deiters, A. *Org. Biomol. Chem.* 2009, *7*, 2506–2508.

Index